Stairway to Empire

Stairway to Empire

Lockport, the Erie Canal, and the Shaping of America

Patrick McGreevy

Published by State University of New York Press, Albany

Printed in the United States of America

For information, contact State University of New York Press, Albany, NY
www.sunypress.edu

Production by Ryan Morris
Marketing by Susan M. Petrie

Library of Congress Cataloging-in-Publication Data

McGreevy, Patrick Vincent.
Stairway to empire : Lockport, the Erie Canal, and the shaping of America / Patrick McGreevy.
p. cm.
Includes bibliographical references and index.
ISBN 978-1-4384-2527-6 (hardcover : alk. paper)
1. Lockport (N.Y.)—History. 2. Lockport (N.Y.)—Geography. 3. Lockport (N.Y.)—Environmental conditions. 4. Erie Canal (N.Y.)—History. 5. Niagara Escarpment—Environmental conditions. 6. Landscape changes—New York (State)—Lockport—History. 7. Nature—Effect of human beings on—New York (State)—Lockport—History. 8. Canals—Environmental aspects—United States—Case studies. 9. Landscape changes—United States—Case studies. 10. Nature—Effect of human beings on—United States—Case studies. I. Title.

F129.L765M38 2009
974.7'98—dc22 2008024766

10 9 8 7 6 5 4 3 2 1

For
Betsy, Blake, and Jonah,
who all lived with it

CONTENTS

Illustrations

Acknowledgments

The state of New York owned and operated the Erie Canal, yet this study would not have been possible had it not also created the New York State Archives and the New York State Library to serve its citizens' thirst for knowledge and understanding of their own history. Whatever laurels the state is due for the former, it is the latter that reveals its commitment to the vitality of democracy. I was privileged to examine countless boxes of original documents, carefully preserved and containing elements of the stories I have tried to tell. They remain there to help in the telling of other stories now and in the future.

I am indebted to numerous other institutions that contribute to public knowledge. These include the New York Public Library, the Syracuse University George Arents Library, the Cornell University Library, the University of Rochester Rush Rhees Library, the State University of New York at Buffalo Library, the State University of New York at Albany Library, the Pennsylvania State University Library, the Clarion University Library, the Vassar College Library, the Lockport Public Library, the New-York Historical Society, the American Antiquarian Society, the Buffalo and Erie County Historical Society, the Niagara County Historical Society, the Niagara County Historian's Office, the Erie Canal Discovery Center, the Erie Canal Museum, and the private collection of Charles Rand Penney.

The people who work at these institutions are numerous, but I must particularly thank Craig Williams, Jim Folts, Dick Andress, and Doug Farley.

I received special help from Tom Grasso, Dave Kinyon, Charles Rand Penney, Dave Kummer, Nancy Batakji, and the mensch Bob Peer. I have benefited from conversations with Yi-Fu Tuan, Michael Curry, Trevor Barnes, Mark Bouman, Terry Young, Joe Wood, Susan Prezzano, Sylvia Stalker, Kay Luthin, Sebastian Ernissee, Herb Luthin, and Stan Green. Jim Boles and David Boles were crucial sounding boards. Helen McGreevy was my most important local informant. My thoughts on Lockport are finally inseparable from my interactions with Kathleen, Marty, Fran, Tom, Joe, Therese, and Michelle.

I was honored to work in collaboration with Cherin Abdel Samie and Yasser Ayad to craft innovative maps that recreate the historical landscape of the study area and mark the canal's progress through the Mountain Ridge.

I was supported in this research by a Larry J. Hackman Research Residency from the New York State Archives, a research residency from the Pen America Society, a National Endowment for the Humanities Summer Institute grant, several professional development grants from the Pennsylvania State System of Higher Education, and two faculty research grants from the Prince Alwaleed Bin Talal Bin Abdulaziz Al Saoud Center for American Studies and Research at American University of Beirut.

Prologue

> How to forgive the world . . . for its illusion of continuing, seamlessly, as the night follows the day, so to speak—whereas in reality life is a series of brutal ruptures, falling upon our defenceless heads like the blows of a woodsman's axe?
>
> —Salman Rushdie, *The Moor's Last Sigh*

In late June of 1824, as workers struggled to cut through the Mountain Ridge, the final obstacle to the completion of the Erie Canal, Reverend F. H. Cuming rose to dedicate the impressive double staircase of locks that would carry the canal to the top of this last ridge. If he could resurrect Washington himself, Cuming proclaimed, and show him "nothing else save this one work," the nation's father would be "satisfied that his country had already done enough to immortalize itself."[1] Four months later, when the waters of Lake Erie reached Lockport—the village that clustered around the locks at the top of the ridge—the state of New York had succeeded in joining the Great Lakes to the Atlantic. The cut through the Mountain Ridge was a rupture that enabled a connection, a connection that would initiate a cascade of further connections and further ruptures. Moreover, the rifts that developed between the settlers and the workers cutting through the Mountain Ridge were a microcosm of the social tensions that would characterize the market world that the canal was helping to initiate.

Exploiting the natural inclination of water to flow downward, the Lockport locks lifted boats and their cargo up into the interior of the continent. Jesse Hawley, who played a key role in both conceiving and promoting the canal, had become convinced of its possibility when he recognized the power of falling water at Niagara Falls.[2] By diverting part of Niagara's flow and channeling it through the Mountain Ridge at Lockport, the builders of the Erie Canal found the power to accomplish some very heavy lifting. Almost immediately upon its completion, citizens began to see the canal as central to a whirlwind of economic, political, and cultural change that was severing them from the world they had known. The Erie Canal helped to make New York City the American metropolis, the North economically ascendant over the South, and the United States dominant in North America. This was heavy lifting indeed. Hence, the serviceability of the Lockport locks and the Erie Canal to a national narrative presenting continental expansion as an inevitable, natural process, like water flowing downhill, or night following the day. The narrative did not deny American agency but characterized its genius as acting in accordance with destiny, conforming to nature.

Almost three decades after the ceremony that marked the completion of the Lockport locks, when both the efficacy of the Erie Canal and the appropriateness of the term "Empire State" were manifest, C. F. Briggs—writing in *The United States Illustrated*, a paean to manifest destiny edited by Charles Dana—took the Lockport locks as a point of departure for a nationalistic reverie. Briggs argued that American "achievements in art, like the Locks at Lockport," were superior to "the great works of antiquity." Theirs are "monuments of human pride" while "ours are made to promote the happiness and welfare of the people who constructed them." The Anglo-Saxons, he continued, "have a patent from Heaven, to improve the earth," to "level mountains, excavate rivers, to bridge impassable chasms, to exterminate the useless tribes that encumber the earth without improving it, to unite together the utmost parts of the globe. . . ."[3] Briggs rendered even the boldest and most violent acts as inevitable—justified by "a patent from heaven" and the purported "happiness and welfare" they created. The Lockport locks became, for Briggs, an icon of American achievement—like the Erie Canal itself, a stairway to empire. In rhetoric and in history, both were implements of power. Connections that rupture.

This book argues against the inevitability of U.S. expansion, the inevitability of the Erie Canal, and the inevitability of the strangely forgotten place where the canal was completed, Lockport. Its subjects are an event (the process of cutting through the Mountain Ridge), a place (Lockport), and the stories through which people have tried to understand them. Its purpose is not to denigrate an achievement or the meaning people have seen in it, but to try to create a more complete picture by attending to the silences and shadows of both. When attempting to grasp what humans have shaped—an event, a place, a story—it is never possible to step outside entirely: there is nowhere to step but into another story. Attending to the largely forgotten stories that highlight what the familiar ones conceal is a task particularly compelling, perhaps, to a child of Lockport severed from its orbit but not from its stories.

1

Leaving Lockport

And Returning to the Canal's Beginning

> [T]he ends and ultimates of all things accord in some mean and measure with their inceptions and originals. . . . [but] all [is] hidden when we would backward see from what region of remoteness the whatness of our whoness has fetched his whenceness.
>
> —James Joyce, *Ulysses*

I became aware at an early age that the place where I lived was cloven by a ponderous, watery pit. Later, I learned to call that pit the Erie Canal and the place Lockport. A flight of locks descends here into a natural gorge, carrying boats down and up the Niagara Escarpment, a tablelike cliff that parallels Lake Ontario. The intersection of these two features, the escarpment and the canal—one carved by ancient glaciers, the other a human achievement—beckoned this town into being in a dramatic series of events over 175 years ago.

I remember people hanging over the bridge railings at Lockport, peering down into the canal, drawn to the water like Melville's "crowds of water-gazers" at the Battery.[1] Of course, Lockport's water-gazers could easily turn around and forget what lay beneath, particularly if they were on the Big Bridge, a structure that covers 399 feet of the canal. It was easy to think of that watery pit as an irrelevant vestige of the distant past. A child who didn't want to forget could follow the old towpath under the Big Bridge and sit, all alone at the heart of the city, and there perhaps ponder another subsurface mystery: the legend of the lost Lockport Cave. Under the center of the city, according to certain early reports and a century and a half of local tradition, lay a colossal cavern. A child could easily imagine that there was something hidden beneath the ordinary streets of Lockport, long lost, but once known.

Such at least were my early perceptions of the city, but later, Lockport's vertiginous terrain seemed to flatten until it assumed for me the form of a hard surface. I had become like one of the teenagers in *Wonderland*, a novel by Lockport's darkest daughter, Joyce Carol Oates. Her teenagers "strolled up and down Main Street, eager to be transformed into adults so they could escape forever the small, maddening confinement of their childhoods."[2] I felt further repelled by the urban renewal demolitions that had begun as I entered high school in 1964 and eventually flattened three-quarters of the city's commercial buildings. This was painful to watch. I did not know then that Lockport had rebounded from similar convulsions several times. Twice, canal enlargements had ripped broad swaths of destruction through the built landscape. Major fires had presented similar challenges; one in 1852 destroyed over thirty buildings in the center of town.[3] I possessed none of this historical perspective in the late 1960s as I watched the buildings come down. Lockport looked to me like a city with its soul ripped out. My escape was college in the Midwest.

It has been almost four decades since I left Lockport, but because of family and friends, I have retained some contact with the place, visiting about twice a year. From this perspective of an exile who was still in a sense an insider, I began to notice, as time passed, certain puzzling facts. In the late 1970s, a friend who lived twenty miles west of Lockport told me about a forty-foot waterfall on the city's western edge. I thought I had explored every square inch of the town's landscape: how could there be a waterfall I had never seen nor heard any mention of by hundreds of local acquaintances?

The crest of the falls was only a few yards from a major state highway, but a high chain-link fence completely blocked access from above. I tried to approach from the south but found myself in the old city dump. Finally, I followed the creek upstream from a road crossing through a wooded ravine that local people refer to as the Gulf. Tires and metal drums lay strewn throughout the valley and creek bed. A large sewage pipe terminated directly in the stream and oozed a putrid and gaudy substance. The valley eventually narrowed to a steep-sided gorge. There, at the head of a box canyon, was the waterfall (fig. 1.1). It had been a dry period, and a small stream of water poured down into a plunge-pool, accompanied by much mist and trickling. On subsequent visits, I learned that the falls could be quite impressive after a rain (fig. 1.2). The exposed rock face was huge for so small a stream; it formed a horseshoe, like the largest falls on the escarpment. A yellow wood-frame house stood adjacent to the crest with a glazed porch extending out on brackets. Building materials and other debris tumbled down from the north side of the canyon. Large pieces of concrete, tires, and other rubbish lay scattered in the canyon and creek bed.

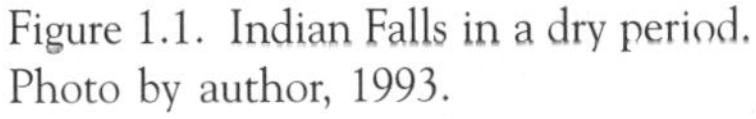
Figure 1.1. Indian Falls in a dry period. Photo by author, 1993.

Figure 1.2. Indian Falls after a rain. Photo by author, 1993.

The state of New York has since declared the dump and landfill a class 2 hazardous waste site. In addition to municipal waste, the site had also accepted toxic chemicals from local industries including PCBs, heavy metals, and acids. The falls itself is just outside the superfund site although the creek downstream is included in it. An even older city dump is closer to the falls on the southeast side. Moreover, it appears that the entire area, unofficially perhaps, had been used for dumping for a very long time.[4]

What sort of process, I wondered, would lead a community to treat with such disrespect this little natural spectacle? Surely not one that was open and publicly articulated. One is tempted to look for a few powerful villains to blame, but when some part of a landscape is hidden, and that concealment is taken for granted, more than a few individuals obviously are involved.

Walking through the contaminated valley that leads to the waterfall, I thought of those people on the bridges in the center of Lockport staring into the canal basin. Did they know there was another valley, a dark twin (fig. 1.3)? In some ways these two valleys seem opposite: one the focus of human activity, the other almost untrodden; one the city's heart, the other a kind of wilderness. Yet it would be a mistake to call one artificial and one natural. No humans carved the Gulf, but the city and its people thoroughly haunt the place.

The fact that the Gulf—with its juxtaposition of natural beauty and toxic waste—had faded from local consciousness struck me in two ways. On the one hand, it resonated with legends long circulating in local popular culture that Lockport's landscape contained a mystery. On the other hand, it seemed to present an anomaly to the usual upbeat story I had always been told about Lockport's genesis and its nature. These considerations, along with my position as a former native who happened to be a historical researcher, gradually compelled me to begin a scholarly investigation.

The story of Lockport we heard at civic events and in local classrooms focused on a single symbolic moment. In the twentieth century,

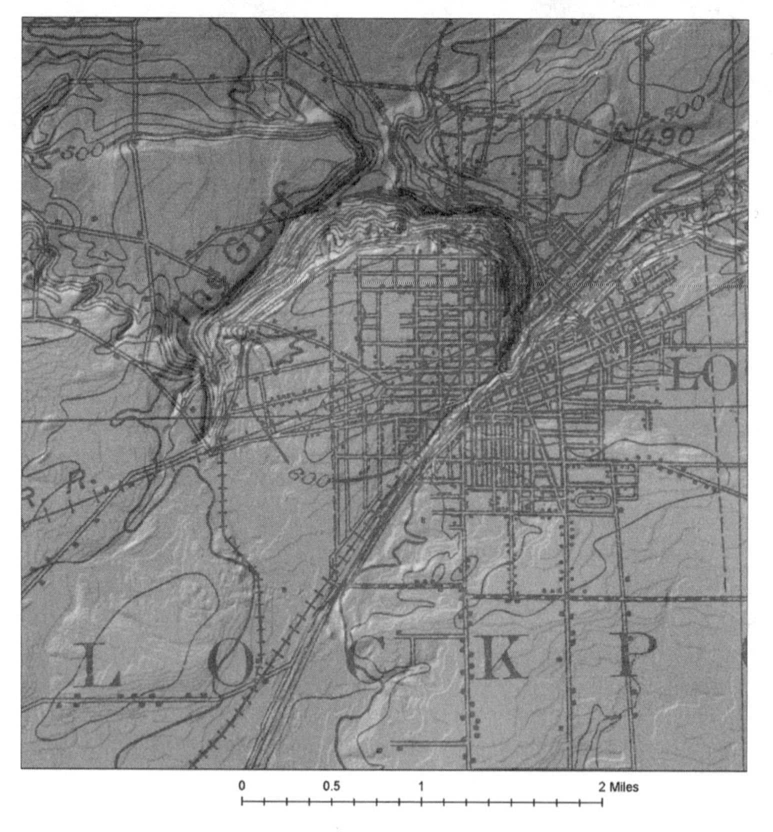

Figure 1.3. Topographic map of Lockport overlain on digital elevation model (United States Geological Survey). Map by Cherin Abdel Samie and Yasser Ayad.

Figure 1.4. *Opening of the Erie Canal, Oct. 26, 1825*, mural by Raphael Beck. Niagara County Historical Society. Used with permission.

many Lockport natives first encountered this symbolic moment through a monumental mural that covered the wall of a downtown bank lobby. The mural, painted by Raphael Beck, a Lockport artist of regional notoriety, is titled *Opening of the Erie Canal, Oct. 26, 1825* (fig. 1.4). The Lockport Exchange Trust Company had commissioned the work at the time of the centennial of the canal's opening and unveiled it at the dedication of its new building in 1928. Beck's mural impressed me every time I entered the bank as a child in the 1950s. Its sheer size and grandiose tone convinced me that the scene depicted must be one of importance. The setting, even to a child, was unmistakable.

The key event Beck's mural depicts is the conquest of the Niagara Escarpment—the final obstacle to the completion of the Erie Canal—and the seemingly instantaneous creation of an urban landscape that, as its name implies, was born of the same event. This was Lockport's moment. Here the Erie Canal had been completed and the waters joined. Even before the canal's opening, many people began to interpret both the event and the place symbolically. From various points of view, they commented on the value and meaning of human (and American) ingenuity and skill, technological and economic advancement and, more generally, progress—always in relation to nature, and in the context of continental power. The moment itself remained the reference point for an ongoing dialogue about Lockport and about these larger issues.

Beck's mural emphasized the nationalistic connotations of Lockport's moment: the flags at the top of the locks point west into the continent's interior. The canal unites that vast undeveloped land to the nation's largest city: Lockport's moment is also a critical New York moment. The flags in the foreground and in the distance show a country, if not yet a continent, knit together by a work of human artifice: an American moment.

Beck's quintessentially progressive view of Lockport's moment is by far the most common perspective expressed in about 180 years of written and pictorial interpretation of the Lock City. Particularly in the three decades that followed the meeting of the waters at Lockport in 1825, the dialogue about the meaning of the event was public and intertextual. It was an exchange that caught up questions about progress, nature, and America's continental destiny—questions that, though of immediate import to New Yorkers of the 1820s and 1830s, continue to resonate today even beyond North America. The presence of Niagara Falls, just eighteen miles to the west, loomed over this dialogue, a natural benchmark from which to assess the town and its artificial river. This was a dialogue among residents and visitors, rich and poor, men and women. Middle-class white men dominated the most visible and prestigious forms of communication. This group included engineers, industrialists, military officers, businessmen, journalists, historians, writers, and artists. They expressed their views not only through public speaking and published writing but also through their power to shape the urban landscape itself. Laborers and other less privileged people participated as well, but it was often their actions, rather than their words, that did the talking.

If Lockport's moment signaled progress to most observers, there were also a number of Jeremiahs who questioned whether the benefits of progress were worth the costs. Many of the laborers who actually dug the ditch and built the locks were killed or maimed in the process. Did the canal bring progressive opportunities to their lives? Two major riots in Lockport during construction suggest the opposite. These events, too, are part of the dialogue on Lockport's moment. The lower right side of Beck's mural depicts a group of Indians and workers with their tools. The picture's narrative structure leaves no doubt about whom Beck thinks deserves credit for the canal: it is not the men who actually built it, but the middle-class gentlemen who imagined and designed it, and, indeed, who would principally profit from it. One worker seems to be waving cheerfully, resigned perhaps to his lower, peripheral, and shadowy position. But, as every Lockportian knows, this dark laborer stands on the slope that leads to Lowertown, a district that, from the Civil War to the mid-1970s at least, was Lockport's closest approximation to an immigrant ghetto, home in Beck's day to an Italian and African American working class. To paint a more honest picture of Lockport's moment, we must attend to the shadows it casts and listen for the voices of those who speak without words or hope of being heard.

In the progressive narrative Beck's mural constructs, the Lockport locks become a stairway to empire. Yet this was not just a story. Completing the Erie Canal helped to secure the country's hegemony in the interior of the continent, and it played a key role in sparking the market revolution that transformed the economic and social structure of the United States.[5] The practical effects of the canal are indisputable—it was a stairway to empire—but the progressive narrative about the canal is built upon exclusions and elisions. For Native American groups and Canadians, the hegemony of the United States was hardly a thing to celebrate. For the largely immigrant common laborers who made up the bulk of the canal construction force, the market revolution was a brutal reality within which they had little leverage or hope. Moreover, as pawns in the new transnational wage economy, they were scarcely more self-identified with the United States—a country that hardly embraced them—than were the Canadians twenty miles to the west.

Then there is the question of the Gulf, the dark twin of the scene Beck's mural celebrates. These two places have a parallel yet paradoxical relationship that prompts me to ask counterfactual questions: if only slightly different decisions had been made, what kind of city, or even continent, might have developed? In a sense, these possible worlds are part of Lockport's moment too. Between the crevices of the actual, we can glimpse the background of possibilities out of which it, far from inevitably, emerged. Because, as Roland Barthes suggests, culture tends to convert history into nature,[6] *what is* can eventually appear *destined to be*, as if no real human choices were involved—in a word, natural—yet it is no more so than the untried possibilities.

This book is an attempt to examine the event and the place Beck's mural depicts. It is the story of a human achievement, but an achievement that has generated shadows and ghosts. Rather than directly fragmenting the progressive story of the achievement, I want to hold it together with what it excludes: the hidden, lost, and counterfactual stories of Lockport. My process is one of excavating beneath the layers of naturalness and inevitability, in an attempt to reveal the human agency, and the human responsibility.

The key events of the 1820s that initiated Lockport's brief moment of prominence form the hub of this book. The next chapter situates these events in a geographical, historical, and cultural context and examines the decisions involved in routing the canal through western New York. Chapter 3 explores the five-year-long struggle to overcome the Niagara Escarpment and complete the canal. Chapter 4 examines the meanings people saw in Lockport's landscape, especially as articulated in words and images during the celebrations that marked the canal's opening and in the following decades of its heyday. The final chapter considers the impact of the canal's completion upon the place where it was completed. Using Lockport as an example that mirrors broader cultural transformations, chapter 5 traces a distinct set of issues and incidents connected to that event.

Had I not been born in Lockport, I would most likely never have stumbled upon the puzzling issues this book attempts to unravel. Yet, despite my efforts to apprehend them as completely as possible, I cannot pretend to stand apart on some neutral ground. Fortunately, one does not have to be detached to be honest. The town that I experienced growing up in the 1950s and 1960s still reverberated with the echoes of what Joyce calls its "inceptions and originals." Yet when we try to discover what is here and why it is here, we find not historical necessities but possibilities: "all is hidden when we would backward see from what region of remoteness the whatness of our whoness has fetched his whenceness."[7]

2

Locating Lockport

The Canal and the Continent's Last Barrier

> The whole world seemed crowded in to render homage to a scheme that God could only bless and make use of himself. Old distinctions and inhibitions were dissolved . . .
>
> —Edward Said, *Orientalism*

The completion of the Suez Canal in 1869, to Ferdinand de Lesseps and his supporters, was an accomplishment blessed by God and the "whole world." There was a certain inevitability about it: in Edward Said's rendering, "nothing, [de Lesseps] repeated on many occasions, could stop us, nothing was impossible." Yet the accomplishment depended on "the combination of old ideas with new methods . . . the genuine imposition of the power of modern technology and intellectual will upon formerly stable and divided geographical entities like East and West."[1] De Lesseps believed that because a bold and confident agency embodied the destined future, it was unstoppable.

In the summer of 1835, Ramón de la Sagra, a Spaniard who had grown to admire the United States during a twelve-year stay in Cuba, traveled across New York State by steamboat, railroad and canal and stood before Niagara Falls:

> On contemplating this immense bulk of water, precipitating itself on the frontier of a happy people who owe to this element their prosperity and prodigies, it occurs to me to consider it as a god for the American nation. Antiquity would have erected to it sumptuous temples. . . . But the American, more industrious and less impressionable, takes advantage of the course of the waters, their falls and their natural deposits, leads them into canals and aqueducts, uniting by their means distant regions; and, not content with dominating it in its liquid form, he transforms

> it into steam, substitutes it for living power, and by it overcomes all obstacles and crosses the greatest distances with the speed of a bird.[2]

It is striking how easily Señor Sagra proceeds from gazing at Niagara Falls to glorifying American technological and industrial progress. He goes on to detail his "amazement" at the canals, steamboats, and industries of the young country and concludes by nearly conflating nature and human action: "and over all this scene of industrial life favored by water presides water itself in the sublime temple of Niagara. It is fitting only for this nation of wonders that it possesses on its frontiers the first wonder of the world!"[3] Here was an apparent paradox similar to the one Said noted: America's success was destined—provided for by ultimate forces beyond human control, blessed by nature, "favored by water"—yet success was equally a product of industriousness, an eagerness to manipulate and dominate nature in all its forms. For Sagra, the two kinds of wonders correspond; there is a unity of destiny and agency. By the time of his visit in 1835, many citizens of the United States were struggling to understand their country's distinctiveness in similar terms. Writers and artists played a key role in articulating such national ideas. Some dealt with the apparent contradictions more honestly than others, but the success of the Erie Canal and its relation to the country's westward expansion provided for many a convincing example of the connection between destiny and agency.

For Said, "the Suez Canal idea" was "the logical conclusion of Orientalist thought, and more interesting, of Orientalist effort" because, "just as a land bridge could be transmuted into a liquid artery, so too the Orient was transubstantiated from resistant hostility into obliging, and submissive, partnership."[4] Before the Erie Canal, it was difficult for anyone in the United States to speak with confidence about a continental destiny for the new republic. The canal's success helped to inspire such confidence—and, as the example of Sagra suggests, not only among U.S. citizens. An American empire required an Empire State, and an Empire State required a canal to pierce the Appalachian barrier with a "liquid artery" that could leash the heart of the continent to its metropolis. Many antebellum observers considered the Erie Canal important precisely because of these continental considerations.

In the section that follows, I address issues of regional and continental hegemony that political leaders and canal supporters thought were at stake in the events surrounding the completion of the canal and the creation of Lockport. I examine in the second section how New York's Canal Commissioners gradually settled on the idea of an Erie, as opposed to an Ontario, canal as the means to effect their continental dreams. Finally, I examine the routing of the canal through the Mountain Ridge—the process, literally, of locating Lockport, the site where the last barrier to the continent's interior would be confronted. Although much of this chapter, and indeed of this volume, focuses on localized events, its more central concern is the interplay of actors and processes operating at local, regional, and continental scales.[5]

The Canal and the Continent

The United States, in the early years of the nineteenth century, was a diverse collection of former colonies with at best an incipient and tenuous cohesion. In addition to the rifts between white and black, native and nonnative, European Americans were a more diverse group than inhabitants of any European country. Citizens were much more likely to identify with their own state than with the United States. The country had become continental, theoretically, with the territories acquired through the wars with Britain and the Louisiana Purchase, yet nearly all its citizens remained east of the Appalachians. Some citizens—notably Washington and Jefferson—glimpsed an immense destiny for the republic in the continent's interior, or even in the far west, but there were also many doubts. For European Americans, a country of continental proportions was difficult to imagine: their main point of reference being the western European pattern of sovereign countries about the size of U.S. states. Even if the thirteen states could be bound together, how would the future settlers of the interior be kept in the fold? "The western settlers," George Washington remarked in 1784, "stand as it were upon a pivot. The touch of a feather would turn them any way."[6] The possible patterns of continental power, both economic and political, were myriad, but one thing at the time seemed certain: a key determinant of that power would be that element to which Sagra would later claim Americans owe "their prosperity and their prodigies," water.[7]

In one sense, the fixation on water routes was a continuation of the earliest European imperial designs in the New World: the goal had been to discover and dominate the right river. Both Jacques Cartier and Henry Hudson had sought a passage that would lead to power. The problem for the United States was that the vast interior of the continent was already penetrated by two major rivers. The cities that dominated the entrances to those rivers, Montreal and New Orleans, seemed most likely to capture the commerce, and perhaps the political allegiance of future settlers. Although the United States had gained nominal control of New Orleans and most of the Mississippi Valley in 1803, the recent demise of France's North American empire underscored the ephemerality of nominal control.

The U.S. interest in water routes was also inspired by the recent success of artificial waterways in Europe. France, Great Britain, and even Ireland had already developed extensive canal systems. Elkanah Watson, a

Rhode Island businessman who had been "constantly impressed" with the importance of canals during his travels abroad, was one of a growing number of Americans who believed that canals could help the citizens of the United States overcome the disadvantages of their location on the eastern side of the Appalachians and compete for the commercial and political allegiance of the interior. When Watson visited George Washington in 1785, he found the Virginian eager "to demonstrate the practicability and the policy of diverting the trade of the immense interior world, yet unexplored, to the Atlantic cities."[8] Here was Washington's hope: that the path of power into the continent would follow neither the St. Lawrence nor the Mississippi but rather an artificial river.

Proponents of internal improvements stressed that canals would help cement the union. "In proportion as the difficulty of communication is removed," wrote Robert Fulton in 1796, "the spirit of enterprise increases, and neighboring associations begin to mingle, their habits and customs assimilate."[9] A canal connecting the east coast with the interior would serve the national interest not only by helping the United States compete with other powers for control of the interior, but also by encouraging that region's settlers to remain loyal.

It was also clear, nonetheless, that canals established lines of power that benefited regions unevenly. There would be winners and losers. Without a national consensus on the location of a canal to the interior, attempts to create a coordinated federal program were repeatedly frustrated. Promoting and building canals became matters of intense regional competition in an era when most identified themselves with designations like "Virginian" or "Pennsylvanian" rather than "American."[10]

George Washington, who had traveled to the forks of the Ohio as early as 1753, was himself an incessant promoter of a canal link between Virginia and Pittsburgh via the Potomac and the Monongahela (fig. 2.1). Thomas Jefferson, in 1785, predicted "a competition between the Hudson and the Potomac rivers for the . . . commerce of all the country westward of Lake Erie, on the waters of the lakes, of the Ohio, and the upper-parts of the Mississippi." He considered the Potomac route superior because it was nearer to the Ohio, less likely to freeze over, and farther from "our neighbors the Anglo-Americans."[11]

In the competition that ensued, it was the business and civic leaders of port cities—Baltimore, Philadelphia, New York, and Boston—who took the leading roles, each group hoping to parlay its transportation advantages into metropolitan dominance.[12] Boston, despite its prominence during the colonial era, lacked any locational advantage. A Massachusetts-commissioned feasibility study showed that a canal through the Berkshires from Boston to the Hudson would require hundreds of locks. But how could commerce be

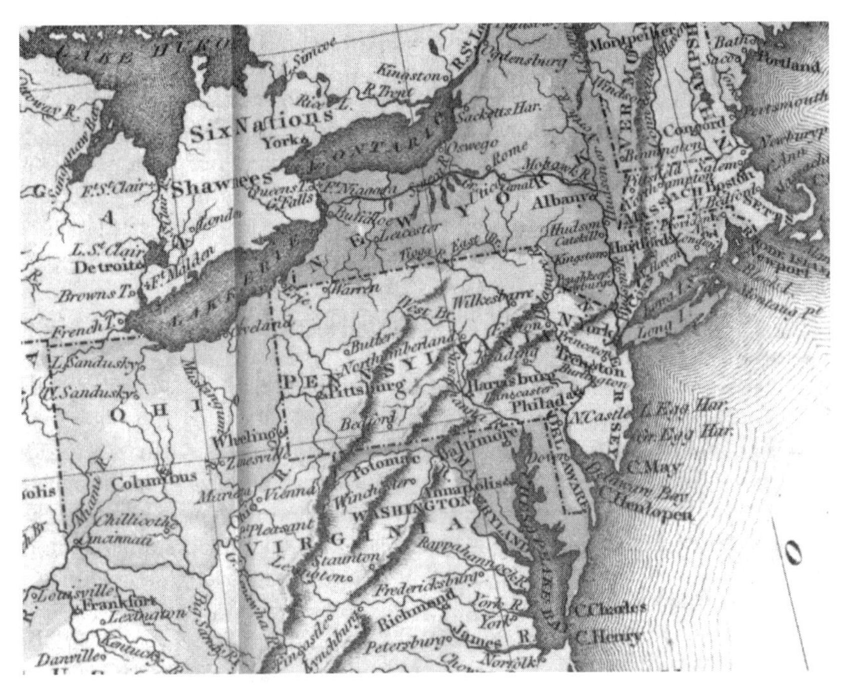

Figure 2.1. Most canal schemes involved connecting coastal rivers with either the Great Lakes or the Ohio River. Section from map of the United States, from Cadwallader D. Colden, *Memoir at the Celebration of the Completion of the New York Canals* (New York: The Corporation of New York, 1825), following p. 4. Print Collection, Miriam and Ira D. Wallach Division of Art, Prints and Photographs, The New York Public Library, Astor, Lenox and Tilden Foundations. Used with permission.

coerced to flow through such a tortuous route to reach the Hudson when that river was already directly navigable from New York City? Baltimore investors aligned their city's destiny with the Potomac. The Chesapeake and Ohio Canal eventually reached Cumberland, Maryland, where it stalled at the foot of the Appalachian's highest ridges. Philadelphia, the continent's largest city at the beginning of the nineteenth century, appeared to be in a better position because of its proximity to Pittsburgh, which many considered the riverine gateway to the entire Mississippi Valley. The chief obstacle was the Allegheny Front, a half-mile-high ridge running north-south through the center of Pennsylvania. The problem of overcoming such a ridge was not only constructing the elaborate lockage but, more fundamentally, securing a reliable supply of water to the highest sections. The Pennsylvania Mainline eventually connected Philadelphia and Pittsburgh with a canal system, but one that required canal boats to be dismantled, loaded on cars, and hauled over the Allegheny Front on tracks. The Pennsylvania Mainline was too cumbersome, too slow, and, moreover, completed too late to play an important role in the contest for metropolitan dominance of North America.[13]

New York's strategic advantage over its Atlantic rivals is a matter of topography. The city dominates the coastal approach to the only significant gap in the Appalachians. To begin with, the Hudson River is itself fully navigable to Albany. Its tributary, the Mohawk, flows directly from the west through a deep valley that severs the Catskills from the Adirondacks. Here is a pass through the Appalachian barrier nowhere more than five hundred feet above sea level.

The Mohawk Valley is a product of North America's last glacial period. As the ice sheet began to retreat about thirteen thousand years ago, enormous lakes of meltwater gathered at its southern margins; these were the precursors of the Great Lakes. Water eventually broke through and began spilling over to the south through two channels, the Illinois and Mohawk spillways. At this time, the glacier was still pressing against the Adirondacks, impounding the meltwater and forcing it to the south. The prodigious discharge of the glacial lakes Algonquin and Iroquois poured through the Mohawk Valley, eroding downward a cleft through the Appalachians. Moreover, a similar process may have begun the cutting during previous glaciations.[14] An artificial river between the Great Lakes and the Hudson was only possible because a natural river had blazed that same path. The Erie Canal would re-create, in broad outline, the very waterway that had made its construction possible.

Although this fluvial prehistory was hardly available to canal promoters, many believed that nature or Providence had laid out a plan that human, and specifically American, artifice must complete. The successful Rhode Island entrepreneur and land speculator Elkanah Watson moved to Albany in 1789 and quickly became an influential agitator for a New York canal. Upon inspecting the region between Oneida Lake and the Mohawk River in 1791, he believed he had glimpsed that natural design. "What a pity," Watson wrote, "since the partial hand of Nature had nearly completed the water communication from our utmost borders to the Atlantic ocean, that Art should not be made subservient to her, and complete the great work."[15] For Watson and his middle-class contemporaries, the word "Art" referred to all human artifice, and particularly to the works of technological and entrepreneurial imagination that were challenging previously unassailable natural barriers. Art was the central term in what historian Paul Johnson has called "the language of progress," and within a few decades, the Erie Canal would be its archetypal example. For workers and artisans, on the other hand, a man's "art" was the specific set of occupational skills and knowledge by which he defined his identity as useful, independent, and equal.[16]

When Watson suggested that "Art" work in harmony with the "hand of nature," he was expressing an American variant of the idea of progress—one that was still very much alive when Ramón de la Sagra visited the United States in 1835. John Seelye argues that this conception of progress had roots not only in the Enlightenment celebration of reason—and its application, art—but also in the Puritan millennial notion of a providential design in which the chosen people were the instruments of destiny.[17] Canals, from this perspective, would overcome rapids and other obstacles, releasing the waters to flow as they were intended, just as the republican form of government released the "natural" energies of people long confined by the barriers of an aristocratic social and political system. Enlightened, neoclassical Americans—like Jefferson and Franklin—shared, in Seelye's words, "a rage for order that, when imposed upon the map, put forth a distinctly rectilinear shape."[18] Jefferson found the continent's interior marvelously configured yet still crying out for orderliness. His geometrical scheme for imposing order, the Federal Rectangular Survey System, was intended to simply assist nature, to perfect nature's own pattern, but it nonetheless required violence in its execution. Like the imperial gaze of Europeans of the colonial era, Jefferson's scheme rendered the continent as a tabula rasa devoid of inhabitants.[19]

A sublime picture of the future served to justify this violence by showing the harmonious pastoral world that progress would create. "My mind voluntarily expanded," Elkanah Watson wrote as he viewed the landscape near Lake Seneca in 1791, "in anticipating the period when the borders of this lake will be stripped of nature's livery, and in its place will be enclosures, pleasant villages, numerous flocks, herds, etc.; and it will be inhabited by a happy race of people, enjoying the rich fruits of their own labors, and the luxury of sweet liberty and independence, approaching to a millennial state."[20] In this image, political and technological progress—the Constitution and the canal—are seamlessly conflated. Though the goal of progress was a world in which nature and art would be in balance, reaching that goal would pit art against a series of natural barriers just as the establishment of republican government had required a war. Also missing from Watson's vision are the Iroquois inhabitants of central New York, recently vanquished in the same war. They are apparently part of "nature's livery" destined to be "stripped."

In the decades following the completion of the Erie Canal in 1825, it seemed self-evident to most Americans that the artificial waterway had allowed New York to outstrip all of its continental rivals in the contest for metropolitan dominance. The most sanguine assertions of the canal's early proponents had become common wisdom. Just before its completion, the *Times* of London had predicted that the Erie Canal would make New

York "the London of the New World."[21] By the mid-1840s, New York had indeed achieved a position atop the urban hierarchy—of the country if not the continent—a place it would never relinquish. As twentieth-century historians reassessed the causes of New York City's rise, a more complex picture began to emerge.

Metropolitan dominance was not simply a matter of connections to the continent's interior. It also turned on New York's success as a port in both coastwise and global trade, its access to information, and its innovative financial arrangements. It is worth noting that New York had already surpassed Boston and Philadelphia to become the country's largest city by 1815, giving it at least a slight advantage in attracting imports, exports, and capital.[22] At that time, however, there simply was no national metropolis, certainly no London that could simultaneously dominate commerce, finance, and information. Moreover, having an advantage hardly determines the outcome.[23] New York's trade during the Napoleonic Wars differed from that of other American ports in its greater proportion of domestic exports, as opposed to transshipped West Indian products. This reflected New York's greater and rapidly growing hinterland that included a third of Connecticut and half of New Jersey in addition to New York State's own expanding frontier—a realm that the Erie Canal would continue to augment. New York developed a particularly strong trade relationship with Britain, partly because its large hinterland made it the preferable market for manufactured goods. A regular packet line to Britain also gave New York first access to information about British demand and prices. These advantages helped New York capture a large share of the cotton trade since cotton transports returned to New York with British manufactured goods.[24]

Innovative financial developments were also crucial to New York's rise, and here the important role of the Erie Canal is not in dispute. After several failed attempts to secure federal funding for the canal, the state of New York decided to borrow money by issuing bonds. Initially, commercial banks and wealthy investors shied away, fearing that they would lose their money if the project failed. A group of canal backers convinced the state legislature to create a new kind of bank, the savings bank, that was only authorized to invest in government-backed securities (like the Erie Canal bonds), while its assets would derive entirely from the deposits of working people, "thus encouraging thrift among the improvident classes and having them bear risks that the rich considered unacceptable."[25] At the same time, new investment banks began to buy large numbers of Erie Canal bonds for resale at higher prices. These new arrangements encouraged the entrance of foreign investors who eventually held the majority of Erie Canal debt. In 1817 a group of New York brokers established the New York Stock and Exchange Board, which initially dealt almost exclusively with canal bonds but soon funneled capital to railroad, land, and manufacturing ventures.[26] Finally, New York banks began to offer farmers and millers as far west as Chicago advances on products to be shipped in the future, a practice that garnered more trade for the Erie Canal.

The success of the Erie Canal was thus one of the "vital factors" in New York City's ascendancy,[27] but only as part of a complex synergistic mix that included not only entrepreneurial innovation but governmental initiative. It is probable, nonetheless, that the canal was the single most important element in New York City's ascendancy. "Above all," conclude Edwin G. Burrows and Mike Wallace in their recent history of New York City, "it was the state-run Erie Canal that secured New York's position as the nation's entrepôt, galvanizing its commerce, its banking, its stock market, and its manufacturing sectors."[28]

Because of its financial success, and because it connected New York with the continent's vast and rich heartland, the Erie Canal played a crucial role in the contest for continental power. It helped to determine that the heartland would be dominated by New York (not Boston, Baltimore, or Philadelphia), by the Northeast (not another section, with dire implications for the South), and by the United States (not another power or breakaway republic). It was partly because so many antebellum Americans had a sense of these immense geopolitical stakes that they would attach so much meaning to the Erie Canal.

Though many New Yorkers (and Americans) saw this alignment of power focused on New York City as a natural, destined development, what is striking in retrospect is its remarkable contingency. Had this contest for the interior taken place two decades later when railroads were available, a different alignment of power could easily have emerged, perhaps favoring Philadelphia with its more direct route to the heart of the Midwest. Had the contest taken place before the political bifurcation of English North America, Montreal might have become the North American metropolis. Had it taken place prior to any major transportation improvement, the continent may have become politically and economically fragmented like South America. Such outcomes, of course, would depend upon not only transport systems, but also complex historical and cultural processes. Moreover, a project like the Erie Canal cannot be explained merely in terms of its feasibility: it required a willingness, indeed an eagerness, to pursue the opportunity that nature, technology, and historical circumstances presented.

An Erie Canal

Before the Revolutionary War, the presence of the powerful Iroquois Confederacy—effectively blocking Dutch, German, and English settlers from moving west of the Mohawk Valley—obscured the topographical advantages of the New York route to the interior. When the war began, the Iroquois, except for some of the Oneidas and the Tuscaroras, sided with Britain. The interior of New York was the site of some of the bloodiest and most strategically important battles of the war. Among these was the Clinton-Sullivan Campaign that utterly destroyed Iroquois resistance, not to mention its orchards, villages, "castles" and many of its women and children. General James Clinton, one of the campaign's two leaders, was the father of DeWitt. The canal that would become the son's obsession and a daring expression of American imperial ambition required first the father's more obvious execution of power. Memories of these gruesome events would live on, particularly in New York.[29] For those who wished to see the canal as the completion of a natural design, the fact that it would pass through territory conquered recently with great difficulty may have been troubling. But as the story of the two Clintons suggests, both war and internal improvements could be viewed as expressions of progress: art overcoming first human savagery, then natural obstacles. Indians could be useful as markers of the past and of nature. In numerous nineteenth-century paintings and lithographs, and even in the ceremonial fleet that would celebrate the Erie Canal's completion, their presence provided a kind of cultural baseline against which to measure the accomplishments of art and civilization.[30]

On the heels of the Iroquois conquest, settlers poured into central New York. By 1800 there were more than one hundred thousand people west of the colonial line of settlement. Within another decade, that number had tripled.[31] New England provided the overwhelming majority of these migrants. The principal thrust of occupation followed a trajectory that skirted the northern tips of the Finger Lakes, spurring the emergence of urban centers at Auburn, Geneva, and Canandaigua, linked by the Geneva Road to each other and to the east (fig.2.2).

These settlers were vocal proponents of a waterway that could transport their products to coastal markets. It was obvious to all that any New York canal must strike through the Mohawk Valley. Beyond that point, however, there were several possibilities. The Western Inland Lock Navigation Company, a private concern headed by Albany investor Philip Schuyler, worked between 1792 and 1820 to connect the Hudson to Lake Ontario. This effort culminated in a waterway that linked the Mohawk and Oswego rivers. Employing a series of short canals and river channel improvements, Schuyler's system allowed a primitive sort of navigation from Lake Ontario to Schenectady, but financial problems prevented the completion of the final section to overcome the drop of more than two hundred feet between that city and the Hudson.[32]

The Western Inland Lock Navigation Company had demonstrated the feasibility of a canal from the Hudson to Lake Ontario, but completing the route west would require a second canal linking Lake Ontario and Lake Erie near the Niagara River. Niagara Falls separates the St. Lawrence Valley into two distinct sections. Above the falls, all the lakes are nearly at the same level. Neither rapids nor other obstructions impede navigation between Buffalo and Chicago. Lake Erie, therefore, was the ultimate goal of all New York canal schemes. A route through Lake Ontario would expose western shipments to the temptation of continuing down the St. Lawrence instead of following a canal to the Hudson. The earliest settlers in central and western New York almost immediately took advantage of the "Quebec Market."[33] Visiting Batavia in 1811, John Melish found that market blossoming. "The principal market is on Lake [Ontario]," Melish wrote, "and it is believed by the people here that it will always continue to be so."[34] The commercial orientation of central and western New York was uncertain at this time. Most settlers were nevertheless aware that, upon passing through the Mohawk Valley, they had left the area of Hudson drainage and entered that of the St. Lawrence. The availability of commercial connections with Montreal was recognized as one of the region's advantages.[35] Indeed, the late twentieth-century demise of the port of Buffalo is due, in part, to the reestablishment of the original shipping conditions by the completion of the St. Lawrence Seaway.

The hostilities that brought so much bloodshed and hardship to the western part of New York—and adjacent Upper Canada—between 1812 and 1814 revealed another ominous problem with using Lake Ontario as part of an American route to the interior. Ontario was a British lake. During the war, attempts to ship supplies to Fort Niagara from Rochester met with disaster, and the British occupied Rochester itself after a naval attack in 1813.[36] Moreover, because the surface of Lake Ontario is more than 325 feet below that of Lake Erie, extensive lockage would be necessary to descend at the eastern end of Lake Ontario and to ascend again near the Niagara River.

The alternative notion of an inland canal from the Mohawk River directly to Lake Erie seemed far-fetched to many when Jesse Hawley of Geneva, New York, first publicly proposed it in a series of anonymous letters to the *Genesee Messenger* in 1807 and 1808. Others had hinted at this idea previously—notably the prominent drafter of the Constitution, Gouverneur Morris—but Hawley was the first to give definite shape to the scheme that would become the Erie Canal.[37] The idea of "tapping Lake Erie,"[38] along with the recognition that the Mohawk Valley pierces the Appalachians at an elevation of less than five hundred feet, were the topographical insights that made the Erie Canal not only conceivable but possible. Hawley was not the only one to conceive of tapping Lake Erie, but his story of the moment of discovery reveals an interesting pattern.

In April of 1805, Hawley, a Geneva merchant, was visiting the mill of Colonel Mynderse at Seneca Falls to prepare a shipment of grain for the New York market. The two men began to complain about "the circuitous and uncertain route then in use." Hawley suggested: "Why not have a canal extend direct into our country?" Colonel Mynderse replied that "it could not be done, for the lack of a head of water."[39] Hawley remembered the moment this way:

> [I] sat in a fit of abstraction for some minutes—then took down De Witt's map of the State, spread it on the table and sat over it with my head reclined in my hands and my elbows on the table, ruminating over it, for—I cannot tell how long—muttering *a head of water*; at length my eye lit on the Falls of Niagara which instantly presented the idea that lake Erie was *that head of water* [emphasis in original].[40]

Like Alfred Wegener—who, while also staring at a map, noticed that the coastlines of Africa and South America curiously matched, and dared to suggest that continents move[41]—Hawley saw something well known in a new way. What he noticed was Niagara Falls, its falling water demonstrating that Lake Erie was a high reservoir with an enormous volume of discharge. By turning some of Niagara's waters to the east, one could tap a seemingly inexhaustible head of water at an altitude above that of the entire canal route. In Hawley's story, Niagara Falls becomes the "source" of the Erie Canal, both conceptually and hydraulically. As we shall have repeated occasion to observe, Niagara Falls and the Erie Canal are multiply linked. The canal would make Niagara the magnet of western travel, a prospect that, in turn, would lend considerable luster to the canal project itself.

Two and a half years after his purported flash of insight, Jesse Hawley found himself in debtor's prison when his business failed. While languishing there, he wrote his famous letters to the *Genesee Messenger*. Although Hawley was a marginal member of the emerging middle class, the form of his story reveals his aspirations. Despite being broke, uneducated, and imprisoned—or perhaps because of this—he sought the high respect he knew would be accorded to acts of technological imagination, bold projections of art. Although even friends initially ridiculed Hawley's proposal, he made a compelling case for the 350-mile waterway. Not only would his route avoid the commercial and military dangers of Lake Ontario, it would eliminate the costs and delays of two additional transshipment points. The detailed route he suggested was very similar to the canal that was actually built. He also presciently argued that the canal should be a public project, under control of the state rather than private investors.[42]

In 1805 President Jefferson proposed devoting excess federal revenue to internal improvements. This prospect, perhaps in combination with the impact of Hawley's persuasive writings,[43] helped induce Joshua Forman to introduce a resolution to the state assembly to explore and survey "the most eligible and direct route for a canal . . . between the tide waters of the Hudson River and Lake Erie."[44] Against Forman's wishes, in 1808 the legislature directed the Surveyor-General, Simeon De Witt, to survey the Ontario route, and left to his discretion any work on an interior route. With a meager budget of $600, De Witt hired James Geddes to conduct preliminary surveys; a judge from Onondaga (later Syracuse), Geddes quickly developed considerable expertise as a surveyor and engineer. De Witt also engaged the support of Joseph Ellicott, the western New York manager of the vast holdings of the Holland Land Company. Ellicott reported that a canal between the Genesee and Niagara rivers was possible via the valley of Tonawanta Creek (see fig. 2.2).[45] After completing his survey of the Ontario route,

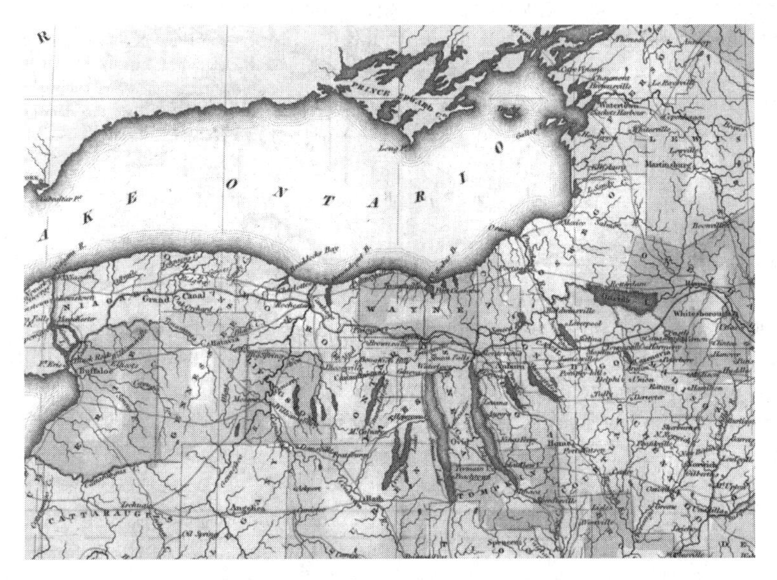

Figure 2.2. Section from map of New York State, from Cadwallader D. Colden, *Memoir at the Celebration of the Completion of the New York Canals* (New York: The Corporation of New York, 1825), following p. 6. Print Collection, Miriam and Ira D. Wallach Division of Art, Prints and Photographs, The New York Public Library, Astor, Lenox and Tilden Foundations. Used with permission.

Geddes made a brief inspection of the interior route between the Oswego and Genesee rivers and found that, contrary to prevailing opinion, there was no high land intervening. Geddes was so enthused by this discovery that his report to the legislature favored the interior route.[46] When Forman traveled to Washington to discuss the report with Jefferson, however, the president replied: "making a canal 350 miles through the wilderness—it is little short of madness to think of it at this day."[47]

In spite of Jefferson's response to the proposal, public support for internal improvements continued to wax, particularly in New York State. In 1810, Jonas Platt convinced a political opponent, DeWitt Clinton, to support a resolution to fund detailed surveys of both routes. The resolution passed both the senate and assembly unanimously, and from this moment Clinton became a vigorous canal advocate. To oversee the surveys, the resolution established a board of commissioners that included De Witt, Morris, and Clinton. Geddes and Benjamin Wright conducted the surveys while Clinton and other commissioners traveled the routes as well.[48] The commissioners' report of 1811 favored the interior route.[49] Under the influence of Gouverneur Morris, the commissioners also proposed an inclined plane, instead of locks, with an average slope of six inches per mile for all but the final descent to the Hudson. This scheme would allow Lake Erie water to fill the entire canal—a direct diversion of the St. Lawrence into the Hudson. The length and depth of intervening valleys rendered this part of the plan almost unimaginable, though the idea of using Lake Erie water would prove crucial to the canal's success.[50] The commissioners sent Clinton and Morris to Washington in 1811 to solicit federal aid, but state rivalries—and now the prospect of war—forced them to return, once again, empty handed.

Had the War of 1812 led to the conquest and annexation of Canada, schemes for internal improvements might have been turned upside down. Because the Niagara Frontier saw the most fiercely contested battles of that war, many recognized that a canal would have greatly facilitated the transport of cannon and other military supplies. These considerations helped to increase support for the canal throughout the state. At the war's conclusion, New York leaders were finally ready to act.

The legislature authorized more detailed surveys early in 1816, and finally passed a law in April of 1817 directing the start of construction on two canals: one from the Hudson to Lake Erie and a second from the Hudson to Lake Champlain. The fact that the commissioners decided to break ground near Rome on July 4 shows that, although the canal was entirely a New York project, they also considered it a nationalistic enterprise. Work immediately commenced on the middle—and, it was hoped, least difficult—section of the canal. By 1821, boats could pass from the Genesee River to Little Falls, leaving a host of thorny problems to be overcome, both east and west of this segment. The Champlain Canal was a more immediate success; it opened in 1822, though the connection with Montreal would be delayed for over thirty years. Construction crews completed the eastern section from Little Falls to the Hudson in October of 1823. Only the western section remained. Obstacles there would prove no less troublesome than those encountered to the east, and it would take another two long years to consummate, in the terms of a contemporary metaphor, Lake Erie's marriage to the Atlantic.[51]

The Mountain Ridge

James Geddes traveled to western New York in 1816 with orders to survey and map a route between Lake Erie and the Genesee River. About twenty-five miles west of the Genesee River, Geddes entered the domain of the Holland Land Company, a consortium of Dutch investors who controlled the entire western portion of New York State—an area today comprising most of eight counties—as well as much of western Pennsylvania. Geddes's task was to map the canal's route, depicting in profile the land sliced into a series of horizontal segments.

Joseph Ellicott, the Holland Land Company's Pennsylvania-born land agent and chief surveyor, had entered western New York twenty years earlier with a similar task of geometrical ordering. Settlement of the region had been long delayed by Native American land claims. In 1797 remaining Iroquois groups (a total population of about seventeen hundred)[52] finally surrendered rights to all land with the exception of a handful of reservations, most of which they would eventually also lose.[53]

Immediately, Ellicott, with a team of twenty surveyors and one hundred laborers, began to survey, mark, and map the grid of six-mile-square townships visible on the company's 1804 map (fig. 2.3).[54] Anchoring this grid to the sun's path were two "Transit Meridian Lines," beginning at the Pennsylvania border and extending ninety-two miles "due North to Lake Ontario." One line marked the eastern boundary of the purchase and the second intersected the lake shore near a creek marked "Quottahongaiga"[55] (fig. 2.4). Such a scientifically exact grid, registered to the cardinal directions, was a quintessential expression of what not only Ellicott but the settlers he hoped to attract would call art. Ordered surveys were already an established tradition in North America. The grid itself first appeared in the street plans of Savannah and Philadelphia, then in federally influenced areas such as central New York's New Military Tract and eastern Ohio's Seven Ranges. Under the provisions of the Ordinance of 1785, it would march across the Midwest and leap to the oases and valleys of the West.[56] To geographer Hildegard Binder Johnson these surveys express "our culture's roots in the rationalism of the eighteenth century."[57] Although the grid represented an imposition of order upon the land, it was an order patterned on nature's

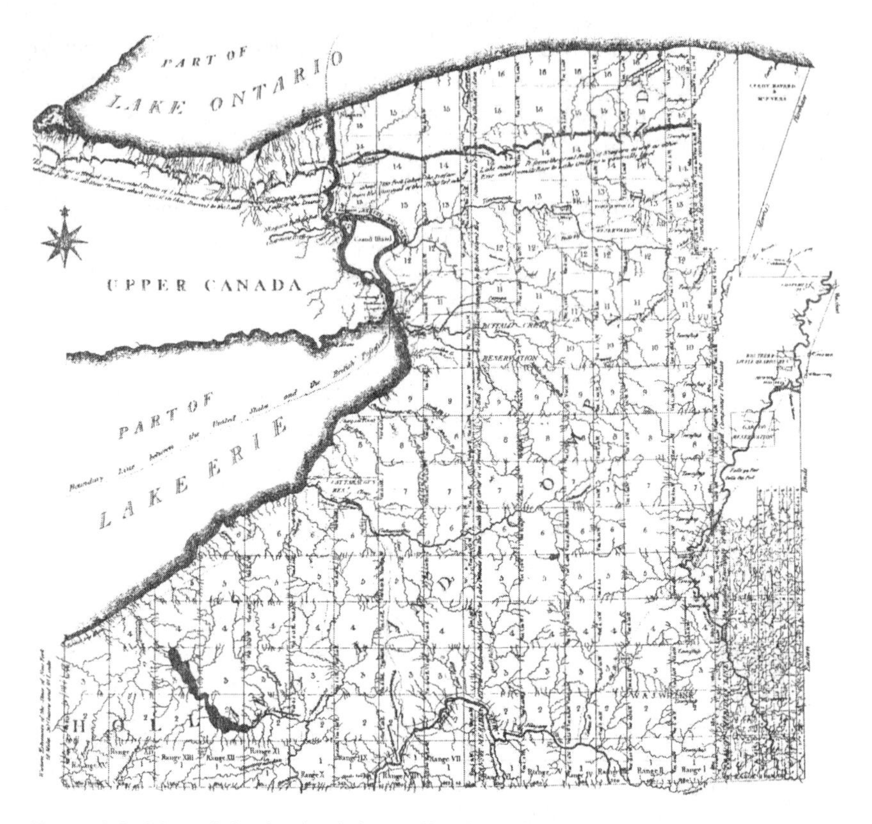

Figure 2.3. Map of the lands of the Holland Land Company. *Map of Morris's Purchase or West Geneseo in the State of New York*, by Joseph Ellicott, with William and John Willink and others, Holland Land Company, 1804. Courtesy of the New York State Library.

own regularities. Apart from their progressive connotations, ordered surveys made good business sense: by eliminating odd-shaped leftover parcels, they helped investors and speculators sell their entire holdings, quickly and unambiguously.[58]

Another artifact of the Holland Land Company's mathematical ordering of the western New York landscape was the renaming of many of the watercourses. Along the shores of both Lake Erie and Lake Ontario, Ellicott simply measured the miles from the Niagara River. Across the northern edge of the company's land, accordingly, we find streams labeled Fourmile Creek, Sixmile Creek, Twelvemile Creek, and near the western Transit Meridian, Eighteenmile Creek (instead of Quottahonyaiga).

As part of his effort to encourage settlement, Ellicott had also created a rational system of improvements. He founded villages and constructed

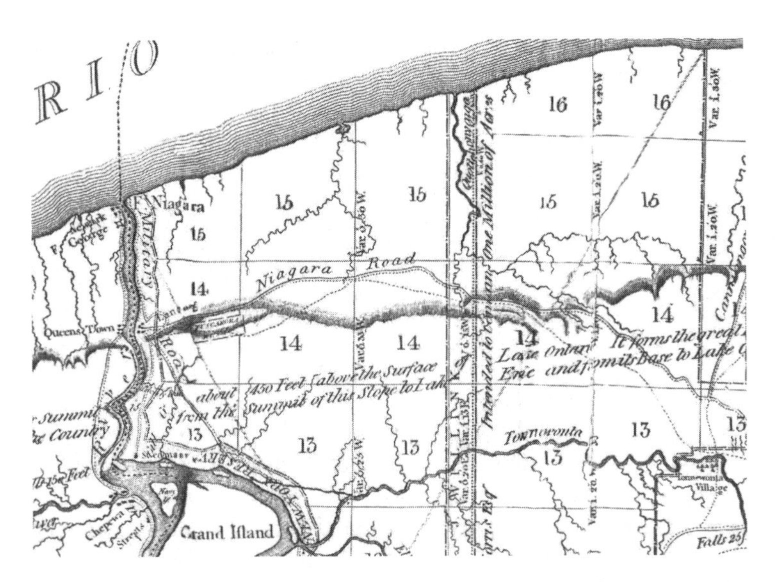

Figure 2.4. Section from map of the lands of the Holland Land Company. *Map of the Morris Purchase or West Geneseo in the State of New York*, by Joseph Ellicott, with William and John Willink and others, Holland Land Company, 1804. Courtesy of the New York State Library.

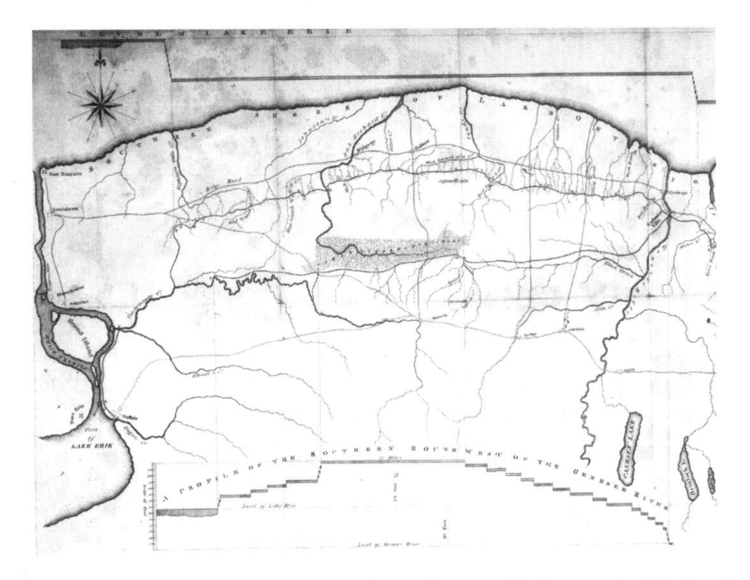

Figure 2.5. Map and profile of the proposed canal from Lake Erie to the Hudson River in the state of New York, Contracted by the Direction of the Canal Commissioners from the Maps of the Engineers in 1817, E. Valentine (Albany: New York State Canal Commissioners, 1817). Courtesy of the New York State Library.

mills, taverns, and, most importantly, roads (see fig. 2.5). From the village of Batavia, the company's headquarters and largest settlement, two main roads led to Buffalo and Lewiston, entry points to Lake Erie and Lake Ontario respectively. Paralleling the Ontario Shore, Ellicott also laid out the Ridge Road, which joined the Niagara Road to a long-distance route from Onondaga. Eventually other roads radiated from Batavia and Buffalo to serve the company's southern townships and provide a route to the west. This was a rationally designed network, yet almost everywhere it followed old Iroquois trails.[59] The Ridge Road, which paralleled the Ontario shore just north of the escarpment following the shoreline of prehistoric Lake Iroquois, was the best of these. The ridge was not only elevated from the surrounding country, but was well drained. Ellicott nevertheless expected the main transport corridor to follow the Buffalo Road. His entire network assumed a trajectory that would extend westward the early thrust of settlement along the northern tips of the Finger Lakes. A shrewd and perceptive manager, Ellicott also recognized the importance of the Erie Canal to the profitable development of his company's lands. He was more than willing to donate whatever lands were needed for its construction as well as contribute the proceeds from the sale of additional lands. Under the 1817 law that initiated canal construction, Ellicott became one of eight canal commissioners.

Ellicott knew that any canal through western New York would benefit his company, but he was by no means impartial on the question of which route was best. When asked in 1810 to provide information on a line through the company's lands, he suggested one paralleling the Buffalo Road from "Black Creek to the Tonnewanta Swamp and down the Tonnewanta Creek" (see fig. 2.5). Geddes reported that the summit level of this route "is estimated by Judge Ellicott only twenty feet or so above the level of the mouth of Tonnewanta Creek," and much of it "may be made almost straight and the cutting very easy."[60] In 1816 Geddes himself focused on the northern route that he had sketched out in 1810. At the same time, Ellicott directed a surveyor named Peacock to map the southern route.[61] Both the northern and southern routes appeared on a special map contracted by the canal commissioners in 1817 (fig. 2.5).[62]

Peacock's team measured the summit section of the southern route at seventy-four feet above the level of Lake Erie. They planned feeders from the upper branches of Tonawanta and Black creeks, but here—unlike central New York—there were no elevated Finger Lakes to serve as natural reservoirs. Serious questions about water supply remained, as is clear from the profiles that accompanied the commissioner's 1817 map (fig. 2.5).

The northern route—in both its 1810 and 1816 versions—was inspired by the idea of avoiding water issues by tapping Lake Erie directly. To appreciate the logic of the northern route, one must recognize how starkly western

New York and Ontario's Niagara Peninsula are bisected by the east-west trending Niagara Escarpment. It appears as the most prominent physical feature on the Holland Land Company's 1804 map (fig. 2.3), where it bears the following label:

> THIS great slope is formed of horizontal Strata of Limestone and the perpendicular Height of its Summit is about 450 Feet above the Surface of Lake Ontario. It forms the great Falls of Niagara as well as other perpendicular Falls in all those Streams which pass it in their Descent to the Lake. The Face of the Country from the Summit of this Slope to Lake Erie and from its Base to Lake Ontario is generally level.

The escarpment divides the landscape into two plains, each slightly higher than the Great Lake it borders. The proposed southern and northern canal routes follow these plains above and below the ridge respectively.

The top of the ridge is generally between 600 and 650 feet above sea level; whereas the surface of Lake Erie is at 572 feet. Geddes knew that if a channel could be opened through this ridge, Lake Erie water would flow toward the east—as much and as far as needed, for the entire canal to the east lay below the foot of this ridge. In his report of 1809, Geddes had already expressed doubts about Ellicott's southern route. "It would be important to know," he wrote, "whether there is not some place in the ridge that bounds the Tonnewanta valley on the north, as low as the level of Lake Erie, where a canal might be led across, and conducted onward, without increasing the lockage by rising to the summit of the Tonnewanta swamp."[63] There was no place on the Niagara Escarpment "as low as the level of Lake Erie," but in 1810 Geddes had identified a notch in the ridge where the elevation was considerably lower than that of surrounding sections. When he returned to conduct detailed surveys in 1816, Geddes focused his attention on this notch. At the point where the western Transit Meridian intersected the escarpment, two declivities or small gorges, etched into the face of the ridge, were now occupied by branches of Eighteenmile Creek (fig. 2.4). Since nature had already accomplished part of the work, this seemed the least difficult spot to channel through the ridge. For almost a decade, this place would simply be known as the Mountain Ridge.

Between 1798 and 1800, the Holland Land Company completed surveys along the two Transit Meridians and the six-mile square township boundaries. The surveyor's field book for the part of the western Transit Meridian between ranges seven and six, township fourteen, depicts the line from south to north through the Mountain Ridge. After descending through steep terrain and crossing a branch of Eighteenmile Creek, the meridian proceeds through some "very stony" level ground before encountering another drop to the Niagara Road and finally descending into a swampy area.[64] The Niagara Escarpment divides into two levels here. Geddes wanted the canal to descend only to the first level and then to hold that elevation eastward to the Genesee River. Later, surveyors completed transects along section and lot lines within the townships and produced township maps depicting lots, major roads, and watercourses.[65]

Because the Holland Land Company's field books also provide detailed information on vegetation, Frank K. Seischab has recently been able to use them to reconstruct late eighteenth-century vegetation patterns for all of western New York.[66] The vicinity of the Mountain Ridge was dominated two centuries ago by a basswood-beech-white ash-elm association with a sprinkling of sugar maple, black walnut, white oak, and hickory.

Unlike some parts of the Holland Purchase, the vicinity of the Mountain Ridge had not recently been inhabited by the agricultural Iroquois, so it was thickly forested. The region had been an important hunting ground, particularly the swampy bottoms of the Eighteenmile Creek below the Escarpment. This was a place rich with beavers, otters, deer, minks, bears, and even panthers.[67] When Geddes arrived in 1816, much of the land had been purchased and a few settlers were trickling in. Some had begun

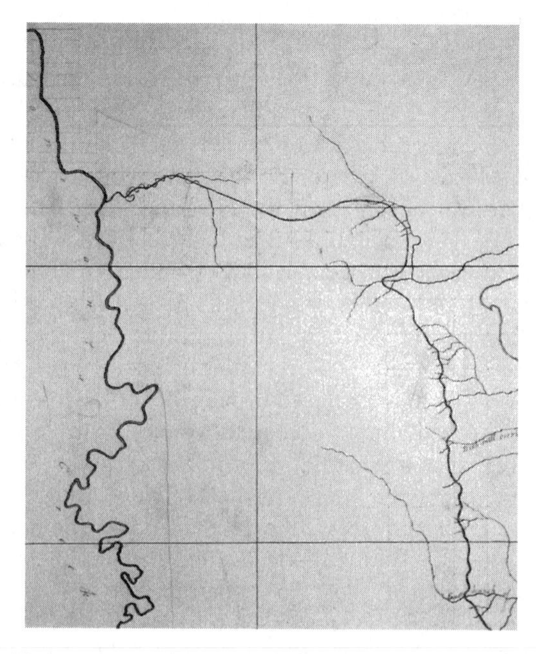

Figure 2.6. Geddes's overview survey map of Mountain Ridge site. North is to the right of the map. From James Geddes, *Original Maps of Surveys for the Erie Canal*, 1817. Courtesy of the New York State Archives.

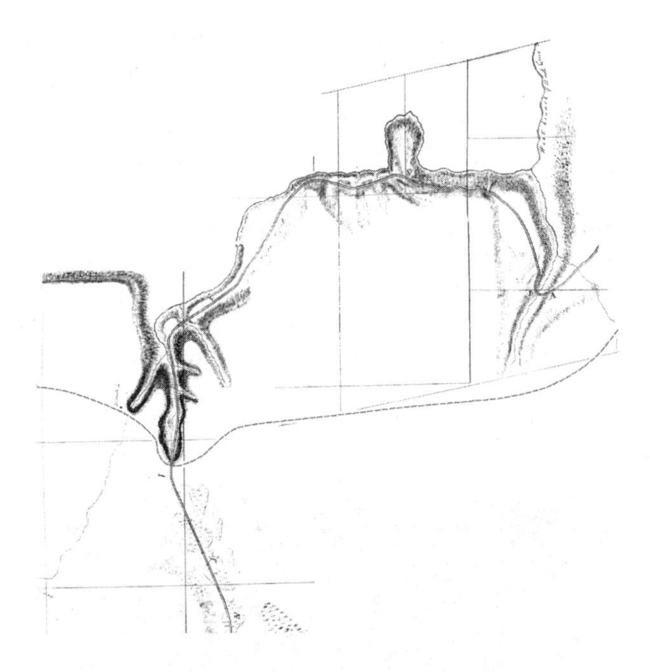

Figure 2.7. Geddes's survey maps of the two gorges at the Mountain Ridge. From James Geddes, *Original Maps of Surveys for the Erie Canal*, 1817. Courtesy of the New York State Archives.

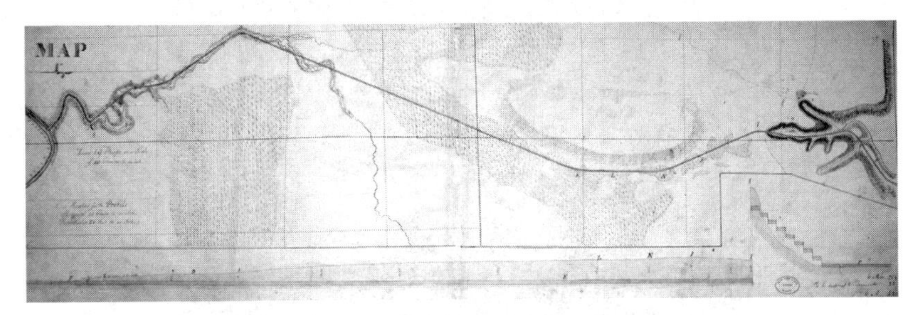

Figure 2.8. Geddes's survey map of the proposed Deep Cut. From James Geddes, Original Maps of Surveys for the Erie Canal, 1817. Courtesy of the New York State Archives.

clearing. Geddes's survey maps (figs. 2.6, 2.7, and 2.8) depict only three log dwellings within five miles of the Mountain Ridge site, two on the Niagara Road, and one near the head of the western gorge.[68]

Geddes was acutely aware of the importance of this place to the entire canal project. Would it really be possible to cut through the Mountain Ridge and divert Lake Erie's water to the east? Success would depend not only upon the state's commitment of resources—and the attendant politics of that process—but also upon local human agency, social dynamics, and topography.

One important local issue faced James Geddes in 1816: through which of the two gorges should the Erie Canal climb the Mountain Ridge? Figure 2.9 is a digital elevation model of the site; since it is based on current conditions, the major alterations to the topography (railroad embankments, quarries, the canal itself) have been removed.[69] It presents a clear picture of the topographical situation that confronted Geddes and indicates the route he eventually chose. Figure 2.10 shows the regional context of the site; it is a combination of a regional digital elevation model and a section of the commissioner's 1817 map. During his 1810 visit, Geddes had noticed that, a little more than five miles north of the Tonawanta Creek,

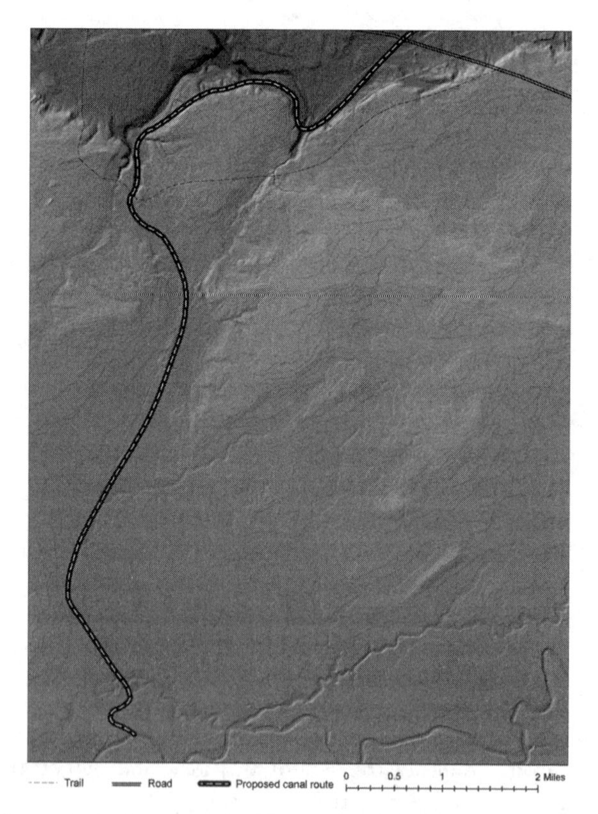

Figure 2.9. James Geddes's 1817 canal route through the Mountain Ridge overlain on a digital elevation model (United States Geological Survey). Map by Cherin Abdel Samie and Yasser Ayad.

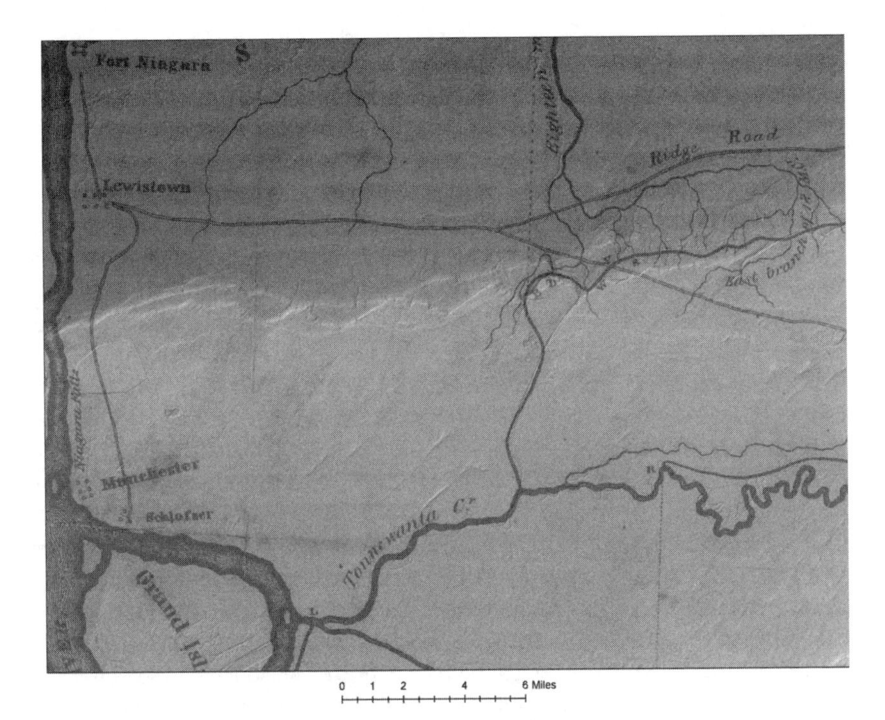

Figure 2.10. A section of the Map and Profile of the Proposed Canal from Lake Erie to the Hudson River in the State of New York, Contracted by the Direction of the Canal Commissioners from the Maps of the Engineers in 1817, E. Valentine (Albany: New York State Canal Commissioners, 1817) overlain on a regional digital elevation model (United States Geological Survey). Map by Cherin Abdel Samie and Yasser Ayad.

> In the middle steep, is the source of one branch of the eighteen mile brook, at a distance of about ten miles from lake Ontario, and about three hundred feet above its surface; consequently, nearly thirty feet below the surface of lake Erie. The greatest elevation of ground between them is twenty-one feet above that surface; it is, however, on an average, for the space of three miles, twenty feet, and the remaining two miles and a quarter, about seven feet.[70]

In 1816, Geddes described his plan in greater detail:

> Tonewanta creek would make a level and wide canal for eleven miles up it. Opposite this place the valley of eighteen mile creek approaches the valley of Tonewanta, so that a cut of 5 miles and 30 chains (course nearly north) will let the canal pass through to the valley of Lake Ontario.

> This five miles will have to be cut in the deepest place twenty-two feet, from the surface to the top of the water line of this canal. About 3 ¼ miles of the surface is on an average, 20 feet higher than the level of Lake Erie. There is great reason to hope that this deep cutting will all be through clay, except a short distance on the north end.[71]

Geddes's map of this crucial section of the canal that would later be labeled "the Deep Cut" (fig. 2.8) shows in profile his estimates of the amount of excavating needed along this section and includes a sketch of the flight of eight combined locks he proposed to bring the canal from the level of the Genesee River to that of Lake Erie.

In both 1810 and 1816, Geddes chose a path through the western gorge. His survey maps of the Mountain Ridge (figs. 2.6 and 2.7) show the canal route entering from the east at an elevation it has held since leaving the Genesee River, then crossing the eastern gorge (by means of an embankment and a ninety-foot culvert). The line then hugs that contour through the Eighteenmile Creek Valley to the western gorge where it climbs sixty-five feet through one of two box canyons with a staircase of eight consecutive locks (inset, fig. 2.8). Deep cutting would commence at the top of the locks and diminish in difficulty to the south. The 1817 Report of the Commissioners discussed both the northern and southern alternatives between the Genesee and Lake Erie. Concerning the Geddes route through the Mountain Ridge, they commented: "As the excavation of the canal, through this distance, constitutes one of the most serious difficulties presented on the whole route, great pains have been taken to avoid all impractical data of calculation relating to it."[72]

Though canal construction commenced a few months after the filing of this report, it would be three years before the commissioners were able to devote significant attention to the western section. Early in 1820 they appointed David Thomas, a Quaker from the village of Union Springs on Cayuga Lake, as principal engineer west of the Genesee River.[73] The commissioners immediately charged Thomas to examine and survey—once again—the route from the "Genesee river to the Tonnewanta Creek, a distance of a little more than seventy-two miles, including the deep cutting through the Mountain Ridge."[74] Although they would not officially announce it for another year, the commissioner's charge to Thomas shows that they had already rejected the southern route. Referring to the Peacock survey of 1816, the commissioners expressed concerns in their 1821 report about the "ultimate deficiency of water, in that passage" and concluded that "Providence would not permit us to adopt any route west of the Genesee River, which should rise above the level of Lake Erie."[75] This decision was disappointing to Joseph Ellicott who had so consistently promoted the southern alternative. Nevertheless, realizing the value that the canal would add to the Holland Land Company's holdings, he deeded 100,637 acres as a contribution to the canal fund.[76]

From May to late November of 1820 David Thomas examined Geddes's route. Thomas was a man of unusually broad interests and talents. In addition to a remarkable career as a canal engineer, he would also become a prominent scientific agronomist.[77] Although his experience as a canal engineer was limited when he traveled to western New York in 1820, he already possessed a keen interest in the theoretical and mathematical issues involved.[78] We know of Thomas's activities during this trip mostly through surviving correspondence from his superiors: Myron Holley, Benjamin Wright and DeWitt Clinton. Holley wrote in April of 1820 authorizing Thomas to hire his survey crew and suggesting he seek advice from Wright and Nathan Roberts, as they were "experienced levelers."[79]

In response to a detailed query from Thomas about the relative merits of cutting versus embankment, Benjamin Wright replied with some revealing advice.

> More security is given by cutting than by embankment so the latter is to be avoided if it can be done without disfiguring the line, *but always regard form and conform to the natural obstacles so as to have persons of judgement see after the canal is formed why curves and sinuosities were made*—this constitutes and exhibits the skill of the engineer and is always beautiful [his emphasis].[80]

The two men were apparently of one mind on the aesthetic nature of their work, for a few months later Wright responded to a letter from Thomas by agreeing: "you are correct as to curves—straight lines are never as handsome as curves."[81] Wright and Thomas believed they were doing something momentous. They saw the exacting mathematical and practical aspects of their task within a broad historical and aesthetic context. They felt the eyes of an enlightened posterity upon their work. How art responded to nature's obstacles—*forming by conforming*—was for them its highest expression. As in the meridians and grids of Ellicott and Jefferson, the special role of art was to give form to the land by conforming to nature's most basic patterns. The coming struggle to overcome the Erie Canal's last major obstacle, the Mountain Ridge, would severely tax this view of art's relation to nature, laying bare some of its contradictions and inviting alternative views.

By the early autumn of 1820, Thomas and his crew were focusing on the Mountain Ridge site. From Wright's October 16 response to a letter from Thomas on October 7, we learn of their plan for a slight slope from the Tonawanta Creek to the Genesee River to help induce an eastward flow.

This would add to the depth of rock cutting. "I pray you to think well of this," Wright cautioned, "and if there can be a plan of avoiding this additional deep and wide cut it is a very desirable one—there is no difficulty in regulating the current so that it will pass around the Locks regularly and feed the canal east of the Mountain Ridge."[82]

Thomas was already thinking carefully about the Mountain Ridge section. On October 11, he wrote to Wright of a "new passage" following the eastern rather than the western gorge of the Eighteenmile Creek. Again, only Wright's response survives:

> Your new passage for the canal appears to me a discovery of importance. The saving in distance is a good deal as well as in the depth of excavation. It will be very interesting to ascertain as far as may be practicable whether your new route presents a greater length of excavation through *rock* than the old route [his emphasis].

Wright was "glad to learn" that Thomas would also be surveying over Geddes's route through the western gorge, "for after both routes are run over and examined by the same mind, a decision may be made with much greater satisfaction." Thomas was particularly fearful of the difficulty of "making the canal over the ground encountered by Geddes this side of the ascent on to the Ridge"—that is, from the foot of the western gorge through the valley of the eastern gorge. Both men recognized the significance of these matters. "Your labors will be so arduous," Wright emphasized, "and they are so important in the vicinity where you now are" that all other problems should be put aside; "it is of much greater moment to do all you can to understand well the line from N. Comstock's [just below and east of the Mountain Ridge] to the Tonnewanta Creek."[83]

This change, along with a few other route adjustments to the east, allowed Thomas to shave six miles from Geddes's line west of the Genesee River.[84] The new route eliminated the winding section from the foot of Geddes's proposed locks in the western gorge to the site of Thomas's locks about three and a half miles to the east. As Wright had advised, the new route exchanged cutting for embankment. Thomas did not know that Wright's cautions about the length of rock excavation were well founded. Geddes's route, although two and three-quarters miles longer, would have required about three-quarters of a mile less of deep rock cutting. Geddes had supposed that most of it would be through clay. In response to a letter from Thomas, dated December 25, 1820, Wright noted: "from your description it appears that the excavation will be heavy," but then he added:

> Your suppositions about the expense of the Rock are correct. It is now pretty much ascertained that any Limestone Rock can be excavated for

> 50 cents or under and if the Mountain Ridge should prove soft as you believe I should say we had got it too high.[85]

Thomas would soon find that even a cost per cubic yard three times this figure would be insufficient to excavate the limestone of the Mountain Ridge.[86] He believed, and hoped, that the rock would "prove soft," but those charged with the actual digging would learn otherwise. The ridge's nearly horizontal top layer—soon to be designated Lockport dolomite[87]—was the geological reason for the very existence of the escarpment. The succession of glaciers that gouged out the Lake Ontario basin, and once stood a mile high above the Mountain Ridge, were unable to grind it down. This was the resistant cap rock that formed the brink of Niagara Falls, eighteen miles to the west. The grueling experience of cutting through this limestone would make Niagara's existence less of a geomorphological mystery.

Diverting the waters of the upper Great Lakes, via an artificial channel, through the Mountain Ridge, seems a bold act of human will. What neither Geddes nor Thomas knew was that those waters had long ago flowed over this exact path. Four successive waves of glaciation, interspersed with periods of stream erosion, had scoured the surface of the Niagara Frontier into three plains, each bounded on the south by escarpments capped by resistant rock.[88] As the last glacier retreated toward the northeast, its face rested near the Niagara Escarpment, and its meltwater formed Lake Dana, occupying the Huron and Erie plains to the south at an elevation of about 825 feet. Lake Dana spilled through the Mohawk Valley and the Hudson to the sea. As the ice sheet continued to retreat, a lower entrance to the Mohawk Valley near Rome became ice free and the lake surface dropped over five hundred feet, forming Lake Iroquois below the Niagara Escarpment (surface elevation about three hundred feet). This event, dated to approximately 12,300 years BP, also marked the birth of Niagara Falls at the escarpment near Lewiston where it began to carve the Niagara Gorge that—twelve millennia later—would be seven miles long. At the same time, the waters of Lake Erie continued to flow into the Huron Plain, forming Lake Tonawanda (fig. 2.11).[89] From Lake Tonawanda, the waters of the upper Great Lakes flowed over the Niagara Escarpment through five spillways. According to geologist Parker E. Calkin, the "outlets at Lewiston [Niagara River] and Lockport carried most of the discharge."[90] The path of the canal would follow "the Lockport branch of the ancient Niagara River."[91] The opening of the spillway through Rome allowed Lake Erie water to help carve that gap through the Appalachian Mountains so essential to the possibility of the Erie Canal. The same event gave birth not only to the Lockport spillway—creating an opportunity for those later seeking to recreate this original flow—but also to Niagara Falls.

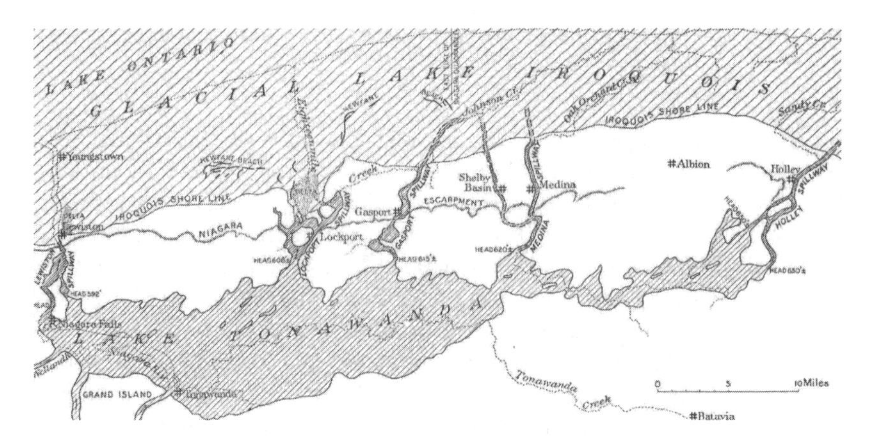

Figure 2.11. Map of Lake Tonawanda. From E. M. Kindle and Frank B. Taylor, *Geologic Atlas of the United States: Niagara Folio* (Washington, D.C.: United States Geological Survey, 1913), figure 10, 19.

Geomorphologist John D'Agostino has delineated the path of the Lockport spillway (superimposed on the digital elevation model in fig. 2.12), a task complicated by the uneven uplift of the land after the departure of the massive ice lobe. The "Lockport branch of the ancient Niagara River" divided into two channels, like the modern Niagara River, before each channel plunged into its gorge, creating twin cataracts and an island. The similarities go further: the channel to the west of the island has a greater discharge of water and consequently eroded farther into the Lockport dolomite. Bringing the Erie Canal through the Mountain Ridge would be eerily close to recreating Niagara's ancient twin.

Lake Erie water poured through the Mountain Ridge for about fifteen hundred years.[92] At first this eastern Niagara River gave serious competition to the western one, as it quickly cut down through the moraine of loose till that the glacier had deposited atop the escarpment.[93] Although the river carved a wide channel leading to its two gorges, the channel became flat-bottomed once it eroded to the level of the Lockport dolomite.[94] Isostatic rebound, or uplift following glaciation, was greater at the eastern end of Lake Tonawanda, leaving the Lewiston spillway lower than the Lockport spillway. This, combined with the latter's inability to erode a deeper channel in the section south of the two gorges, eventually caused the abandonment first of the eastern channel, and later, the western, leaving only a ghost river and ghost waterfalls. Later, the northward retreat of the glacier opened a channel through the St. Lawrence, and the Mohawk spillway also dried up.[95]

The Lewiston spillway had another advantage over its eastern rival: before the last glaciation, a river about as large as the modern Niagara flowed

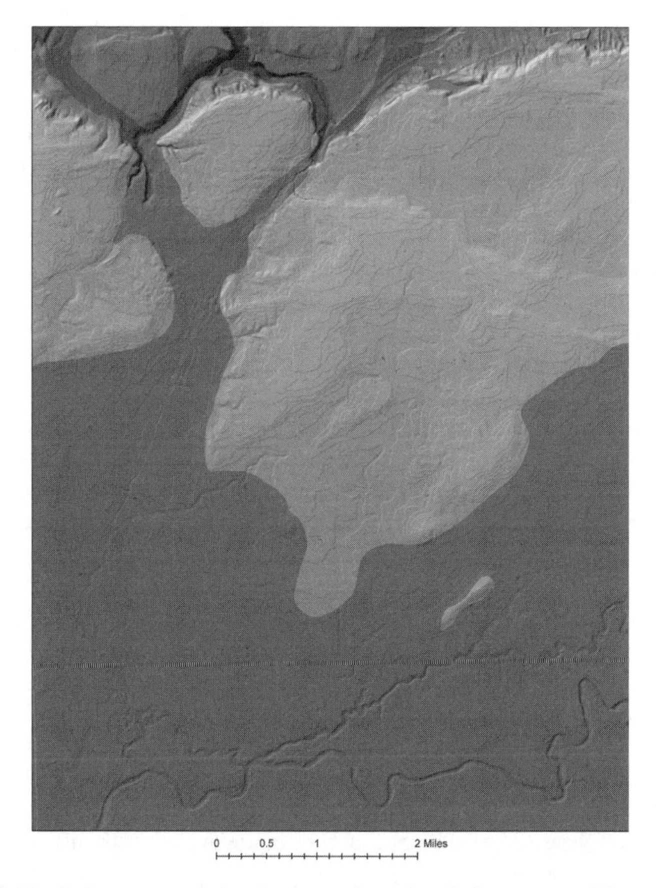

Figure 2.12. Delineation of the Lockport branch of the ancient Niagara River superimposed on a digital elevation model (United States Geological Survey). Adapted from John P. D'Agostino, "Lake Tonawanda: History and Development," MA thesis, University of Buffalo, February, 1958. Map by Cherin Abdel Samie and Yasser Ayad.

in the same general course but plunged over the escarpment about two miles west of Lewiston. It eventually eroded a wide and deep gorge about three miles into the escarpment. The last glacier entirely buried this gorge, but the interglacial river had created a slight trough even above its cataract that would help channel the flow after the glacial retreat.[96] Investigations by Parker Calkin and Carlton Brett have revealed that "major spillway cutting at Lockport" also "occurred prior to the last ice advance," so the interglacial Niagara Falls had a twin at this site as well.[97] The Mohawk Valley, similarly, saw fluvial erosion prior to the work of its most recent spillway.

We can specify the birth of Niagara Falls as the initiation, about 12,300 years ago, of the flow of Lake Erie water over the escarpment, but that Niagara was just the latest falls to cut the escarpment in that vicinity. At Lockport, the reestablished channels and waterfalls occupied exactly the same sites as their interglacial predecessors. Glaciers advance and retreat; the land sinks and heaves. In the geological imagination, the earth is as seething as the sea. Yet those little tiltings and changes in flow can have enormous significance in human affairs.

In his "Private Canal Journal" of 1810, DeWitt Clinton noted that there were many ancient forts south of the Niagara Escarpment but almost none to its north and concluded that "the ridge was the ancient boundary of Lake Ontario." He also reported that "some suppose that Lake Erie formerly discharged itself by the Tonnewanta Valley into the Genesee River."[98] Progressive middle-class Americans like Clinton believed that they were at the forefront of modernity. With the groundbreaking work of the Scottish geologist James Hutton, and the widely read work of John Playfair, a dynamic sense of the earth itself was beginning to evolve.[99] To some extent, Clinton seems to have been influenced by this new thinking. What, after all, was the natural way for the waters of the interior to flow? Downhill surely, but changing circumstances—ice sheets, uplift, erosion, and indeed human intervention—could alter the flow. If there is no "natural" state of things, then interventions of art become hard to distinguish from nonanthropogenic change. The American Revolution had convinced many that humans could act in history, that change need not equal desecration. People like Clinton were hardly loose from their moorings, but they were deeply preoccupied with—and unsettled about—the proper boundary between art and nature.

Between May and November of 1820, David Thomas and his survey crew worked exclusively on the proposed northern route from the Genesee to Lake Erie, although the commissioners had yet to make an official announcement in favor of this route. Since the Geddes and Peacock surveys of 1816, the commissioners had kept open both possibilities, partly out of deference to Joseph Ellicott and partly because the northern route, in historian Noble Whitford's words, "seemed to stagger the canal commissioners and the engineers, on account of the heavy rock cutting at Lockport."[100] The commissioners finally made public their decision in the report to the legislature of March 12, 1821.[101]

Although it would not be announced officially until the following year,[102] Wright and Thomas had already convinced the commissioners to accept the new route through the eastern gorge at the Mountain Ridge. By the spring of 1821, the route of the canal through western New York had been settled, leaving only the question of whether the canal should terminate at the village of Black Rock on the Niagara River or continue to the Lake Erie port of Buffalo. After a decade of investigation, debate, and political wrangling, contingencies had been reduced to apparent certainties. Now what had been marked on surveys and maps would have to be shaped in earth and stone. The ancient spillway had blazed a trail through the ridge: now humans, with their animals and tools, would have to transform that trail into an artificial river. The engineers and commissioners knew that overcoming the Mountain Ridge would be the principal construction challenge west of the Genesee River. They knew it would be difficult, but no one in 1821 imagined just how difficult.

3

Cutting Lockport

Laborers, Citizens, and Five Years at the Mountain Ridge

> [T]he canal that runs through it . . . seemed an object of utter ineffable beauty. (It must be remembered that beauty does not mean mere prettiness but something more brutal, possessed of the power to rend one's heart.)
>
> —Joyce Carol Oates, *(Woman) Writer: Occasions and Opportunities*

Overcoming the Mountain Ridge, the Erie Canal's final barrier, became a key element in progressive narratives that developed during and after the canal's construction. In chapter 4, I will investigate attempts to define the meaning of the events at the Mountain Ridge and to weave that meaning into larger narratives about progress, America, continental power, and the relation of art to nature. Here, I will focus on the construction period itself, paying particular attention to the stories that have been lost, in part because they were difficult to fit into those narratives. To do so, it will be necessary to analyze a sequence of events, and hence to create a narrative of my own, or perhaps a counter-narrative. There is an inherent tension in this chapter. On the one hand, there is the need to examine the construction process diachronically, piecing together what the surviving evidence can tell us, while recognizing the gaps, uncertainties, and ambiguities that persist. On the other hand, there is a need to step back from the unfolding events in order to analyze the remarkable social dynamics that were emerging between the so-called citizens and the common laborers thrown together for five long years at this remote site. The Mountain Ridge offers us a snapshot of developing class formation during a moment of transition to a market economy. While cutting a channel through rock, the people at this site were also creating a "deep cut" that bifurcated their social world. Without diminishing their accomplishments, the most important lost story I want to recover is the violence involved in both kinds of cutting. Like Joyce Carol Oates, I can still see the canal at Lockport as an object of beauty—not only because of its origin in human imagination, intelligence, and perseverance, but also because of its origin in violence, for it is this that gives it "the power to rend one's heart."[1] Rather than submerging the tension between the chapter's two goals, I will make it visible by interrupting the chronological structure in the year 1824. I will then examine the violence of the social bifurcation at the Mountain Ridge before returning to complete the narrative of construction.

Prelude

In their report of 1822, the canal commissioners discussed in great detail the scheme—conceived by Geddes and given final form by Thomas—to carry the canal through the Mountain Ridge at the level of Lake Erie. There were two main challenges. "Through that ridge," the commissioners warned, "occurs the most extensive deep cutting, which we have anywhere to encounter." At the brow of the ridge, where a small stream ran through a narrow valley before falling into the eastern gorge (visible on Geddes's 1816 map of the two gorges, fig. 2.8), only thirteen feet of excavation was needed. From this point, "the ground rises, as the line extends southerly, for a mile and a half, where the excavation is required to be thirty feet six inches." A little more than seven miles south of the ridge, where the canal joins Tonawanta Creek, twelve feet of cutting would be necessary. At the northern end of the cut, the commissioners cautioned, "the excavation for about three miles, must be through limestone rock."[2] The ancient spillway had blazed a trail, but left much to be done.

In addition to this deep cutting, a second challenge was to construct locks to carry the canal sixty feet down from the level of Lake Erie to the canal's longest level, extending from the foot of the Mountain Ridge eastward to Rochester and continuing over the Genesee River by means of the famous 802-foot aqueduct.[3] Geddes had proposed a staircase of eight combined (or consecutive) locks to climb the escarpment. Although Thomas changed the location of the locks, he retained Geddes's idea of combined locks. "In almost all cases," the commissioners reported,

> the locks of a canal should be separated by spaces, or pound reaches, of at least forty rods in length, in order to save water, and to prevent injurious delays in the passage of boats. At the brow of the mountain ridge, from the almost perpendicular descent of the whole sixty feet, such a separation would be impractical, without enormous expense.

In almost every important European canal, conserving water had been a crucial issue, but here Lake Erie provided an "inexhaustible fountain."[4]

The problem of "injurious delays" could not so easily be dismissed. Combined locks cause delays because, while a boat is descending, boats wishing to ascend must wait until it passes through the entire flight before they can enter the first lock. The staircase locks of Europe, the precedents of the Lockport locks, were bedeviled by this problem. The magnificent Sept Ecluses (Seven Locks) at Rogny on the Canal du Briare (1604–1642) that joined the Seine and the Loire rose twenty meters. The Ecluses de Fonserannes on the Canal du Midi (1667–1681) that connected the Mediterranean and the Atlantic climbed twenty-one and a half meters.[5] David Thomas invented an entirely new solution. In a letter to Benjamin Wright on December 25, 1820, he proposed two side-by-side flights of locks, one for ascending and one for descending.[6] Nathan S. Roberts, the engineer who designed and directed the construction of the Lockport locks, has usually been awarded most of the credit for their renown,[7] but Geddes and particularly Thomas provided the concepts that he implemented.

Before the Geddes survey of 1816, only a few settlers had begun to clear land in the vicinity of the future locks. They clustered along the Niagara (or Lewiston-Queenston) Road, about two miles east of the locks, and on the Ridge Road, three and a half miles to the north.[8] Beginning in 1816, a new group of settlers began to arrive at the Mountain Ridge site specifically because of its projected importance to the success of the Erie Canal. Among these were a group of relatives, all named Comstock, who purchased land along the presumed canal route. They came most recently from the vicinity of Canandaigua in Ontario County, the last outpost of commercial connection to the east. Zeno Comstock, learning of Geddes's recommendations, sold land near the eastern gorge to his brothers and invested at the head of the western gorge. He was so disheartened when Thomas altered the route that he immediately departed.[9] East of the Transit Meridian, the Comstocks controlled most of what was to become the village of Lockport. Eseck Brown purchased the adjacent land west of the meridian.

As work on the canal was about to commence in the spring of 1821, Dr. Isaac W. Smith arrived, also from Ontario County, just in time to attend an impromptu meeting. Jesse P. Haines, who was surveying village lots on land owned by Jared and Darius Comstock and Otis Hathaway, convened a meeting with them, Nathan Comstock, and the new physician to discuss a name for the place. Haines suggested Locksborough. Isaac Smith proposed Lockport, which, after more discussion, they agreed upon. All of these men were Quakers, as were the majority of the village population that had reached nearly one hundred.[10] Although many of them had previously experienced frontier conditions, they were literate people of some means. The first place of worship in the village was the Quaker Meeting House, erected in 1819. Canal engineer David Thomas was another Ontario County Quaker. Joseph Ellicott was a Pennsylvania Quaker, as were Jesse Haines and a few other Lockport settlers.

Eseck Brown, another Quaker, opened the village's first tavern on his land west of the Transit Meridian on the path that followed the top of the escarpment from the Niagara Road westward. Jesse Haines designated this path Main Street (see fig. 2.10). Brown hung a wooden sign, naming his tavern the Lockport Hotel.[11]

Morris H. Tucker, an aspiring merchant who arrived in the summer of 1821, observed that "there were some half dozen families residing in unfinished log houses, and a number of men were building small houses, expecting to bring their families as soon as they could finish the tenements." Altogether there were about one hundred acres of cleared land, but already "Jared Comstock and Eseck Brown were selling village lots on Main Street." Almost nothing was cleared north of the future locks, and "the land from the head of the locks, around the ravine, embracing all of Lower Town . . . was a dense forest."[12] Throughout the coming decades, travelers and local citizens would measure the remarkable growth of Lockport against a presumed wild baseline. The obvious persistence of the region's previous inhabitants—there were three Iroquois reservations within twenty-five miles—provided for most not a contradiction but a reinforcement of this wild baseline story, despite the fact that local Quakers were on remarkably friendly terms with the Tuscaroras and Senecas.[13]

Edna Deane Smith, the wife of Dr. Smith, published her memoir, "Recollections of an Early Settler," as a newspaper series in 1873. Although perhaps colored by nostalgia, it provides a vivid image of the early scene.

> From the present aspect of the basin below the locks, crowded as it now is with mills and manufactories, one can scarcely imagine how wild and picturesque an appearance it presented in early times.
>
> Before the canal was commenced there was a pond in the ravine fed by numerous springs and rivulets flowing from the hillside. The slopes and sides of the surrounding precipices were covered with trees and shrubs, mossy rocks, ferns and wild flowers, reaching to the water's edge, which was gay with ivy and cowslips. Late in the season the foliage was completely beautiful, owing to the profusion of sumach [*sic*] and maples, whose autumnal brilliance of crimson and gold mingled with the darker green and brown of oaks and evergreens made all the hillside like a gorgeous carpet.[14]

Although Aunt Edna, as she was affectionately known, was respected by almost everyone in the village, her views on a number of things—as we shall see—were hardly common in Lockport. Most of her neighbors displayed an unrelenting urge to combat the wildness of the site.

Morris Tucker reports that, in the summer of 1821, "the rattle snakes were so numerous that they occasioned much alarm to the villagers."[15] They congregated in the "rocky borders and sides of the ravine or basin," according to Edna Smith. "Everyone considered it a duty to dispose of as many of these reptiles as possible, and a young man would return from a ramble in the woods dragging a dead rattlesnake behind him, or bringing in the rattles as an Indian warrior does his scalps as trophies."[16] One man, known as Rattlesnake John, actually earned a living by killing rattlesnakes and selling their oil as a medicine; he did his hunting in the basin and the lower gorge of Eighteenmile Creek, an area still known as Rattlesnake Hill.[17] Apparently all the rattlers were gone within about ten years.[18]

The first wave of settlers prosecuted a similar war on trees. Overlooking the site of the locks stood a colossal black walnut tree. Some suggested it be spared until the completion of the canal when it could be made into a gigantic canoe and taken on a celebratory cruise through the locks. Instead it was felled, buried beneath piles of stone, and forgotten for years.[19] "There was so much public spirit," Mrs. Smith informs us, "that evenings, the clerks and young professional men, would all turn out with their coats off, and roll logs, burn stumps, fill mud holes and kill snakes."[20]

The canal commissioners, in their 1822 report, also noted the difficulties presented by the site.

> When the ridge sections were contracted for, their entire extent, except a small part of the line at the northern extremity [cleared by Eseck Brown and the Comstocks], lay through an unsettled wilderness, very heavily timbered, and much of it under water. It was not therefore, without great difficulty at first, that the contractors erected their houses for shelter, opened roads for the delivery of their provisions and tools, and collected their hands.[21]

Most of those who came to Lockport before 1824 traveled from Rochester on the Ridge Road that followed the well-drained ridge of an ancient beach and paralleled Lake Ontario north of the Mountain Ridge. Because there was no road from there directly north to Lockport until 1823, travelers either tramped through the woods or zigzagged via the Niagara Road to Cold Spring and then eastward on Main Street. A daily stage began to ply this route in the summer of 1822.[22] The isolation that early white settlers faced in western New York is captured in Orasmus Turner's picture of the first stage of settlement from his *Pioneer History of the Holland Purchase* (fig. 3.1).[23] Turner was the long-serving editor of Lockport's first newspaper.

Figure 3.1. "The Pioneer Settler Upon the Holland Land Purchase, and His Progress," a poem introducing a series of engravings. This view, "No. 1, introduces the pioneer" who "has plunged within the forest, there to plant / His destiny." From Orasmus Turner's *Pioneer History of the Holland Land Purchase* (Buffalo: Jewett, Thomas and Company, 1849), following p. 562. Courtesy of the New York State Library.

1821

In May of 1821, David Thomas entered into contracts for the canal line between Tonawanta Creek and the Mountain Ridge, including the locks. To facilitate the discreteness of contracts, engineers generally divided contractors' sections at intersections with watercourses, allowing independent draining and testing.[24] Such a scheme was impossible in this long rock excavation known as the Deep Cut. The commissioners decided to make special arrangements.

> After mature deliberation the whole seven miles was divided into four separate contracts. The two sections commanding the drains at each end, are required to be excavated one season before the other sections. And it is specially stipulated, that the latter shall have the privilege at all times, of draining through the former.[25]

Drainage in the old bed of Lake Tonawanda would prove a recurrent problem, and each of the four cutting sections would prove too much for individual contractors (fig. 3.2).

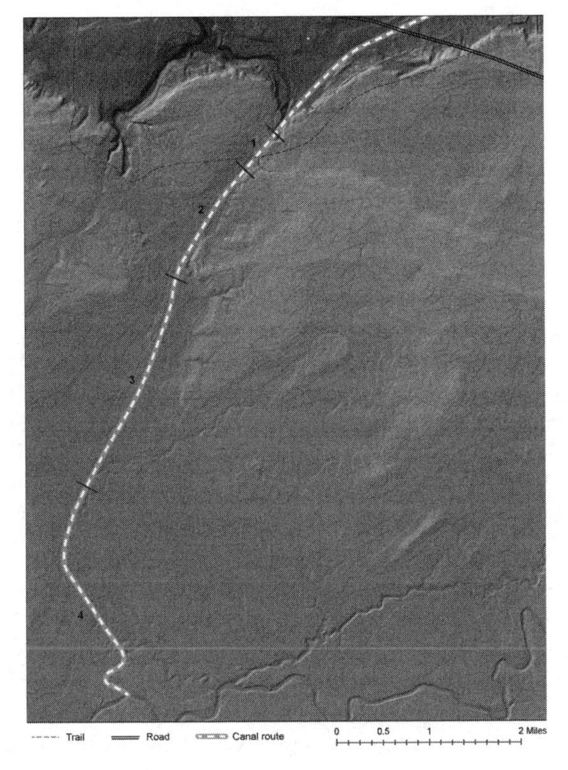

Figure 3.2. Mountain Ridge contracting sections of 1821 overlain on digital elevation model (United States Geological Survey). Map by Cherin Abdel Samie and Yasser Ayad.

The most northerly section was only a half mile long, but it included the construction of the double combined locks. Local landowners Joseph Comstock and Otis Hathaway secured the contract for section 2; although the second shortest of the four sections, it would require the heaviest rock excavation, the experience of which would soon lead to a crisis in the contracting system at the Mountain Ridge. Because of the southward tilt of the Lockport dolomite, section 4 and the southerly part of section 3 required less rock excavation. The four contracts, all entered on May 21, 1821, provided advances of $2,000 to $3,000, and stipulated that the commissioners would pay only twenty-five cents per cubic foot of rock excavated, and that contractors were expected to complete their sections between October 1, 1822, and September 1, 1823.[26] A table of the Mountain Ridge contracting sections and their year-by-year subdivisions and changes is provided in the appendix to this volume.

The contractors now needed a veritable army of common laborers. Local historian Clarence O. Lewis writes that

> in January 1821 there appeared in most newspapers in New York City advertisements inserted by the contractors of Niagara County stating that they needed canal workers at "Twelve Dollars a month and found [room and board]." About twelve hundred men, largely Irishmen were hired by the Niagara County contractors.[27]

From other sources we know that at least one thousand and eventually as many as fifteen hundred laborers descended on the Mountain Ridge site.[28] Job seekers from easterly sections of the canal did not merely rely on newspaper advertisements: word of mouth and letters mailed back to family and friends made them wary of the sometimes overinflated estimates of the number of workers needed.[29]

The early 1820s was a period of general labor shortage in the United States, and there was almost no surplus farm labor west of Rochester. Apart from workers already on the canal line, the impoverished residents of New York City's Five Points district were the obvious labor pool. In the republic's poorest neighborhood, free blacks, Irish, and others coexisted in a dense, squalid, and colorful environment.[30] Most people with steady employment would not consider traveling to what they could only consider a distant wilderness for indefinite seasonal work. Only the dregs of the wage-labor workforce—immigrants (largely Irish and Welsh) and blacks—could find such a prospect attractive. Twelve dollars a month was slightly more than established laborers' rates in New York City.[31] Because of labor shortages, North American canal wages were marginally higher than those in Europe. As Peter Way points out, immigrant workers were available in the United States because they had been dissevered from the land. People willing to come to an isolated, densely forested, incipient frontier village were obviously making a clean break from the immemorial world of agriculture where one's means of subsistence was always at least partly in one's own hands. Here there would be near total dependence on selling their time and effort for cash—now the only means of acquiring the necessities for survival.

In a sense, canal work represented a transition between traditional paternal work arrangements and the subsequent free labor of the industrial workforce. Contractors provided food, shelter, and whiskey, and worked alongside their men and boys. New labor and class dynamics, however, were rapidly evolving. A canal contractor with a hundred men to oversee could hardly sit them down at the family table as artisans and shop owners traditionally had done. Paternal arrangements could ameliorate exploitation, but they also facilitated control. On the Erie Canal, we can see the development of class antagonisms characteristic of the coming industrial economy. Since contractors generally were paid per cubic foot of excavation or embankment, they had a strong incentive to exploit their workers. Moreover, the size of their individual workforces removed some of the disincentives to exploitation that the intimacy of traditional arrangements had fostered. The sheer size of the total workforce that toiled for five years at the Mountain Ridge was also something new. The middle-class contingent of Yankee and Yorker engineers, surveyors, contractors, preachers, and merchants was confronted by an army of largely immigrant common laborers who were not only cut loose from the land but also from direct middle-class control.

The workforce of laborers, contractors, and engineers immediately began to dramatically transform the landscape of the Mountain Ridge. During the first season of work at Lockport, assistant engineers oversaw the contractors and reported to David Thomas, chief engineer for the section west of Rochester. Thomas, in turn, reported to the chief engineer for the entire canal, Benjamin Wright. Canal Commissioner Myron Holley topped this hierarchy and took an active interest in the practical issues involved in overcoming the Mountain Ridge.

Apart from engineers, contractors, and land speculators, other middle-class people began to arrive at the Mountain Ridge in 1821, most no doubt attracted by the economic opportunities surrounding the great construction project there. Canandaigua served not only as a staging area but also as a center of information about canal work to the west. It was common to use personal connections and insider information in ways that today might be considered inappropriate. In June of 1821, for example, Canal Commissioner Myron Holley wrote from Canandaigua to his father in Connecticut that his father-in-law, William House, "will set out tomorrow morning for New York to lay in a store of goods, which he intends to sell out on the canal-line, in Niagara Country, on the Mountain Ridge."[32]

Lockport's first merchant was Morris H. Tucker. When he arrived in the summer of 1821, the contractors for section 1 were clearing and grubbing from the head of the locks westward. Tucker brought with him an "old store of goods" from Batavia, which he stored at Eseck Brown's tavern.

> When it became known to the women that I had good tea stored at Brown's, no excuse would answer, have it they would, and I was obliged to open shop. In two or three weeks I moved my goods into a new framed store, an imposing building at the time, twenty-two feet square, a story and a half high. Here for several weeks I had no opposition in trade. Soon however House & Boughton got their new store finished and Lilbeus Fish brought on goods from Batavia, and Lockport began to be a place of no little importance.[33]

Dr. Isaac Smith, who had arrived in April of 1821, just in time to suggest the town's name, brought his wife Edna to the Mountain Ridge in July. Like so many settlers in the vicinity of Canandaigua, Edna's family had moved from Massachusetts to Ontario County less than a decade earlier. She married Dr. Smith in 1815 and became a Quaker.[34] Although there were "a few Irish shanties," the Smiths built "the first house, in what was then the village proper. It was of hewn logs and boasted *three* rooms on the ground floor, quite a luxury for a new country where one was the rule."[35] According to one contemporary, "this dwelling was neat in every respect, and in perfect keeping with Dr. Isaac and Aunt Edna. It was built of small logs, nearly equal in size, with the bark peeled off, and whitewashed both inside and out."[36]

Joseph Landon of Buffalo built what Aunt Edna called the "first public house." He waited until the building was well underway before bringing his family to town in August of 1821. It was an extensive log structure that became known as the Cottage. Many of the canal engineers and contractors boarded there.[37]

We have no direct reports of laborers' experience during the first season of work, but Edna Smith wrote that almost as soon as her house was complete, "three men were blown up and seriously injured by a blast, and they were brought to us to be nursed and taken care of. We had no place for them but a low loft or chamber, where two were put, the other in a cot in our common room."[38]

There were hints of an impending crisis in the canal commissioners' annual report of February 1822. Although the contractors suffered a number of setbacks, the commissioners wrote,

> still, they went on with resolution, increasing their means as they supposed they could apply them most advantageously: and although from the difficulties attending their commencement, and the long course of unfavorable weather in the fall, they have not performed so much as we hoped they would, they are continuing their application of their labor though the winter and promise a great increase in hands in the spring.[39]

In the 1821 contracts, the commissioners had required excavation only to the surface level of Lake Erie, hoping that local streams might provide enough water to avoid the additional four feet of rock excavation. Although they knew this was unlikely, they thought that adding on the additional work later would be easy enough. By 1822, the commissioners knew they would need the extra depth, but they decided to contract for this work when the original contracts expired in September of 1823, and they concluded that the extra work would be completed a year from that date.[40]

1822

In March of 1822, the canal commissioners reappointed David Thomas "chief engineer of the line from Rochester to Buffalo."[41] The commissioners also decided to convene a special meeting at Buffalo in June of 1822 to obtain "further local knowledge" concerning the controversy surrounding the canal's western terminus.[42] The rivalry between the villages of Buffalo, on Lake Erie, and Black Rock, on the Niagara River, had become deeply embroiled in state politics. Supporting Black Rock's case was one of DeWitt Clinton's most powerful enemies, Peter B. Porter. Joseph Ellicott was a key supporter of Buffalo. Ellicott's elaborate radial street plan for Buffalo, which has often been compared to that of Washington where he had worked as a surveyor, is one indication of his aspiration for the site's future.[43]

Increasingly, partisan politics divided commissioners and engineers as well, with Clinton, Myron Holley, and David Thomas supporting Buffalo, while William Bouck, James Geddes, Nathan Roberts, and David Bates sided with Porter and the so-called Bucktails, arguing for Black Rock. Geddes and Thomas were particularly bitter rivals.[44] A month prior to the Buffalo meeting, the commissioners had declared that the question of the canal's terminus should be made on technical rather than political grounds and decided to pass the responsibility to the canal's leading engineers. It happened that Wright, Geddes, Thomas, Roberts, and White had all been in Albany, and all but Geddes agreed to a resolution to extend the canal to Buffalo.[45]

At the special Buffalo meeting, representatives of both villages made presentations to the commissioners, who decided to terminate the canal at Buffalo but also encouraged both villages to begin to develop harbor facilities with the prospect of state support for harbor development if, after a year, the commissioners judged it wise. "So long as construction of the deep cutting and locks at Lockport delayed the opening of the whole line," historian Ronald Shaw argues, the commissioners "saw no reason to close the question."[46] Indeed the construction difficulties at the Mountain Ridge were an important subject at the Buffalo meeting. The most powerful argument in favor of a Buffalo terminus was that Lake Erie water could be taken at a higher level than at Black Rock, thus reducing the amount of excavation necessary at the Mountain Ridge.

The commissioner's assessment, in February of 1822, that the Mountain Ridge contractors "have not preformed as much as we hoped they would" was a great understatement. After a season of work, they had completed only about 6 percent of the estimated rock excavation in the "worst two miles" of the ridge. At that rate it would have taken seventeen years to complete just those two miles.[47] Progress during the spring of 1822 must have been equally disappointing. Nowhere on the canal line had engineers encountered rock like Lockport dolomite. "The rock is shelly," the commissioners reported, "and more difficult to remove, by blasting or otherwise, than was anticipated."[48] At the Buffalo meeting the commissioners apparently got an earful from the engineers. "And it was perfectly apparent," they concluded, "that the work at that place, where it was of the greatest importance to advance it with all possible celerity, would soon fail entirely if a new course was not adjusted."[49] Here was the commissioners' response:

> The work was to be conducted in the following manner: The contractors were to employ as many hands as the commissioners should require; and a minute account was to be kept, by an assistant engineer, of all the expenses necessarily incurred in the prosecution of the work. The assistant engineer was to be constantly on the contracts, and to inspect the whole course of operations.[50]

This arrangement effectively turned the contractors into overseers who were paid wages of about $3 a day. "Here was the first assumption of public works by the state on The Erie Canal," Ronald Shaw comments, "and perhaps the first such project in the history of the United States."[51] The commissioners attempted to keep the new system flexible by reserving the right to continue paying the contractors per yard of excavation. Instead of the $0.25 they had offered in 1821, they now hoped to excavate for $1 per yard. They actually paid more than this; Darius Comstock, for example, received $1.50 per yard for rock excavated in 1822. Great flexibility was built into this new scheme as all determinations of payments were to be left to the engineers.[52]

The commissioners also divided the "largest" Mountain Ridge section into three parts with separate contractors.[53] By "largest" they meant section 2, which had been under contract to Joseph Comstock and Otis Hathaway; it was actually the second shortest of the original four sections but it contained the heaviest rock cutting. Recognizing the enormity of their task, Thomas had given these contractors a second advance in 1821. Engineers now renumbered the six Mountain Ridge sections from north to south and, on June 15, issued new contracts reflecting the new system to the four original contractors and two new ones (fig. 3.3). The original contracts had required completion as early as October 1, 1822, but under the terms of the new contracts the date was changed to June 1, 1824. In addition, the commissioners advanced each contractor $2,500.[54]

It might seem that the crisis at the Mountain Ridge was good news for common laborers. Instead of the twelve dollars a month originally promised they generally earned seventy-five cents a day (potentially as much as twenty-two dollars a month).[55] As day laborers, however, they were

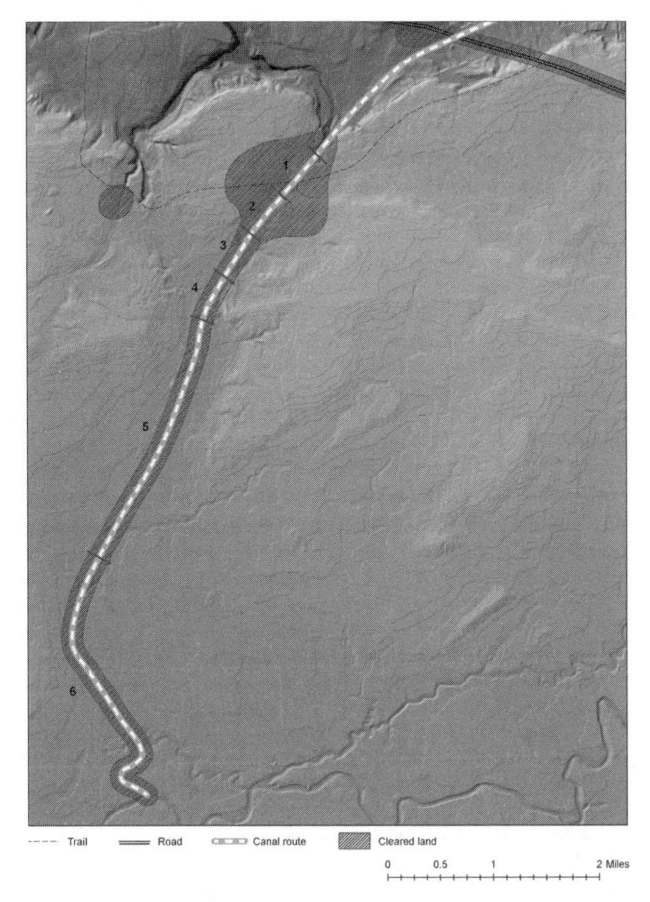

Figure 3.3. Mountain Ridge contracting sections of 1822, with cleared land indicated, overlain on digital elevation model (United States Geological Survey). Map by Cherin Abdel Samie and Yasser Ayad.

now responsible for their own food and shelter. The sense of panic among the commissioners and engineers probably meant that laborers were being pushed more forcefully to perform. Moreover, because the engineers could now dictate the number of workers contractors must oversee, the crew sizes increased, further eroding the moderating effects of paternal intimacy. As we shall see, there are indications that the common laborers on the Mountain Ridge were under considerable stress. In an additional response to the local labor shortage, the commissioners redirected workers from the sections just west of the Genesee River—deliberately slowing work there—because "it was desirable that a large portion of the laborers to be hired in the country, should be employed at Lockport, and in its vicinity."[56]

By the summer of 1822, a much larger workforce was engaged on the six contracts. The commissioners reported at the end of the season that under the new system, effective prices per yard of excavation varied "from about eighty cents to more than double that sum, giving an average not far from nine shillings and six pence" ($1.19),[57] notwithstanding Wright's estimation to Thomas that "any Limestone Rock can be excavated for 50 cents or under."[58] The commissioners concluded:

> Under this system, the work has been conducted with much more vigor than it was before, and the passage though The Mountain Ridge, which will doubtless be the last labor to be done in the construction of the Erie Canal, will be all cut and completed before the end of the year 1824.[59]

It is difficult to discern the basis for this optimistic prediction that the ridge sections (and the canal itself) would be done in two more years of work. During 1822, workers had excavated 38,226 cubic yards on the "worst two miles" of rock sections. That was 2.5 times the amount excavated in 1821, but even at that rate it would take another six and a half years to complete those two miles. Another hint of concern was the commissioners' decision to construct a feeder for the long level between Rochester and the Mountain Ridge by diverting the upper Tonawanta Creek into Oak Orchard Creek. They contracted this work "with a view of accomplishing the desirable object of extending the canal navigation as far west as Lockport, in the spring of 1824."[60] Given the fact that this feeder would become entirely redundant once Lake Erie water flowed through the Mountain Ridge, which the commissioners claimed was to occur before the end of 1824, did it make sense to incur this expense? Perhaps they suspected that the canal might be stalled at Lockport for many years. Maybe some even secretly despaired of ever overcoming the Mountain Ridge. Such a scenario played out in the 1830s when the Pennsylvania Mainline Canal stalled at the Allegheny Front, an obstacle it could surmount only by dismantling the canal boats and hauling them over it on tracks.

During the summer of 1822, as more engineers, contractors, and workers—as well as more resources—converged on Lockport, the village began to develop into a distinctly urban place. George Boughton's store became the first post office; until 1823 mail would arrive from the Ridge Road by horseback.[61] The old Seneca path that followed the top of the ridge west from Cold Spring was transformed into Main Street. A bridge carried this road across the canal line; it consisted of two log stringers with split logs laid across bark side up. It was wide enough to accommodate only one team. As might be expected in a village so dominated by Yankees, a log schoolhouse also opened this year.[62]

In 1821, Niagara County was divided at the Tonawanta Creek. Buffalo, the original county seat, now became the seat of Erie County. A fierce rivalry ensued between Lockport and Lewiston over which town would become the county seat of the reduced Niagara County. Lewiston, at the northern end of the portage around Niagara Falls, was by far the older settlement, and still served as the Niagara Frontier's principal Lake Ontario port. The first commissioners designated by the state to settle the question became deadlocked, delaying the decision until the summer of 1822. As a second set of commissioners were deliberating, Mrs. Edna Smith notes, Lockportians realized that Lewiston had the advantage of a newspaper "by which they could send out their side of the question all over the county." After learning that "the printer was not very well supported, some of the more enterprising Lockport citizens met and appointed a committee, consisting of Dr. Isaac Smith and Otis Hathaway to go to Lewiston and purchase the paper, press and printer."[63] The 1874 *Illustrated History of Niagara County* embellished the story:

> William Carney was engaged to transport the press and accompaniments. The venerable pioneer is still living to relate the difficulties under which he labored in the undertaking. The road was in a most wretched condition. Deep mud was prevalent in a partially frozen state [it was July], and such obstruction as logs, brush and stones were plentiful. With two yoke of strong oxen and a heavy lumber wagon he was enabled, after two days of diligence and hard labor [Edna Smith remembers it taking less than twelve hours], to accomplish the task. On his arrival the interested inhabitants gathered about to gaze on the scene as if a caravan was approaching. The advent of this old printing press marked an important era.[64]

By noon on the day the press and printer arrived, the committee "had a paper out on their side," Edna Smith recollects, "with fiery and convincing articles, and blazing with exclamation points." They called it the *Lockport Observatory*.

Mrs. Smith describes reports that, when the commissioners finally decided in July to make Lockport the county seat, Dr. Smith mounted his pony and proceeded "on a gallop the whole length of Main Street across the "Big Bridge" to the Washington House shouting the glad news at the top of his voice.[65] Mrs. Smith's memory was not perfect—the Washington House would not be built until 1823—but she was correct in pointing out that a great many people had confidence that Lockport would be more than a construction site. Although the village was only one year old, it not only had become the county seat but had established a school and a newspaper. Middle-class people like Isaac Smith, though he too had only arrived the previous year, were already deeply committed to the place. Indeed, Smith would live the rest of his life in Lockport. He also must have felt a strong sense of agency. Smith had taken the initiative to lure the newspaper and had also given Lockport its name. Laborers at the Mountain Ridge clearly had less agency, and in particular less freedom to choose where to live. Local historians have tended to give these workers short shift, perhaps partly because they seemed less committed to the place.

The contractors for the locks and rock section 1, Culver and Boughton, advertised in the *Lockport Observatory* on August 1, to purchase "pork, flour, whiskey, oats and hay for cash."[66] Before the arrival of the canal, the scattered farmers of the Niagara Frontier could find a market only for concentrated products like potash for the Quebec market. There was, as yet, little excess food production locally, but the growing cash market would soon fundamentally transform the lives of local farmers.[67]

In August of 1822, Orasmus Turner arrived from Palmyra, purchased the *Lockport Observatory*, and issued his first paper on September 26. He would remain an editor in Lockport until his death in 1855. In his *Pioneer History of the Holland Purchase* (1849) Turner recalls that, when he arrived, Lockport was "a rough and primitive village." It had "advanced considerably in one year . . . yet there were log heaps and huge piles of rocks in the principal streets. There were not over a dozen or fifteen frame buildings, and but one of stone."

> Culver and Maynard were clearing the timber from the slopes of the mountain, around the ravine, and excavating the first rock section. . . . the dense forest between Lockport and Tonawanda creek looked as if a hurricane had passed through it, leaving a narrow belt of fallen timber, excavated stone and earth; and that, to complete the ragged scene, log boarding houses and Irish shanties had been strung along the whole distance. The blasting of rocks was going on briskly, on that part of the canal located upon the village site; rocks were flying in all directions; framed buildings, and the roofs of log buildings were battered by them, and huge piles of stone lay upon both banks of the canal with a narrow opening to admit the passage of teams over a log bridge, on Main street.[68]

The deforested portion of the central part of Niagara County resembled a tadpole with its head at Lockport and its tail following the line of the canal southwestward (fig. 3.3).

Turner's *Lockport Observatory* maintained an upbeat stance on the progress of excavation. In October 1822 the paper reported that "the excavating of the canal through the Mountain Ridge at this place, is progressing in a manner highly advantageous to the state and creditable to the contractors. There are now employed on the canal in and near this village about eleven hundred men; and the amount of money daily expended by the contractors is rising of $1,000."[69]

On December 22, 1822, Lyman A Spalding came to Lockport from Canandaigua to open a store. Although only twenty-two years old at the time, Spalding would make Lockport home for the remainder of his eighty-five years. As an entrepreneur, industrialist, journalist, and abolitionist, he would quickly become an influential citizen. Spalding was also among the most articulate of early settlers in western New York.[70] Like most of the early nineteenth-century settlers in the former Iroquois lands, Spalding had family roots in New England. His father, Erastus Spalding "was a very prominent man in Cayuga County" who gave young Lyman his first gun at age seven. In 1810, Erastus moved his family to the mouth of the Genesee River, where they narrowly escaped a British naval attack in 1813. Religion was of central importance to the Spaldings, but like so many others in central New York, their religious commitments were in turmoil. Originally a Presbyterian, Erastus and his entire family joined the Society of Friends in 1814.[71]

Lyman Spalding managed to complete about six years of schooling. Beginning at age 11, he tried his hand at clerking, bookkeeping, legal studies, and farming. By 1820, he had accumulated enough resources to join in partnership with Canandaigua merchant William Cromwell. That same year Spalding began to travel to New York to buy goods for the store. Spalding was attuned to news of commercial opportunities along the canal line.[72] "I had seen Addison Comstock," he writes, "son of Davis C., who was one of a company having a large contract for cutting the canal through the mountain ridge, and employed a large number of men, and he encouraged the idea."[73] As the end of 1822 approached, Spalding observed, "we had purchased a larger stock of goods than we feared we could dispose of in Canandaigua, and we concluded it best to send part of it to Lockport." Cromwell visited Lockport first and secured a building for the new store. They sent "as many goods as two three-horse-teams would draw" while Spalding "took the stage" on the Ridge Road and "had to walk from Wright Corners to Lockport . . . carrying my valise." The building his partner had chosen "was nothing but a square of hewn logs about seven feet high, more like a hog pen than a building." He found a small house, "hired the occupants to vacate," and opened for business. Sales were so brisk that the partners brought the rest of their goods to Lockport and closed the Canandaigua store.

Spalding recalled the scene in Lockport in late 1822:

> At this time, Lockport was principally log buildings, and very few of them. Work on the lock pits had not progressed but little, and the excavation probably about eight feet below the surface, and for about three miles of the line of the canal, large numbers of men were blasting and excavating rock. The explosions (cannonading) were almost continuous during the day. The buildings within reach of the falling stones were protected by trunks of trees, about six inches at the butt, and long enough when set around the buildings at about 45 degrees, to meet at the top. Any stone falling could not go through the roof, though occasionally a stone would drop through. In walking near the work, on hearing an explosion, one would look upward so as to dodge any falling stones. During the work, several men were killed by falling stones and carelessness in blasting.[74]

In a separate description, Spalding provided more detail:

> The atmosphere was murky with the smoke of burning powder. There was the din of battle and yet but the peaceful pursuit of enterprise overcoming the most formidable barrier to the construction of The Erie Canal. Beyond the deep rock cut through low wet, heavy timbered land the track of the canal had been grubbed. There were uprooted trees, slashings, log shanties, a succession of log cabins, boarding houses, and groceries.[75]

On December 7, 1822, just two weeks before Spalding's arrival, Samuel Jennings placed an advertisement in the new newspaper for an establishment that he hoped would appeal to the aspirations of Lockport's middle class: a "public house for the reception of ladies and gentlemen. From his long acquaintance with the business of entertaining genteel company. . . . He keeps on hand at all times the choicest of liquors; and will spare no pains to afford comfort and satisfaction to those who favor him with their company."[76] Perhaps the rough frontier conditions, the presence of Indians, and especially the huge, largely foreign born, corps of common laborers, made the accoutrements of middle-class distinction seem more necessary.

We have no statements—corresponding to those of Jennings, Spalding, Turner, and Mrs. Smith—that express the laborers' perspective on the situation at the Mountain Ridge. But words are not the only means of expression. The violent events of late December 1822 indicate that laborers' experience at the Mountain Ridge was a world apart from that of their middle-class counterparts. A disturbance, which began when a fight broke out at a tavern, quickly turned into a battle between canal workers and other local citizens. The Rochester *Telegraph* reported the story under the heading

"Riot at Lockport." In a condescending tone, the *Telegraph* identified the source of the violence as the laborers "getting fairly into their cups" while "keeping Christmas Eve." The "riot" "excited considerable uneasiness among the citizens." The "rioters," the *Telegraph* reported, "attempted to raze several buildings. Stones flew as thick as blackberries, and bludgeons were brandishing in every direction. Two persons (citizens) were mortally wounded, and several others very seriously injured."[77] One man had a stone embedded in his fractured skull that "penetrated to the brain." The nearest surgeon was in Batavia, but before he was able to get to Lockport the stone was "dug out with a jack-knife."[78] After the smoke cleared on Christmas morning, fourteen laborers had been arrested, and another twenty fled to Canada.[79] A large group of workers then gathered and attempted to rescue the detainees. In response, the sheriff called out Captain Howell's rifle company "for the protection of the court."

This was a short-lived intifada, facilitated by an abundance of very hard rocks of every dimension. About three weeks later, the county court, which had not yet moved from Lewiston, indicted eight men—most with Irish surnames—for the murder of John Jennings, "a victim of the Lockport Riot."[80] Ten months later, the trials began in Lockport. They were presided over by Judge William B. Rochester, who had traveled by coach from the village that bore his name.[81] Most of the men were quickly convicted of manslaughter and sentenced to several years in the state penitentiary. One of the defendants, James Kelly, managed a hung jury and gained an acquittal in the retrial.[82] Within three weeks of Kelly's acquittal, the *Niagara Sentinel* (Lewiston's new paper) was advertising a fifty-two-page pamphlet called "Report of The several trials, held at Lockport, for the alleged murder of John Jennings in the unhappy Riot of 24th of December 1822." The announcement claimed the pamphlet was of more than local interest because it recounted "the arguments of some of the ablest lawyers in this state, on the several points of riot, murder, manslaughter, and justifiable homicide."[83]

In a later article, the *Telegraph* said that, after the disturbance had begun, it "soon changed from a pitched battle to a riot."[84] It is hardly surprising that the court's response should basically support the "citizens" interpretation of the events. Although the term "pitched battle" seems to indicate considerable violence on both sides, all of the indictments were on one side. It is unfortunate that neither the court records nor the fifty-two- page pamphlet have survived; the discussions of "justifiable homicide" might be particularly revealing. Laborers did not fit into the category "citizens." Like the Tuscaroras and Senecas, they had no standing. Yet, apart from these remaining Iroquois, *there were no locals here*, no established community for the laborers to disrupt. Both canal workers and "citizens" had come at about the same time. There was no precanal Lockport; as historian Blake McKelvey put it, at Lockport the Erie Canal "achieved immaculate conception."[85] Taken together, the press accounts suggest that violence, drunkenness, and irrationality were inherent characteristics of laborers, perhaps imported from the Old World. Nowhere is there even a hint that the conditions under which laborers were operating—their current experience at the Mountain Ridge—had anything to do with their behavior. We do know that when the powder keg of resentment exploded on December 22, 1822, workers were willing to defy the local establishment and even raze buildings. Their immediate world was one in which they had little stake.

1823

Near the end of their report of February 1823, the canal commissioners made a surprising admission. Among a long list of reasons why the canal would cost over a million dollars more than originally estimated, they noted that some of the alterations to the canal line increased rather than deceased expenses, "particularly that at The Mountain Ridge."

Although "at that place, the alteration shortened the line more than two miles," it also meant "near a mile more of rock than would be found on the original line." The commissioners also "confessed" that "the expense of excavating the rock is much greater than we expected. This unforeseen expense will probably amount to $200,000."[86] Thomas's decision to alter Geddes's line through the Mountain Ridge, although entirely rational given the information available to him, was a costly mistake. On "rough examination" it had seemed that Geddes's route through the western gorge would be vastly more difficult and expensive. Although Thomas's route was shorter, it saved neither expense nor time. After two full seasons of work, it was too late to undo the change. Lockport might be in the wrong place, but the inertia of events meant that it was here to stay.

Given the longstanding animosity between Thomas and Geddes, one might speculate that his decision at the Mountain Ridge was politically motivated. Indeed, one might farther speculate that, since Thomas was a Quaker, he probably attended the log meetinghouse erected in 1819 by Hathaway, the Comstocks, and the small group of Quaker settlers who had arrived just before his surveys. We know that the Comstocks purchased all the land east of the Transit Meridian that included the future site of the locks. Perhaps they somehow learned from Thomas that the Geddes plan of a route though the western gorge was being reconsidered. We know Thomas resurveyed both routes before the end of 1820. He could hardly have prevented the Comstocks from observing his crew at work on the eastern route. If most of the Comstock brothers guessed that Thomas would choose the eastern gorge, one brother, Zeno, guessed wrong and invested in land near the western gorge. It is probable, therefore, that Thomas acted professionally and proffered no insider information to the local Quakers.

What is striking, in retrospect, is that nearly everyone involved in planning the western section of the canal focused so intently on the openly political battle between Black Rock and Buffalo that they hardly considered a question of much more practical significance: which route should be taken through the Mountain Ridge? Had these engineers and commissioners chosen the western gorge the canal would not only have been less expensive but probably would have been completed a year earlier.

Because the question of the canal's terminus so deeply divided both the commissioners and the engineers, it would eventually have direct effect on the work at the Mountain Ridge. The terminus decision had many twists and turns. When the commissioners visited Black Rock in June of 1823, they found the Black Rock harbor experiment an apparent success. The commissioners agreed, with Clinton alone in opposition, to fund the construction of Black Rock harbor and to make that port the canal's principal transfer point with Lake Erie for both freight and passengers. Engineers created the harbor by running a berm from the Niagara River bank to the north end of Squaw Island. The canal entered directly into the harbor, but then an extension canal linked this harbor with Buffalo. At the mouth of Tonawanta Creek, a lock provided a third entrance to the canal. A final reversal came in 1825 when the legislature voted to allow the commissioners to build an "independent canal" directly to Buffalo, bypassing Black Rock harbor.[87] When damage by ice and spring floods in 1826 forced Black Rock Harbor to close, all the forwarding business moved to Buffalo, and the "independent canal" became the main line.[88]

The tensions surrounding the terminus controversy, together with the developing antagonism between Thomas and canal engineer Nathan Roberts, had important ramifications at Lockport. In February of 1822, DeWitt Clinton wrote that "David Thomas called on me to signify his intention of resigning the post of principal engineer." Clinton then described the incident that prompted the resignation:

> It appears that Mr.________, a sub-engineer treated Mr. Thomas with great rudeness, recently in Albany, and that his unaffected meekness shrinks from collision with such a rough and rude temper. I have written to Mr. Thomas that he must not resign. The report has excited great alarm among the friends of the canal.[89]

Three weeks later, Clinton informed Thomas that the Canal Board had unanimously voted to continue him as "chief engineer of the line from Rochester to Buffalo with the full control of the line in that capacity"[90] Whether Thomas ultimately resigned or was "removed" is unclear, as is the exact timing of the event. Thomas retained some responsibility at the Mountain Ridge until December of 1823. Nathan Roberts then assumed the post of chief engineer of the western section.[91] By 1823, Clinton's own statewide political support had substantially eroded. He had decided in 1822 not to seek reelection as governor, though he retained his position as president of the Canal Board.[92]

Reporting on progress at the Mountain Ridge in 1823, the commissioners began by reciting a litany of difficulties and unforeseen obstacles that were increasing the cost of the work. The high water table in the floor of old Lake Tonawanda "even in the driest season of the year, very much embarrassed the operation of drilling and blasting." Contractors sunk a ditch "below their work" for two miles to "the brow of the mountain." Still "much pumping" was necessary. As excavation "approached the summit," the rock became "a mix of flint and lime, in irregular layers, dipping in different directions, which is found hard to drill, and after it is drilled and charged, no ingenuity or skill in blasting, is at all times sufficient to produce much effect by the explosion." And finally the remoteness of the site, the lack of local food sources, and the great "number of hands" made it necessary to "provision from a great distance."[93]

Then in the very next paragraph, the commissioners announce that excavation in the northerly four ridge sections, which they had referred to as the "worst" two miles in the 1822 report, totaled 144,000 cubic yards for 1823—that is three times as much rock as was removed in the first two years combined.[94] How was this "miracle" at the Mountain Ridge accomplished?[95] A combination of factors, some local and some emanating from Albany, were responsible for the turnaround. Among these factors were a dramatic increase in resources, important technological innovations, and further subdivision of the work.

In 1823, a great deal more money began to flow into Lockport. The commissioners had focused efforts on the northern part of the Deep Cut for all of the first three years. This included the locks and about two miles of heavy rock cutting. After dividing original ridge section 2 into three contracts in 1822 (new sections 2, 3, and 4), the commissioners spent about $72,000 on this 1.44 mile stretch that year. In 1823, they doubled expenditures in these sections to about $142,000.[96] Presumably such increases in expenditures translated into a corresponding increase in workers; in 1822 the Lockport *Observatory* had estimated eleven hundred laborers at work in the vicinity of Lockport.[97] Perhaps now that number swelled closer to fifteen hundred since, under the contracting system initiated in 1822, engineers could force contractors to increase the size of their crews.

In 1823, prodigious amounts of blasting powder arrived at the Mountain Ridge. The reports of Commissioner William Bouck listed almost $16,000 for powder—that is about 3,100 kegs, or more than two kegs per laborer.[98] A recurrent pattern in the industrial history of the United States is for shortages of labor to spur technological innovation and the substitution of capital for labor.[99] Indeed the blasting powder used at Lockport was itself a recent innovation. The ingredients of the powder had been known for hundreds of years, but the French immigrant Irénée DuPont had recently perfected the process of manufacturing the powder with just the right proportions to create a much more effective and consistent product, a great boon to those excavating the Deep Cut through the Mountain Ridge.[100]

While the number of workers and the amount of powder used both increased in 1823, and total expenditures doubled, yards of rock excavated more than tripled. In part this success was due to technological innovations developed locally in response to specific problems that delayed the work. One of the thorniest problems was drilling the holes for the charges. This was a two-man job, with one man holding the drill—a long steel rod sharpened to a diamond shape at one end—and turning it after each swing of the other man's hammer. The rock was so hard that drills immediately became dull. Engineers offered a $100 reward for a drill that could penetrate the rock, and a man named Batsford from Niagara Falls, who was skilled at tempering steel, produced a new drill hard enough to do the job.[101]

Another difficult task was hauling the blasted rock out of the steep-sided canal and onto the growing spoil bank. The first attempt to improve this process was the development of a special "dumping wheelbarrow."[102] Orange H. Dibble and Walter Osborn, two engineers working at the Mountain Ridge, developed a new stationary device that made the removal of debris much easier and faster. Pictured in a George Catlin lithograph (fig. 3.4), this horse-operated crane allowed buckets of rock to be quickly lifted, swung around, and dumped on the towering spoil banks.[103] To some extent, these innovations may have been reinventions, but they became part of local folklore among the largely Yankee middle class of Lockport and proved a harbinger of the inventiveness that would characterize the entire canal region of New York in the coming decades.

Another important change in 1823 was an intensification of oversight. There were six canal commissioners in 1823. DeWitt Clinton, president of the Board, and Stephen Van Rensselaer attended meetings and helped to scrutinize fiscal and practical decisions. The other four commissioners—Samuel Young, Myron Holley, Henry Seymour, and William C. Bouck—were designated acting commissioners; they took direct charge of construction, performing the duties of engineers and superintendents.[104]

Bouck succeeded Holley in the summer of 1822 as the acting commissioner with direct oversight responsibility for the western section of the canal.[105] He took seriously the directive to keep detailed records, especially

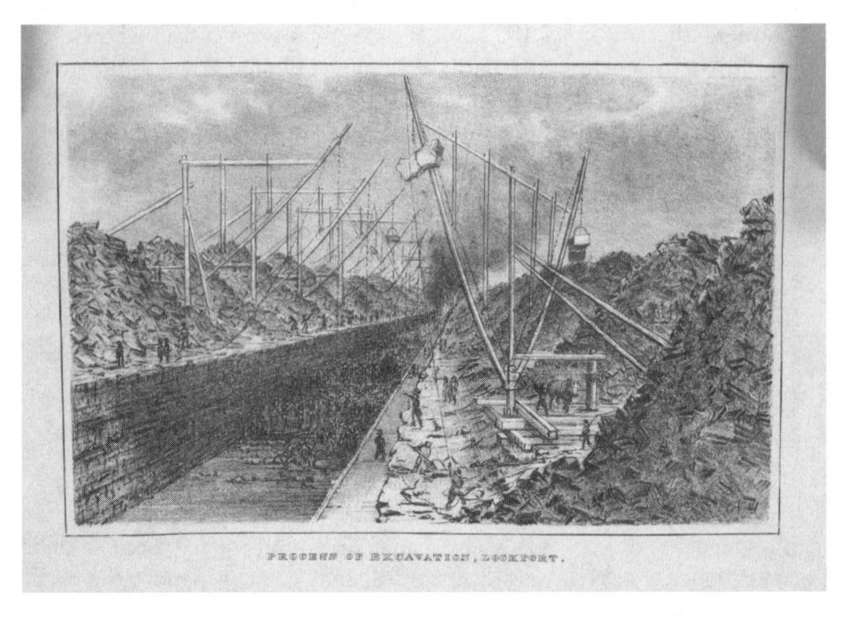

Figure 3.4. George Catlin, *Process of Excavation, Lockport*, lithograph, in Cadwallader D. Colden, *Memoir at the Celebration of the Completion of the New York Canals* (Albany: State of New York, 1825), opposite p. 298. Print Collection, Miriam and Ira D. Wallach Division of Art, Prints and Photographs, The New York Public Library, Astor, Lenox and Tilden Foundations. Used with permission.

under the stipulations of the revised contracts of 1822. He sent monthly as well as yearly reports listing all expenditures and funds released to individual contractors and engineers. Bouck was an important presence at the Mountain Ridge, making Lockport his principal place of residence for four years. Orasmus Turner, the editor of the Lockport *Observatory*, describes Bouck as "indefatigable . . . in the discharge of his duties," and remembers him

> in those primitive canal times, traversing the forest on horseback and on foot, from the log shanties of one contractor to those of another; sleeping and eating where emergency made it necessary, in quarters no matter how rude or humble; or in his room at the old "Cottage" in Lockport, coolly and good naturedly resisting the fierce importunities of the dissatisfied contractor; yielding to exigencies here and there, when public interest demanded it, or strenuous and unyielding when it did not; pressing on the difficult work upon the Mountain Ridge, amid great difficulties and embarrassments, persevering to the end, until he had seen the last barrier removed that prevented the flow of the waters of Lake Erie through their long artificial channel.[106]

As acting commissioner, Bouck had the authority to settle most disputes, make adjustments to contracts, and even offer special inducements.

The change to direct state control of the work at the Mountain Ridge in 1822 gave Bouck considerable flexibility. This system effectively spread the risk that otherwise might induce a contractor to be cautious, for instance, about adding more workers to his crew. In their 1824 report, the commissioners announced their intention to return to the standard arrangement of "a definite price per cubic yard."[107] The advantages of the flexible system, however, were so apparent to Bouck and the engineers that they continued to apply many of its elements for the remainder of the construction period, funding many of the contractors' expenses over and above the price per yard of excavation, including expenses for blasting powder, whiskey, board, extra laborers, and especially pumping.[108]

Another kind of intensification that hastened the excavation work in 1823 was a further subdivision of the contracts. The four original sections had become six in 1822, but by the end of 1823 there were seventeen much shorter sections on the same seven miles (fig. 3.5; see also the appendix to this volume). Contractors themselves initiated additional changes. For example, Boughton, Lobdell, Smith, and Maynard—the contractors for the most southerly and longest section—with Bouck's approval, divided most of their work into eight sections of ninety-six rods each (.3 miles), and also divided oversight responsibilities among themselves.[109]

The first state-initiated subdivision of ridge contracts in 1823 was a curious one. In July, the commissioners requested Oliver Phelps of Tompkins County to attend a special meeting in Albany and then "as soon as possible to go to The Mountain Ridge and take possession of all that part of Section no. one which is South of the road bridge in Lockport." As in the 1822 contracts, they agreed that "he should be paid all the necessary, actually incurred [expenses] together with three dollars per day for his services."[110] In his first voucher for this work, Phelps included the expense of traveling both to Albany and to Lockport—a perk no other contractor or laborer received.[111] Was Phelps a relative of the influential financier and speculator from Canandaigua who bore the same name?[112] The commissioners gave Phelps a target price per yard, but they also offered him a special incentive to economize, to wit: "that if he can reduce the said expense so as to make the average cost below twelve shillings per yard ($1.50), he shall be paid the amount of the difference between the actual cost, and twelve shillings per yard."[113]

The commissioners initiated most of the 1823 subdivisions in the autumn. Bouck began with a preliminary step: he marked off the entire Deep Cut from the Main Street Bridge to the Tonawanta Creek (6.9 miles) into 184 sections of three chains each (198 feet). He retained the current contractors, albeit some with reduced sections. In addition to

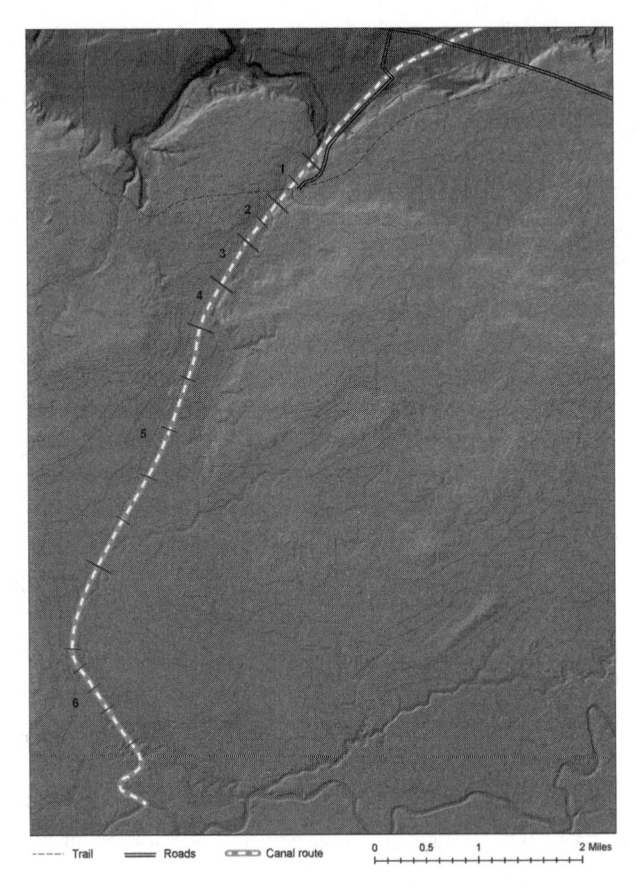

Figure 3.5. Mountain Ridge contracting sections of 1823, overlain on digital elevation model (United States Geological Survey). Map by Cherin Abdel Samie and Yasser Ayad.

dividing section 1 between the original contractors and Phelps, he also split Child and Hamlin's section, giving half to each. The commissioners had estimated in 1822 that the first four sections (1.94 miles) would require 260,000 cubic yards of rock excavation, but in 1823 they corrected this upward to 323,346 cubic yards. Still, the improved progress in 1823 left only 126,100 cubic yards in those sections—less than had been excavated that year. There was still "much to be done" on section 5 and 6. Moreover, there remained considerable uncertainty about these two southerly sections, which together were over five miles long. No one knew how much rock lay under the muck, nor just how "embarrassing" drainage issues might prove. Consequently, the commissioners decided to expedite their work "by subdividing these sections into eleven parts, and placing them into the hands of energetic and experienced contractors."[114] Only six of these contracts were actually entered into by the end of 1823.[115] Most of the progress in 1823 had taken place on the first four rock sections (sections 1 and 2 in 1821). Most of the new contracts on the southerly two sections were only signed in December. The progress that had been made in 1823 on those sections was due to the original contractors' efforts and, in particular, to the intensification of work on the most southerly section by the contactors' own subdivision of the work.

The terms of the 1823 contracts combined elements of standard contracts with those of 1822. A price per yard, from $1.56 to $1.76 was subject to "such reduction or increase of price" as Bouck and Roberts "may deem just and equitable."[116] The contracts also required contractors to "render an account" of expenditures monthly, to give access to all accounts and records, and in most cases to complete the work by the following autumn. The commissioners offered John Gilbert, who held the contract for section 4, an added inducement: "if the said Gilbert excavates a greater quantity than 12,000 yards by the first of April he is to receive one dollar and eighty-eight cents per cubic yard."[117] These were the highest definite prices offered at the Mountain Ridge, and like the incentives given to Phelps, they were calculated to entice him to press his workers as hard as possible. When David Thomas signed these autumn contracts, it was among his last official acts as chief engineer of the western section.[118]

From the beginning, the commissioners and engineers had given a high priority to work on the double locks by which the canal was to descend from the Lake Erie to the Genesee River level. Geddes's original idea of combined or staircase locks—that is, locks without gaps separating them—had the advantage of greatly lessening the amount of rock excavation necessary to prepare the site. Nonetheless, this excavation work was much more costly and time consuming than the lock construction itself. It was not until July 9, 1823, that workmen laid the first stone of the locks.[119] By this time, all the necessary plans, materials, and personnel were on hand, and engineers were optimistic about an early completion of these complex stoneworks at the northern end of the Deep Cut.

In 1823, a crew finally succeeded in cutting a road to Wright's Corners (see fig. 3.5). For the first time there was a direct connection between Lockport and the Ridge Road, three miles to the north. The Ridge Road was still the principal connection between Niagara County and points east. The new connecting road facilitated the transport of powder, provisions, animals, and workers, all of which arrived from the East.

During this year, the Washington House opened at the corner of Main and Transit (the perfectly straight street that followed Ellicott's meridian line).

A large framed structure, it immediately became Lockport's most prestigious hotel and stood for the next forty-two years.[120]

The increasing human efforts at Lockport in 1823 did not make the place seem less chaotic or more orderly. The sights and sounds of the village made decidedly the opposite impression on Marcus Moses who arrived in the autumn of 1823. Timber stood "close about" Main Street, which was studded with "great stumps." On both sides of the canal near the Main Street bridge, were "piles of stone so high that the buildings on the opposite side of Main Street were hidden from sight." Moses's account—like those of Spalding, Mrs. Smith, and Turner—was a retrospective, in his case published in 1878. His description proceeds up one side of Main Street and down the other, listing and commenting upon every business establishment. Perhaps the most numerous of these were the taverns and groceries that provided alcohol to thirsty workers and residents. Most people at this time believed that "business would be done west of the canal."[121] The placement of the county buildings several blocks west of the bridge coincided with this perception. Moreover, the large workforce and its shanties scattered along the line of the canal tended to draw business toward the west. After the construction period, the business center of Lockport would migrate eastward toward the head of the locks.

For those with sufficient resources and knowledge, the chaotic scene at Lockport presented financial opportunities. Near the end of 1823, Holmes Hutchinson wrote to Lyman Spalding, his cousin and close associate: "Lockport must be at present a very money making place and could I make arrangement with you to employ a considerable sum so as to realize as much as 20 percent it would suit my pocket extremely well."[122]

By 1823, the project at the Mountain Ridge was attracting widespread attention. Newspapers across the country followed the progress of the canal. William Hamilton Merritt, the father of Ontario's Welland Canal, had been impressed by the energy of his neighbors across the Niagara River. "An enterprising people can effect wonders," he noted in his diary in 1823. Merritt was the leader of a group from St. Catherine's who were preparing to submit a canal proposal to the Upper Canada legislature. The group charged him to visit Lockport to observe the progress and gather information. He arrived at the Mountain Ridge on July 19th. "Lockport bids fair to become a large and flourishing city," he noted; "the canal progresses as fast as it can, from the slow progress of blasting the rock." So committed was he to the benefits of "the Age of Canals" that he proffered one more comment at odds with nearly all the surviving evidence: he observed "no intemperance, and much cordiality, directions given in a mild unassuming manner."[123]

1824

The year 1824 began badly for DeWitt Clinton and his political ally Myron Holley. When the commissioners submitted a list of expenditures in February, Holley was unable to account for over $30,000. The ensuing political investigation showed that he had used public money for private purposes, and he immediately resigned. Clinton's enemies, known as the Albany Regency, wanted to weaken his power before the coming gubernatorial election. Clinton had declined to enter the 1822 campaign, but he retained his position as president of the Canal Board. On April 12, the last day of the legislative session, Regency legislators introduced a bill removing Clinton from the Canal Board, a position he had held, without pay, for fourteen years. A huge public outcry followed the bill's passage, sweeping Clinton back into the governorship in November. His opponent in that campaign was fellow canal commissioner Samuel Young. The canal, in 1824, was at the center of New York State politics.[124]

The commissioners reported that contractors at the Mountain Ridge had hoped to work during the winter but flooding and harsh weather made this impossible. They did take advantage of the frozen roads to stock up on "stores of subsistence" for the coming year. High water in the spring made it necessary to construct ditches on both sides of the canal. In addition, horse-powered pumps "were introduced on almost every section to throw water on the towing path, on which it was conducted for more than three miles, into the Tonawanta Creek." It was late May before any work could begin.[125]

At the beginning of the 1824 season, Bouck, Roberts, and the other engineers on the western section worked under an added pressure. In 1823 their counterparts on the eastern section had finally overcome the narrow parts of the Mohawk Valley, a place the commissioners said was "obstructed with a greater complication of difficulties than . . . any other part of the canal."[126] In addition, crews had completed the descent from Schenectady to the Hudson River at Albany, a section commissioners dared not even attempt at an earlier stage. This last descent required two aqueducts and twenty-nine locks.[127] Finally, workers at Rochester had finished the "great aqueduct across the Genesee River," a jewel of hydraulic engineering featuring nine stone arches, each spanning fifty feet. The famous aqueduct carried the canal into western New York at a level that would extend to the foot of the Lockport locks.[128] By May of 1824, the canal was navigable from the Hudson River to Brockport—twenty miles west of the Genesee River and forty miles east of Lockport. Engineers expected the level section between Brockport and Lockport to go quickly and smoothly. Now all eyes turned toward the tiny part of the line between Lake Erie and the Mountain Ridge.

The Buffalo *Journal* reported that on May 9, 1824,

> the laboring Irish upon the canal between Black Rock and the Tonawanta Creek, had a misunderstanding, the origin of which cannot be ascertained, but which ended in a very general conflict. A number of persons were severely injured by the various missiles used by the combatants, and one man, who was missing at the end of the affray, has yet to be found. Strong fears are entertained that he was killed.—*Twelve* of the principal actors were secured and . . . committed to await their trials.[129]

Two months later, violence on a larger scale exploded at Lockport. The trouble began on the morning of July 12 when Irish Protestant canal workers gathered in the center of the village to celebrate the anniversary of the Battle of Boyne, the decisive turning point in Irish history loathed by their Catholic countrymen. "The Catholics, in order to prevent an observance of the day," the Lockport *Observatory* reported, "assembled to the number of more than 300, armed with guns and cudgels."[130] The outnumbered Orangemen, quickly "repaired to their work."

> The Catholics, after procuring a drum and fife and marching several times through the village, started up the canal line, frequently attacking those who were peaceably at work, and threatening to put to death every Orangeman that came in their way. In several instances, severe wounds were inflicted. Towards evening, the mob presented an appearance so alarming that civil authorities deemed it necessary to call out the militia.[131]

Before the militia arrived, "our citizens had for the most part procured arms and organized themselves into a body." Confronted with this better-armed group, the Catholics dispersed. "Much credit" was due to the militia, the article concluded, since its appearance, "if it had no other effect, must have convinced the rioters that the means are at hand at all times to punish their outrages." This account places the blame for the "outrages" on the Catholics rather than the Protestants, whom it describes as eager to return to work. Yet even the Protestant workers were hardly to be classified among "our citizens." Indeed, from the point of view of *Observatory* editor Orasmus Turner, the cause of the violence could only be a dark irrationality imported from the Old World. Both Catholics and Protestant Irishmen, were "cultivating even here a hatred towards each other, which is universally prevalent in their own country."[132]

Lewiston's *Sentinel* also covered the story, praising the sheriff's prudence. As the militia approached "the camp of the enemy," he ordered "the music to appraise them of their danger.—It had the desired effect, and nothing was to be encountered but the cursings of deserted women and the screams of little children."[133] The *Sentinel* was also sympathetic to the Protestants, calling the Battle of Boyne "that great event." Yet the condescension toward all the Irish laborers was even more striking. The events of July 12 were similar to "all like occasions in the land of Potatoes" when "Pat meets Pat." "Shelalies, brick-bats and stone missiles were in high requisition"—a "state of things which held the sober Yankees of Lockport in terrorem."[134]

Historian Peter Way lists the 1824 Lockport Riot as the first example of military intervention against canal laborers in North America and perhaps it's first use against laborers of any kind.[135] Apparently, he was unaware that the militia had also intervened at Lockport in 1822. What is clear is that, in the subsequent three decades of North American canal building, there would be many more labor uprisings of various kinds.[136]

How are we to understand the Lockport Riot of 1824? The most common explanation is the one suggested in the local press accounts; as historian Ronald Shaw put it, "the Irish laborers engaged in lusty combat over old country loyalties."[137] That this interpretation is at least partially true is bolstered by the fact that a similar Orange Day riot took place in New York City in 1824. The so-called Greenwich Village Riot also began when a group of Irish Catholics confronted the assembled Orangemen. Which side started the violence is unclear, but, like the Lockport Riot, it resulted in many serious injuries on both sides. Authorities arrested thirty-three Catholics. The ensuing trial revealed a growing rift between Protestant and Catholic Irishmen. Many early leaders of New York's Irish community had been refugees from the failed 1798 Rebellion. DeWitt Clinton, who was himself part Irish, had helped to formulate the secular united Irish political vision that was anticolonial, egalitarian, and republican. After 1798, the spirit of unity had disintegrated in Ireland, replaced by an intense religious factionalism. In the intervening decades, Irish immigrants to North America were increasingly poor, illiterate, and imbued with animosity toward their countrymen of a different religion. Judge Richard Riker, who presided over the New York trials, allowed discussion of the entire history of Ireland. He then issued sentences, condemning both sides for importing their religious bigotry "to this land of freedom." "It was irrational," he admonished, "for Irishmen to go back so far into history for causes to perpetuate quarrels and bloody affrays with each other."[138] Burroughs and Wallace point out that, with the swelling tide of poor and fractious Irish immigrants in the decades ahead, authorities would rarely react so leniently, concerned as they were about a new and fundamental issue: "the growing numbers of masterless men."[139]

The fact that Lockport's middle-class citizens had intervened on the part of the Orangemen might be seen as an indication that they considered only Catholic workers as dangerous and irrational. Indeed, since July 12 was a Monday, perhaps contractors and even engineers had sanctioned the celebration by permitting laborers to leave their jobs. In 1823, a group of prominent citizens had established Lockport's first Presbyterian Church, an institution that would remain an important center of middle-class community life.[140] The Orangemen, of course, were also Presbyterians, yet church records indicate that not a single Irishman was admitted to the congregation during canal construction.[141] Lockport's crucial lines of social demarcation in 1824 were not based on religion. Nor were they a simple matter of the town—with its bakers, merchants, and doctors—versus the canal—with its engineers, contractors, and laborers. Clearly, the important division was between the canal's common laborers and just about everyone else. There was a sharp contrast between the world common laborers experienced at the Mountain Ridge and the one the "sober Yankees" of the middle class knew. The divergence between these two worlds merits a closer examination. To do so, it is necessary to interrupt the diachronic structure of this chapter with a synchronic interlude. Before returning to examine events at the Mountain Ridge in the remainder of 1824 and in 1825, I analyze the divergent worlds of laborers and citizens in the following two sections.

The Laborers' Lockport

Throughout the peak period of canal construction in North America (1817–1840), canal labor was almost universally considered the most degrading of all work. Accordingly, slaves formed the majority of the canal workers south of Maryland,[142] and in response to the labor shortage on the western section of the canal, Governor Clinton pardoned convicts from the Auburn State Prison in 1821 and 1822, providing that they agreed to work on the canal.[143] It is not surprising that desperate Irish immigrants—the objects of scorn and bigotry—should find themselves in such company, accepting work and living conditions others universally shunned. Since most Americans considered canal work beyond the pale, those who accepted it entered a world separated from the mainstream of society. They found themselves in a box, not of their own design. Because canal work in North America, by its very nature, often took place far from major cities and populated areas, laborers were also geographically marginalized. Moreover, laborers' shanties were usually separated even from villages.[144] At the Mountain Ridge, shanty settlements were strung out for seven miles, but because of the enormous labor at the locks and the northern few miles of cutting work, laborers' shanties clustered along the canal right into the village of Lockport. This situation, combined with the fact that the work continued for five full years, created distinctive social relations resembling those of future industrial cities. At Lockport, laborers were a presence impossible to ignore.

While slaves had no choice, and convicts had an offer they could hardly refuse, immigrant laborers presumably volunteered for canal work. Indeed, to justify the lack of compensation for illness, injury, or even death, contractors and canal companies argued that laborers knew of the inherent risks when they accepted canal work.[145] How much choice did they really have? Peter Way argues that

> the vast majority of canal workers did not have the power to choose to be a nonmanual worker, say a merchant, clerk, lawyer or even a farmer. . . . They did not make a choice to be canallers after some objective evaluation of a multitude of alternatives nor of the dangers of accidental injury or exposure to all manners of ailments and infectious diseases. Most moved into canalling by accident and as a result of having precious few options.[146]

Irish canal workers, in a process that sometimes took more than one generation, faced "gradual separation from the land, transfer to wage labor and migration to where the market dictated."[147]

At the recommendation of the canal commissioners in 1819, the state legislature—in the same act that authorized the completion of the entire canal—exempted canal laborers from militia duty.[148] The commissioners were motivated entirely by labor force issues: they wanted to prevent delays and interruptions of canal construction. An effect of the law, nonetheless, was to increase laborers' marginalization. By making it clear that laborers were not part of the community of citizens, the militia exemption facilitated the social bifurcation of communities like Lockport into two separate spheres and exacerbated tensions whenever and wherever the two spheres intersected.

Contractors at the Mountain Ridge offered laborers twelve dollars a month plus food, lodging, and whiskey. Such arrangements represented a transition between traditional paternal labor systems and the free labor of industrial capitalism. Although contractors no longer included workers in their own households, they often maintained their stable of workers in shanty settlements under their indirect control, a condition not dissimilar to that of indentured servants.[149] This shielded workers from the cold blast of free labor capitalism, perhaps allowing them food and shelter during work stoppages and illness. It also facilitated surveillance of the workers, rendered them dependent, and perhaps deterred them from shopping around for a better situation. Of course, it was at Lockport that the canal contracting system first broke down. When the state assumed control of construction at

Lockport, contractors became mere overseers and the shield of paternalism presumably weakened, a trend that would only accelerate on North American canals.[150] In the debates over the morality of slavery in the decades preceding the Civil War, defenders of the "peculiar institution" would often argue that slaves were treated better than the "wage slaves" of the North. Indeed, when immigrant labor became available, slaves were often removed from canal work in the South, partly because slaves represented an investment, whereas laborers—in the days before insurance and workers' compensation—shouldered their own risks. In this wage-labor system, an employer need not own the bodies of workers when he could merely rent them.[151]

Twelve dollars a month was slightly more than most Erie Canal laborers earned and substantially more than wages on subsequent canal projects.[152] For a month with no work stoppages due to weather or other factors—that is, about twenty-four to twenty-six workdays—this amounted to fifty cents a day. A young healthy worker with no dependents, self-discipline, and a great deal of luck, had a chance to save enough to extricate himself from canal work. For a host of reasons very few laborers were able to follow this path.[153] To begin with, contractors paid wages only for days actually worked. When work ground to a halt in the winter or was delayed in the spring by drainage problems—as in 1824 when most workers were not on the job until late May—there was simply no pay. Even if contractors continued to provide food and shelter during periods of work stoppage, workers continued to spend money on everything they needed beyond mere subsistence. This included "luxury" items such as tobacco, whiskey, and newspapers. Hence, whatever savings workers managed could easily disappear during the winter. Nor did workers receive allowances for transportation. For those who made their way to Lockport from New York City, this would have been a substantial expense. When work ended on a particular job, laborers also needed resources to get to the next work site.

Those who were not healthy or had families to support often found themselves in desperate straits. Illness and injury struck many workers at the Mountain Ridge; without wages, their only recourse was charity. Since contractors offered no food or shelter to the dependents of workers, they too were forced to work, usually for lower wages. Although the site was isolated, the fact that workers continued for five years probably induced many laborers to bring their families to Lockport. Women and girls worked as housekeepers, cooks, menders, laundresses, and—given their desperate situation and the preponderance of young men at the Mountain Ridge—probably prostitutes.[154] Boys, of course, were an integral part of the workforce, often delivering whiskey rations to other workers and driving horses on carts, pumps, and cranes. Some families threw up their own shanties and took in boarders.

The flexible contracting introduced in 1822 probably increased the proportion of day laborers, who received seventy-five cents a day but no meals or lodging. Since all provisions had to be carted in from a great distance, high prices put pressure on all workers but especially day laborers.[155] To put it simply, most laborers at the Mountain Ridge were under considerable financial stress and some—particularly those too sick or injured to work—were utterly desperate. The *Niagara Sentinel* reported in March of 1824 a growing problem of paupers at and near the canal that available resources were inadequate to meet.[156] In addition to products of all kinds, a human flotsam was beginning to ebb up the new channel.

The conditions of shanty life are difficult to reconstruct. Laborers' material culture was ephemeral: abandoned log shanties in the wilderness soon crumbled. Nor did workers leave us writings describing their experience. The few middle-class descriptions of shanty life are condescending glimpses of an appalling hidden world. Although these accounts say as much about their authors as they do about laborers, they are all we have. In the woods near Troy, New York, the British visitor Frederick Marryat stumbled upon some huts that looked "more like dog-kennels than the habitations of men."

> In a tenement about fourteen feet by ten, lived an Irishman, his wife, and family, and seven boys as he called them, young men from twenty to thirty years of age, who boarded with him. There was but one bed on which slept the man his wife, and family. Above the bed were some planks, extending half way the length of the shealing, and there slept the seven boys, without any mattress, or even straw, to lie upon.[157]

During his travels of the 1840s, Charles Dickens encountered workers "hovels" on the Pennsylvania Canal.

> The best were poor protection from the weather; the worst let in the wind and the rain through wide breeches in the roofs of sodden grass and in the walls of mud. Some had neither door nor window; some had nearly fallen down and were imperfectly propped up by stakes and poles; all were ruinous and filthy. Hideously ugly old women and very buxom young ones, pigs, dogs, men, children, babies, pots, kettles, dunghills, vile refuse, rank straw, and standing water, all wallowing together in an inseparable heap, composed the furniture of every dark and dirty hut.[158]

In such descriptions, it is difficult to separate the actual degradation from the nightmarish middle-class shock, perhaps reflecting a fear of falling in the harsh new world of market dynamics. Shanties usually had dirt floors and a hole in the roof to serve as a chimney. The crowded, unsanitary conditions in Mountain Ridge shanties facilitated infestations of human parasites such as lice and bedbugs. There was no escape from the swarming mosquitoes in the poorly drained bottom of ancient Lake Tonawanda.

Darius Comstock and his Quaker associates, who held the contract for a rock section just outside the village, built unusual barracks for their workers. "Some of them," Mrs. Smith recalls, "were a hundred feet long all in one room, and had the beds hung on the sides like berths in a canal boat."[159] The sleeping arrangements on canal packets may indeed have inspired these barracks, as they undoubtedly helped inspire George Pullman of the western New York canal town Brockport, the inventor of the railway sleeping car.[160] An intriguing possibility is that the model for these barracks is the Iroquois longhouse. The Comstocks were among the first to arrive at the future site of Lockport, and they were on intimate terms with many of the local Tuscaroras and Senecas.[161] If this were the case, it might indicate again just how marginal contractors considered canal laborers: it was appropriate to house them like those the Comstocks undoubtedly considered savages.[162]

At the Mountain Ridge, alcohol was more than just a part of the environment; for many laborers, it *was* the environment. They imbibed it all day long at their work. In the evenings and on Sundays, they continued drinking at groceries, taverns, and shanties. Alcohol assumed a pivotal position in the working-class culture free laborers were in the process of forging. It was one constant in a new and difficult world.

While the broad outlines of alcohol's role at the work site are clear, the details are obscured by the contracting system. Whiskey rations, like most other aspects of laborers' conditions, were entirely under the control of contractors. While commissioners and engineers often maintained detailed records, they generally disavowed any knowledge of common laborers and their conditions. After the state assumed control at the Mountain Ridge in 1822, a few illustrative details began to appear in contractors' expense vouchers.

The quantity of alcohol a contractor issued to his laborers is generally hidden beneath the threshold of written records. Many contractors, even where listing work by the number of days of labor (rather than, say, yards excavated), remain in a middle position between paternalism and free labor—household production and the factory—because they are still boarding their workers and paying them on a monthly basis. Otis Hathaway, for example, on a job leveling the towpath just east of the locks early in 1825, received from the commissioners one dollar per day per laborer. Since this price included "board, whiskey and use of tools" (but not powder, which is listed as a separate expense), it tells us little about expenditures on alcohol and quantities consumed.[163]

Contractors Boughton and Lobdell, in a voucher from December 1824, list an expense of $16 for one barrel of whiskey "for the pumpers" who worked 222.75 days. Dividing gallons by workdays reveals that each pumper consumed about eighteen ounces of whiskey (or twelve shots of 1.5 ounces) per day, at a per-day cost of $0.07.[164]

Late in 1825, Seymour Scovell, another Quaker contractor, submitted a remarkably detailed voucher for "cutting back drains." He listed all fifty-four of his workers by name. They were overwhelmingly Irish. Many could only sign their pay receipts with a mark. These were day laborers who had to find their own boarding arrangements. Scovell requested $768.46 for 1,005 days of labor at $0.75 a day. He then added $125.63 for "furnishing whiskey and tools." Since Scovell already owned the tools and was simply adding this charge as a kind of rental fee, most of this additional expense—which amounts to fourteen percent of the job cost—would likely have gone for whiskey. If we assume that proportion to be 60 percent, and further assume a cost of $0.50 per gallon for whiskey, each of Scovell's fifty-four workers would have consumed nineteen ounces a day. If the proportion were 75 percent, the consumption rises to twenty-three ounces a day.[165] Such estimates are the only direct evidence we have.

Peter Way estimates that canal workers, on average, consumed from twelve to twenty ounces of whiskey on the job each day. Under certain circumstances, George Condon claims this figure could rise to as high as thirty-two ounces.[166] Given the multiple intensifications of work at the Mountain Ridge, it is probable that whiskey rations were very high there, perhaps twenty ounces (over thirteen shots of 1.5 ounces) per day. Surviving records are not detailed enough to pinpoint either the total amount or the total cost of whiskey rations. By making some simplifying assumptions, it is nonetheless possible to estimate these figures. If we assume a workforce of five hundred in 1821, one thousand in 1822, and fifteen hundred for the next three years, and if we further assume a daily per capita consumption of twenty ounces and a two-hundred-day work year, the total consumption for five years would be 187,426 gallons (6,046 barrels). Almost all of this would have arrived by cart through the torturous roads from Rochester. The wholesale price per barrel at Lockport varied between $0.28 and $1.50.[167] Assuming a price of $0.39 per gallon yields a total whiskey expenditure on the Mountain Ridge contracts of $73,196.14. These figures probably represent less than half of the total consumption at the Mountain Ridge during the construction period because they do not take into account workers' drinking off the job or the considerable consumption of "citizens."

After work and on Sundays, workers would have had no trouble finding as much whiskey as they wanted. For about $0.25 they could buy a quart.[168] More likely they would buy it by the gill (four ounces) from the innumerable taverns, groceries, and boarding houses both in the village of Lockport and along the line of the Deep Cut. The economics of whiskey, with its complex system of large and small dealers, resembled the low-technology fresh water delivery systems still operating in developing-world shantytowns. After many years of work at Lockport, the infrastructure was well developed. The profit margins of large dealers exceeded 400 percent; those of small dealers were much lower. Apart from canal work itself, alcohol sales—taken as a whole—must have been about the largest and most profitable sector of the local economy.

At the work site, a small boy known as a "jigger boss" delivered whiskey rations at regular intervals. When the section of the canal between Black Rock and Buffalo neared completion early in 1825, the contractors tried a new approach. "Along the line of the canal, at convenient distances, was to be found a barrel of whiskey, pure old rye, with part of the head cut out and a tin dipper laying by and all were expected to help themselves."[169] Without direct evidence, many have repeated the legend that workers were expected to refrain from partaking until the excavation reached a barrel, at which point they were allowed to stop and drain the barrel.[170]

How are we to account for the prodigious levels of alcohol consumption at the Mountain Ridge? Some have suggested an Old World explanation: heavy drinking "came from Ireland with the immigrants."[171] How, then, are we to understand the almost equally heavy drinking of nonimmigrants? During the decade of the 1820s, per capita alcohol consumption for the entire United States reached its historical peak.[172] An enormous amount of alcohol was already moving through the Erie Canal; in their reports of 1823 and 1824, the canal commissioners list a total of 11,464 barrels of alcohol passing through Rome. Of this, 92 percent by volume was whiskey.[173] Perhaps Americans were frightened of their drinking water, but could it have been any worse than in previous decades? William Rorabaugh proposed an alternative explanation for the excessive drinking of this period: it was a by-product of rapid social change.[174]

In 1824, the Erie Canal was almost the epicenter of change, facilitating national expansion, geographic mobility, and market dynamics. At the Mountain Ridge, the incipient class structure of industrial capitalism was already beginning to emerge. In the name of industrial discipline, alcohol would soon be eliminated from the workplace. It was an immemorial fixture of the paternal system where it functioned as a solvent, permitting a measure of cross-class male camaraderie. Canal laborers drank more than any other identifiable group, but they did so not merely because they were free (for part of every day) from direct oversight, but also because they lived and labored under conditions that were not only hellish but almost entirely beyond their control.[175]

The effects of laborers' drinking are multiple and evident. Although both laborers and contractors viewed whiskey rations as part of the latter's payment to the former for time on the job, these rations were neither an unalloyed benefit to the laborers nor a useless expenditure for the contractors. For workers toiling in the hot sun, the immediate effect was dehydration. A chronic effect, combined with inadequate shelter, was susceptibility to disease. Then there was the danger of inebriated men operating cranes and handling explosives. Issuing large amounts of a habit-forming substance at regular intervals was bound to cause, in many workers, a deep dependency upon the contractor. To the extent that alcohol rendered the workforce sedated and dependent, it served the aims not only of the contractors but also of the engineers and commissioners. Under the new conditions of free labor, however, the social dynamics of alcohol in the workplace were evolving in ways that must have disturbed many contractors. Instead of sedating workers, at times it inflamed them.

Even with the large crews, whiskey breaks still provided a measure of camaraderie between workers and their overseers. Among workers, alcohol could enhance or undermine solidarity. Drinking was a central aspect of a developing oppositional workers' culture, but it could also fuel divisive and violent rivalries. Alcohol certainly played a part in the Lockport Riot of 1824, which began with violence among opposing groups of workers, then evolved into a confrontation between one group of workers, villagers, and eventually the militia. Although alcohol consumption may have been very high at the Mountain Ridge compared to other parts of the canal, it was not in any direct sense the cause of the riot.[176] Alcohol flowed freely on every section of the canal. Yet the *only two* significant riots during the construction of the Erie Canal both occurred at Lockport. If alcohol consumption intensified, it was merely one more response to the intensified pressure focused on the workforce toiling in the Deep Cut.

While drinking on the job linked canal laborers with the paternal past, other aspects of working conditions presaged the work discipline of industrial production. Laborers heard the wake-up call a half hour before sunrise and worked from sunup to sundown—fourteen hours in the summer, about ten in the winter. In addition, Peter Way points out that "canallers were driven at industrial pace" by contractors paid by the yard—or, at times on the Mountain Ridge, by engineers and Commissioner Bouck, themselves compelled by a strong incentive to speed up and finish the work.[177]

The multiple dangers inherent in the work at the Mountain Ridge turned the laborer's day into a kind of Russian roulette. The navvy who managed to avoid dehydration and illness might be maimed or killed in a number of other ways. In April of 1824, on the Tonawanta Creek, the wind blew down a "partly grubbed up" tree, killing "two laborers on the canal."[178]

About a week later, the Lockport *Observatory* reported that David Gilroy, "a laborer upon the canal, was killed" when a box lifted by one of the horse-powered cranes, "amounting probably to 1,600 weight," after "being raised to the height of thirty feet, directly over the head of the man, the chain by which it was suspended broke, and the box with its contents fell upon him and killed him instantly."[179] In the Russian roulette of work in the Deep Cut, David Gilroy got the bullet.

Because of the contracting system, there simply are no records of construction casualties on any section of the Erie Canal. All that remains are anecdotal stories like the aforementioned two. At Lockport, even this record was mostly lost in a series of fires that almost entirely erased Lockport's antebellum newspaper archives.[180] Indeed, the story of the crane accident survives only because a Lewiston paper reprinted it.

All the anecdotal evidence indicates that the greatest danger to human and animal flesh came from blasting. Lyman Spalding observed that "the explosions (cannonading) were almost continuous during the day."[181] Orasmus Turner reported that "rocks were flying in all directions."[182] The sky above Lockport was full of stones. Not even middle-class citizens in their buildings with second roofs of timbers were entirely free from danger. Edna Smith reports that

> stones several inches in diameter were daily thrown over into Main Street. When the warning cry of "Look Out" was sounded for a blast everyone within range flew to a place of shelter. The small stones would rattle down like hail, and were anything but pleasant, particularly when one was caught with uncovered head. One stone weighing eighteen pounds was thrown over our house and buried itself in the front yard.

Mrs. Smith's home was more than seven hundred feet east of the Deep Cut. She also describes the "narrow escape" of Judge Ransom, whose Main Street office was only three hundred feet from the canal.

> He was one day sitting in the front room, his chair tilted back, and his feet resting on the table, when crash came a stone weighing twenty pounds just within the doors, rolled in, hit the legs of the chair, and down came the young counselor, in a very undignified manner and a surprised state of mind, at his unexpected fall.[183]

If eighteen-pound rocks could fly more than seven hundred feet, workers, overseers, and animals toiling in the artificial canyon of the Deep Cut were particularly vulnerable. George Catlin's lithograph, the *Process of Excavation, Lockport* (fig. 3.4), depicts a remarkable scene: a thousand men and boys, together with hundreds of horses, oxen, and other animals, toil in the perfectly orthogonal artificial canyon, while an explosion at the western vanishing point sends rock into the air.[184] During World War I, armies learned to construct zigzag trenches to blunt the linear enfilade effect of shrapnel from exploding shells. But if "a straight trench is a death-trap,"[185] what kind of trap was the Deep Cut, a channel as straight as a Parisian boulevard?

Thomas Nichols, a British traveler who visited Lockport in 1837, described a passage "through miles of deep-cut ravines, where hundreds of lives were lost in making them." This was a

> deep, dark cutting through the hard blue limestone, where tons of gunpowder were burnt, and the Irish laborers grew so reckless of life that at the signal for blasting, instead of running to the shelter provided for them, they would just hold their shovels over their heads to keep off the shower of small stones and be crushed every now and then by a big one.[186]

Presumably this story was circulating at Lockport where Nichols heard it twelve years after the end of the construction. A manly pride in facing their arduous and dangerous work was an element of the "rough culture" canal laborers were forging in response to their difficult situation.[187] Yet the implication that Irish laborers had no concern for their own lives and limbs and were therefore uniquely suited to canal work is perhaps a story middle-class people tell themselves to rationalize simple exploitation. Such stories about the Irish, in Peter Way's view, are "reminiscent of the assertion that blacks were built to work under a broiling sun or in fetid rice swamps."[188]

Workers' carelessness, bravado, and tipsiness undoubtedly contributed to their own danger. Sometimes, as Edna Smith reports, they strayed "too near the blast," or "if the fuse went out or burned slowly, they would rush back recklessly to see what was the matter, often blowing [on] them to revive the dying fire. Many a poor fellow was blown into fragments in this way."[189] Middle-class observers, even those like Edna Smith who recognized the human costs of excavating the Deep Cut, attributed almost the entire responsibility for injuries and deaths to worker negligence: their own free choices, and their own stupidity, seemed the only available explanation. Such accounts fail to take notice of the compulsion imposed by need and constrained options. Workers had little influence over working conditions. Even tipsiness was no simple choice.

"On some days," Edna Smith tells us, "the list of killed and wounded would be almost like that of a battle field."[190] Although such a human disaster could not be completely ignored, it seemed to cause most citizens little anguish. Like patriotic soldiers, these men might have been lauded for making the ultimate sacrifice in a crucial conflict that helped secure the expansion of the United States. From a cultural standpoint, however, canal laborers—particularly immigrant canal laborers—had no standing. Their status, in fact, came close to that of *homines sacri*, those who could be killed without the commission of homicide—literally, "the killing of a human"—because they had been written out of humanity.[191] There was certainly no place for them in the stories that middle-class Americans—those who did count—were telling themselves to make sense of their experience.

By the twentieth century, the directors of large construction projects built in the "costs" of injuries and deaths from the start, but in these opening days of the wage-labor era there were no insurance polices for the dependents of those killed, and little help for the injured or ill. Indeed, for those maimed by blasting, accidents, or even riots, the only recourse was charity. At Lockport, charity was represented by the person of Mrs. Edna Deane Smith. Upon her arrival in 1822, Mrs. Smith had taken in three laborers "blown up and seriously injured by a blast."[192] Along with her husband, the Quaker physician Isaac Smith, she opened her home as a makeshift hospital where the injured could "be nursed and taken care of." Whether motivated by her religious beliefs or simple compassion, Mrs. Smith seemed more willing than most middle-class Lockportians to extend the boundaries of community in the basic sense that, for her, even Iroquois and Irish laborers deserved kindness. Yet, despite her remarkable ability to stretch the boundaries of middle-class culture, even Edna Smith, as we shall see, remained firmly rooted in that culture with its aspirations of a refinement that seemed at once achieved and natural. A cynic might even label her concern for injured workers an emblem of middle-class refinement, as some have interpreted the wave of reforms that swept through the Northern middle-class in the ensuing decades.[193]

Workers were not passive in the face of economic and social marginalization, long exhausting days, and the ever-present dangers of injury and death. At the Mountain Ridge, fifteen hundred workers were massed together, many for a full five years. At times they must have sensed the potential strength of such a body. The conditions of wage-labor allowed them an unprecedented measure of freedom. When they clambered out of the Deep Cut at the end of a day or week, their time was their own. In the space provided by this freedom, they created a culture of sorts, independent of and oppositional to the culture of contractors, engineers, and local citizens. To some extent, this culture unified a heterogeneous group of Irish, Welsh, blacks, and the most desperate of Yorkers and Yankees. Canal workers had one thing in common: according to the assumptions of the ascendant middle-class culture, they were the most degraded specimens of humanity. Navvies arrived carrying elements of this culture, garnered from previous experience, but it was a culture forged in the process of canal work, in response to current experience and to the middle-class culture that encompassed their world.

Peter Way's insightful analyses of the "rough culture" of North American canal laborers brings some clarity to a neglected issue. It was a culture workers constructed from materials at hand. Since they felt valued only for their brawn and willingness to accept the appalling conditions and dangers of canal work, laborers turned manliness and bravado into central values. Drinking enhanced a "camaraderie founded on male bonding," and at the same time provided a means to demonstrate one's manliness. The emphasis on physical power and courage led directly to other contests such as gambling, arm wrestling, and brawling. Even their "exclusion from the rest of American culture" added to the "sense that they were outside the pale, a social threat, which gave them a status that swelled their male pride."[194] All of this served to further marginalize the wives and other women who lived among the workers. In a cascade of injustice, their subordination "served the psychic needs of men who were themselves held down within society."[195]

At times the culture of laborers could encourage explosions of violence and even uprisings, like those at the Mountain Ridge, that struck out against middle-class people and property. More often, these episodes pitted groups of workers against each other along lines of ethnicity or religion, as in the Lockport Riot of 1824, at least in its initial stages. Ethnicity and religion, particularly for socially excluded groups like blacks and Irish, were among the materials at hand for the creation of an independent culture. While such loyalties could serve as bastions, uniting groups of workers in familiar ways, they also laced workers' culture as a whole with deep fractures. That these divisions represented something more than simple ethnic bigotry is supported by the fact that ethnic feuds "were virtually unknown away from public works construction sites."[196] Particularly telling was the division between Irishmen from the south (Corkians) and the north (Connaughtmen). The feuding between these groups "crystallized on British and North American public works" where workers from these regions were "in direct competition for jobs."[197]

In Way's view, labor historians have tended to "overstate the integrity of working-class culture" partly because they have focused on the greater organizing success of artisans, as opposed to common laborers, and partly because they were reacting to earlier accounts that accorded no agency to workers. To glorify their rough culture is "to downplay the difficulty of workers' lives and the forces changing them."[198] The sexism, violence, and self-destructive qualities of this culture did little to alter workers' conditions or their powerlessness.

The middle-class citizens of Lockport were appalled at the behavior of laborers. "The Irish element largely predominated in the army of laborers employed in digging the canal," Edna Smith observed, "and with the hot blood of that nation, away from many of the restraints of a settled town, it needed all the force and vigilance at command to prevent wild work now and then." Violence generally originated "from quarrels among themselves" but "the inhabitants were always alarmed at such times, not knowing when the spirit of mischief would end, when once loosed from restraints."[199] She also reports numerous brawls and Donnybrooks in the shanty settlements touched off by "a religious or political discussion."[200] A newspaper account tells of an incident involving two young middle-class women who were spending the night together because the husband of one was out of town. Terrified of the rough laborers, they barricaded themselves in.

> About 10 p.m. they heard men quarreling outside. Stones were thrown, some hitting the house. Shrieks and groans could be heard as if somebody was being killed. . . . They climbed out the window at the rear of the house. . . . They ran in the darkness along the rock strewn Canal Street to the escarpment.

There they found safety with George Levelley who was tending his lime kiln. He "prepared a kettle of hot lye to use if necessary in their defense." They remained with him until the following morning.[201] Although the fears of these women may have been exaggerated, canaller violence was a great deal more than a stereotype born of middle-class hysteria. It certainly makes little sense to glorify workers' defiance when it turned criminally violent, regardless of whom its victims were.

The story of an 1823 murder in Lockport provides a revealing illustration of the pattern of violence among workers. The Lockport *Observatory* reported on February 6 that "*David Gilbreth*, a foreigner, and a laborer on the canal, was murdered in this village on Saturday evening last, by *William Smith*, a man of color" (emphasis in original). The incident began when

> Gilbreth, in company with five other canal laborers, went to the house of Smith about 7 o'clock in the evening, and after remaining a short time commenced playing cards. Smith, who had been before lying upon a bed, arose and forbid their playing, Gilbreth and his companions refused to obey him, when Smith pulled off his coat and made some attempts to put them out of doors, a scuffle ensued, in which Smith with the help of two other black men, succeeded in getting them out: and while one of the blacks was endeavoring to fasten the door with an axe, Smith caught it out of his hands, sallied out into he the dooryard, and coming up with Gilbreth, gave him two or three violent blows with the axe upon his head which killed him instantly.[202]

Despite the invasion of his house, and the intense harassment and provocation of three black men by six Irishmen, only Smith was arrested. This shocking incident provides a glimpse of the dynamics at the bottom of North American society. Whatever terrors middle-class people like the two young women who fled into the darkness faced, they paled in comparison to everyday life among the shanties, particularly for blacks who were essentially without protection.

It would be misleading to characterize the relation between Irish and blacks in the early 1820s as one of continual conflict. At the Mountain Ridge, as well as in New York City's Five Points, the two groups lived, worked, and socialized together. A separate incident underscores this dynamic. When two slave hunters from Kentucky seized a local, a free black man who worked as a barber, contractor Darius Comstock, "noting the excitement on West Main Street, brought his men up from the canal, seized the men and sent the barber back to his shop." They disarmed the slave hunters "and drove them out of town."[203] Irish workers, although led to this noble rescue by their Quaker boss, apparently felt some sympathy for the black man who almost lost his freedom, a feeling that would become increasingly rare among Irish Americans in subsequent decades. In a process that required nearly a century and a half to complete, Irish Americans slowly climbed out of that ditch that defined social and economic exclusion in the 1820s. Blacks, for the most part, did not. The Irish took advantage of the one opportunity that American society afforded them: they became white. It may have been the easiest path out of the ditch, but it was an ugly one. In the ensuing decades, Irish Americans increasingly defined themselves against blacks.[204] Daniel O'Connell—the man who most embodied Irish hopes and who led the successful campaign for Catholic emancipation in the British House of Commons—admonished Irish Americans to remember their own history of exploitation and oppose slavery. Instead, they largely opposed abolition and many openly attacked black communities.[205] The fact that Irish Americans eventually achieved economic success and social acceptance underscores the malleability of cultural categories. The very qualities other Americans cited to justify the exclusion of the Irish—their ethnicity and religion—became neutral markers of identity, yet the system of exclusion remained at least partly intact.

On September 28, 1960, Lockport's large and still partly segregated Irish community came to the Big Bridge to join one of the largest crowds ever assembled in the town in order to witness an event that seemed to signify the completion of the journey that had begun, among other places, in the ditch beneath that very bridge. In the late afternoon, a motor caravan arrived at the speaker's platform and out stepped presidential candidate John F. Kennedy. His speech urged Americans to move "forward in the Sixties" by "proving that through freedom we can outproduce the communists." Accompanying him was Robert F. Wagner, the mayor of New York City, who, Kennedy quipped, had "come along to see what a big city looks like."[206] Apparently neither Kennedy nor Wagner were aware that one of the key moments in the rise of New York—arguably *the* key moment—had taken place in Lockport 125 years earlier when canal laborers and their overseers had overcome the last barrier and completed the Erie Canal. Although some of Lockport's Irish Americans were direct descendents of canal laborers, they were not drunk that day. They neither rioted nor terrorized the "sober Yankees." Had the "hot blood of that nation" undergone a genetic mutation? No. Their adoption of middle-class norms had been a decision, facilitated in part by the presence of black Americans against whom they could distinguish themselves. This, along with their economic success, makes one thing clear: rioting, drinking, and other rough behaviors were not entirely of foreign origin, nor were they indications of an innate barbarity; rather, they were, at least in part, responses to their exclusion itself and to the hellish conditions of their lives and work.

The Citizens' Lockport

Although Lockport's citizens lived in a different world from that of the laborers, these worlds intersected at the work site, in the streets and shops, and occasionally in violent confrontations that clearly demarcated the boundary between them. The two groups evolved together, an oppositional pair, each constellating in the other a sense of their own group as a body.

In the world of Lockport's citizens, there was less deprivation and violence, but still considerable anxiety and insecurity. Like the laborers, they suffered from what Dr. Smith's wife called "agues, fevers and other diseases incident to a new country,"[207] and of course the rocks raining from Lockport's sky respected no social boundaries. Joseph Comstock who, together with Otis Hathaway, had secured the contract for section 2 in 1821, died in 1822.[208] Amy Comstock, daughter of contractor Jared Comstock, in 1821 married George B. Rogers, a skilled blacksmith who made "a large part of the iron works for the locks." She died one year later, perhaps in childbirth, and was the first person buried in the Quaker cemetery.[209] Isaac J. Thomas, an assistant engineer and son of Chief Engineer David Thomas, died in 1824. Resident engineer Davis Hurd wrote on the voucher that "in consequence of his sickness and death, his account was settled with his father, David Thomas."[210] Of course, the names of laborers who died are much less likely to have left a historical trace.

Because Lockport was, in Holmes Hutchinson's words, "a money-making place," merchants, artisans, doctors, and ministers flocked to it almost as rapidly as laborers. The Buffalo *Journal* reported early in 1822 that Lockport already had 337 families exclusive of canal laborers.[211] If there were four persons per household, the total population of citizens would be 1,352, a number about equal to that of the laborers.

In the days before the temperance movement, alcohol played an important role in middle-class life. While contractors and overseers shared whiskey with laborers when the arrival of the "jigger boss" signaled a break from work, citizens and workers rarely imbibed at the same establishments after work hours. Samuel Jennings was clearly appealing to citizens when he advertised his "long acquaintance with the business of entertaining genteel company" and promised at his "public house" to keep "on hand at all times the choicest of liquors."[212] At times it seemed perfectly acceptable to consume large quantities of alcohol. As we shall see, after public celebrations of the canal's completion, dignitaries at canal towns would retreat to an indoor venue to continue celebrating with prominent local citizens. On such occasions, the social elite would drink to countless toasts in swift succession.[213] Middle-class people, as much as laborers, were experiencing rapid social and economic change, and like their neighbors in the shanties, they too craved community. When Otis Hathaway completed Lockport's first gristmill on the Eighteenmile Creek northwest of the locks, he invited all "his friends" down to the mill to help celebrate. "In those days," Edna Smith observed, "everything ended in a jollification." They suspended a large potash kettle over a bonfire and made "a mighty dish of mush." The group "had a great frolic, threw Mr. Hathaway into his own meal chest, and came near to smothering him to death." Obviously, they had already been drinking.

> There was some snow on the ground so they lifted the kettle onto an ox sled, and drove up town to "the cottage," a log tavern on the corner of Main and Cottage Sts., carried it into the dining room, set it in the middle of the floor, poured in wine, brandy, molasses and butter and each swallowed a plateful of that compound beside some glasses of the liquid ingredients, in an unadulterated state, and no one went "home till morning."[214]

Apparently, Lockport's Yankees were not always sober.

If both laborers and citizens experienced danger, death, and drunkenness—albeit to different degrees—a fundamental difference remained. Citizens had the palpable sense that they were building a community that would endure. One indication of this was the founding of churches. By 1824, four denominations had completed church construction: the Quakers, Presbyterians, Methodists, and Baptists.[215] Meanwhile, Lockport's largest religious group, the Roman Catholics, not only had no church, but there is no evidence even of visiting priests. The citizens who secretly moved the first newspaper press to Lockport and campaigned for their town to become the county seat felt a stake in the place no laborer could possibly have felt.

Indeed, Dr. Smith, Lyman Spalding, Orasmus Turner, and many other early middle-class leaders would make Lockport their permanent home. Partly in response to the flux around them, citizens embraced endurance and solidity as markers of middle-class identity. Although they too were recent migrants to the place, and many would eventually move again, citizens strove to distinguish themselves from the transients in the shanties.

Citizens' sense of community, like laborers' solidarity, was heightened by the forces surrounding it: the primitive frontier conditions, rapid economic change, and the unruly army of workers. Hence, middle-class retrospective evocations of community life often depict winter scenes. Taverns were important centers of community because they offered the warmth of both brandy and the hearth. They were bastions against the cold and chaos. Marcus Moses, writing fifty-five years after his arrival in 1823, praised Joseph Langdon, proprietor of the Cottage, because "he always kept a good fire when needed. In cold weather, in the old-fashioned Dutch fireplace would be seen a large back log and on the andirons a large fore stick and heaped upon these about a quarter cord of wood."[216]

Edna Smith felt a sense of community most intensely in the confusing and difficult "first few years of our settlement in Lockport." She recalled that, upon her family hearth,

> many a winter evening before a flaying fire, stood dishes of apples, dough-nuts, mince pies, and pitchers of cider, warming for the delectation of guests gathered from the village—a social circle, that for good feeling, wit and fun, is seldom surpassed.[217]

Such descriptions afford a glimpse of an incipient bourgeois sensibility that would soon propel the revival of the cult of Christmas as a celebration of the private family home, a "haven in a heartless world."[218]

Edna Smith described a remarkable incident in her *Recollections of an Early Settler* that reveals a great deal about middle-class consciousness. Edna's husband, Dr. Smith, told her of "a case of suffering which he had just left" and asked her "to do something to relieve it." Her description is the only detailed picture we have of the inside of a Mountain Ridge shanty:

> In a shanty of the poorest description, no floor but the bare ground, for a door a board set up, for a bedstead, some poles with one end fastened to the side of the house, the other resting upon stakes driven into the ground, and pieces of wood laid across, upon which was a straw bed with very little bedding.

On this "miserable pallet" was Dr. Smith's patient,

> a young, beautiful, refined English woman, with her first-born infant in her arms, no one in attendance, altogether in about as forlorn a condition as one can well imagine. The poor girl was sobbing and almost in despair, and you may be sure my tears flowed as freely as her own to see the wretched state she was in. Upon questioning her she poured forth the old and not uncommon story of love opposed by "cruel parents" on her side ending in an elopement, and emigration to Canada. Then the money they brought with them was soon expended in vain search for employment. Hearing that public works were in progress on this side, the poor young things drifted over here, the same ill luck followed them and finally with starvation staring them in the face, he went to work as a common laborer on the canal.

Edna, with the help of some of her "kind neighbors" quickly had the couple "comfortably housed." Soon after, when the woman's husband, "an Irish gentleman, uncommonly fine looking with a noble frank face," met with his contractor to receive his wages, the contractor

> noticed the extreme delicacy of his hands, that were blistered and bleeding from the unusual handling of the shovel. When called upon to sign the receipt, his pride at his situation was such that he took the pen and made his mark. The contractor looked at him a moment and said "a man with such a face and such hands can write his name I know." With a flush of shame, he wrote his name in a most graceful and beautiful manner.

The contractor immediately hired him as a bookkeeper, a position he retained until the canal's completion. He then moved to Rochester where he became a successful businessman and eventually was elected mayor. Edna met him on a packet boat years later and observed that he "never forgot his old friends and remembered enough of his old sorrows, to have his hand and heart always open to the wants and sorrows of others."[219]

The fact that Dr. Smith was tending to people among the shanties, and that he and Edna took injured laborers into their own home, shows that they recognized the humanity of laborers and sympathized with their suffering. Yet the instant class solidarity the Smiths felt for this particular "case of suffering" is telling. If other laborers were human, why was it any less tragic that they should have to put up with inhumane conditions? Why was this couple more deserving? Perhaps because she was a "beautiful, refined English woman," and he was a proud, "intelligent" "gentleman" with hands of "extreme delicacy" that rendered a "graceful and beautiful" script. If these were the qualities that made the couple seem out of place among the shanties, what does that imply about the other common laborers?

That the men were clumsy, illiterate, heavy-handed, and stupid while the women were coarse, ugly, and Irish? Did the couple's qualities mean that they deserved to suffer less than the laborers? Were laborers born to sleep in draughty, insect-ridden shanties and to work fourteen-hour days with rocks raining upon their heads? Certainly, Edna Smith could see herself in this English woman; their tears flowed together. In this changing world, it was possible to fall through the floor of respectability. Edna expresses more than her own anxiety when she recounts the gentleman's "shame" at admitting he was an educated man.

This incident raises other questions. Was the instant solidarity the Smiths felt for the couple a middle-class solidarity? Since the couple's economic status was no different from that of other laborers, we can categorize them as middle-class only if we define that class in cultural rather than economic terms. Does the sharp boundary at Lockport between laborers and citizens mean that the latter were all middle class? Clearly they were not all bourgeois in the structural sense of controlling some means of production. Many owned no land. Some worked for relatively low wages as clerks for merchants or in the shops of bakers and blacksmiths. Most citizens probably shared an appreciation of middle-class respectability, an aspiration to achieve it, and more fundamentally, so deep a fear of falling to the status of common canal laborers that they would avoid it until starvation was "staring them in the face." Although rags-to-riches stories of upward mobility during this period are largely mythology, "booming smaller places"[220] like Lockport could be exceptions. If such opportunities existed in Lockport, they would have helped to fuel middle-class aspirations among the lower echelons of citizens that, in turn, could enhance overall citizen unity.

Class exploitation was a fact of the entire canal enterprise with laborers subsisting on wages that barely secured their ability to reproduce themselves, while landowners, merchants, speculators, and others garnered colossal rewards. In this changing world, historian Charles Sellers argues, "a middle class of consciousness encompassed people of whatever class" who were willing to embrace a mythology that posited "a middle class of the effortful" in a society atomized by "a marketplace rewarding each according to effort." Because of this, he argues, the "so-called middle class was constituted not by mode and relations of production but by ideology."[221] Henry Clay's campaign manager, the Reverend Calvin Colton, who introduced the term "middle class" to the American scene, proclaimed that "ours is a country where men start from an humble origin and . . . rise gradually in the world, as the reward of merit and industry." Failure to rise is, therefore, a moral failure since it implies a lack of either merit or industry. Accordingly, "the idle, lazy, poor man gets little pity in his poverty."[222] The gradual ascendancy to middle-class economic status of the gentleman and English woman in Mrs. Smith's account underscores the reality that effort and merit alone were no guarantee of success. They rose because of the accumulated cultural capital of their education and upbringing, sheer luck, and the kindness of strangers.

Although the developing class relations at Lockport were in some ways a harbinger of things to come, middle-class ideology was only beginning to take hold in 1824. Indeed Sellers contends that middle-class ideology arose along with the market revolution, the "decisive phase" of which was ignited only at the canal's completion.[223] Hence, the most self-repressive aspects of the ideology—ratcheted up in the following decades by the temperance movement, among other things—were not yet fully in evidence.

What is laid bare by the tumultuous events of Lockport's construction period is the fundamental connection between the rise of middle-class consciousness and the challenging presence of an army of defiant free laborers. Here was something no one had seen before or expected: laborers, "loosed from restraints"[224] and armed, marching three hundred strong through the center of the village with fife and drum. In direct response to this presence, as the Lockport *Observatory* account put it, "our citizens organized themselves into a body." The citizens' coming together is presented as a rational effort, but the use of the term "body" implies an entity whose boundaries are given by nature, not choice.[225] The citizens become a body in response to a circumambient environment of apparent chaos—chaos embodied in the defiant bodies of laborers. Ironically, what for the citizens was chaos, was actually one of the rare moments when a large group of laborers also formed a body. Indeed, this was the central threat, an abyss that needed to be covered over, a task that would be recurrent for the panicked Northern middle class throughout the antebellum period.[226] Although contemporary middle-class accounts dismissed the 1824 riot as a bizarre aberration disconnected from local realities, it is thus possible to see it as a moment of clarity that revealed the new dynamics of the commercial order the canal was so instrumental in initiating. In the riot and the citizens' response of joining together and arming themselves against the unruly workers, the boundary between the two groups was clearly articulated. Just as the worker's defiant presence forced citizens into the recognition of themselves as a "body" wholly separate from the workers, so too middle-class consciousness and ideology formed partly in response to this same presence. "By the time of the Erie's completion," Carol Sheriff concludes, "the project's sponsors no longer claimed that the new commercial era could be launched without class tensions."[227] Moreover, the place where the Erie was completed—the Mountain Ridge—and the way it was completed—with the canal's only two major labor riots—underscored this fact.

The laborers' oppositional culture, and even their violent uprising, did little to improve their own situation. Their actions, nevertheless, had the profound effect of driving the citizens toward an anxious unity and concern for their own status. These actions also established laborers as an undeniable collective presence. In the decades following the canal's completion, at industrial places like Lockport, Utica, and Rochester, they would remain a challenging presence that would continue to affect the development of the Northern middle class.[228]

The day after the riot, as if to underscore its futility, work resumed as usual. At the end of the 1824 season, the commissioners reported that the combined locks were nearly completed.

> This is a work of the first magnitude on the line, and one of the greatest of the kind in the world. The superior style in which it is executed—its situation at the brow of a perpendicular precipice of about seventy-six feet, overlooking a capacious natural basin, with banks on each side of an altitude of more than one hundred feet, connected with the deep rock excavation, renders it one of the most interesting points on The Erie Canal.[229]

The commissioners also noted that excavation work had proceeded at a rapid pace, similar to that of 1823. By the end of the season, rock sections 1, 2, and 3 (1822 numbering) were finished, leaving only 56,000 cubic yards of rock in the three southerly sections of the Mountain Ridge.[230] The contracting situation had become very complex. At the close of 1823, commissioner Bouck had divided the two southerly sections (5 and 6) into eleven contracts. Some of these were further subdivided in 1824. Surviving documents show that the four original sections of the Deep Cut were now divided into twenty-two separate contracts, with indications that even these may have been further subdivided (fig. 3.6; appendix).[231]

In late summer, crews completed the remainder of the long level between Rochester and the foot of the Lockport locks. The Rochester *Telegraph* announced that "the water was let into the canal as far as Lockport" on August 30th and that "boats could reach there in about a week."[232] For the next fourteen months, Lockport would serve as the canal's western terminus, the head of a navigation system that reached to New York city. The *Telegraph* also noted that a "*Daily Stage* now runs between Lockport and Buffalo." During the spring and summer of 1824, stages had plied the

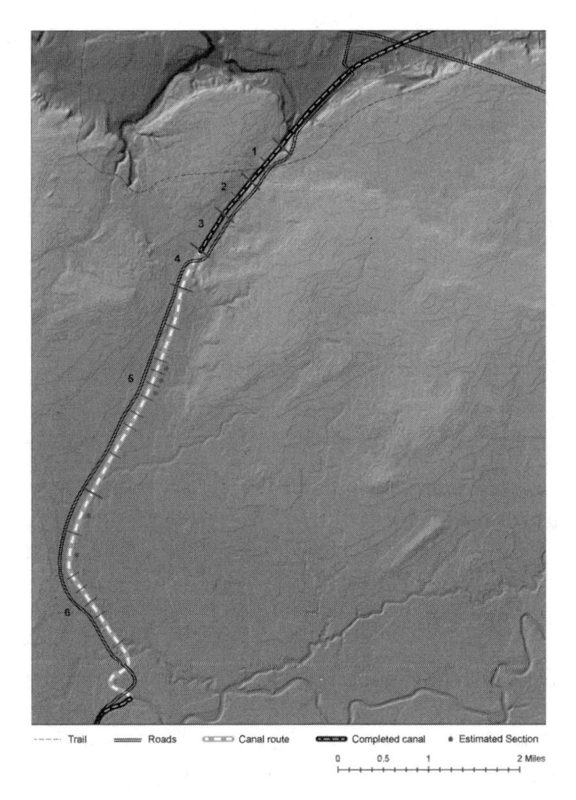

Figure 3.6. Mountain Ridge contracting sections of 1824, overlain on digital elevation model (United States Geological Survey). Map by Cherin Abdel Samie and Yasser Ayad.

long portage between Brockport and Buffalo via the Ridge Road, avoiding Lockport and its still notorious road connections. The site of Lockport, a rocky ridge forming a steep cliff, surrounded north and south by swamps, possessed few natural transport advantages. Nor, without the canal and the Lake Erie water it channeled there, did it possess industrial prospects. Without the canal, there would be no town, but with the canal, Lockport suddenly became the key transshipment point for goods and passengers bound not only for Buffalo and points west on the Great Lakes but also via stages to Niagara Falls, Lewiston, and Upper Canada.

Months before the canal reached Lockport, the Utica and Schenectady Packet Boat Company announced plans to run as many as two daily boats in each direction from Schenectady, Utica, Weed's Basin, Rochester, and Lockport. The company emphasized that "stages would be provided at all times" to connect these canal ports with the "principal villages" near the canal and beyond.[233]

There were additional improvements in 1824 to the portaging route west of Lockport. Contractors and their crews completed the canal section from the junction of the Tonawanta Creek and the Deep Cut to the creek's mouth. This involved the construction of a dam and a lock that connected the canal to the Niagara River, opening the first of the canal's three western termini. By the close of the 1824 season, the commissioners reported that "boats have passed the whole way."[234] The Mountain Ridge was now the last obstruction, but overcoming it would require another whole season.

In the meantime, commissioner Bouck contracted with two men to build a stage road "between Lockport and the Tonnawanta Creek . . . , the clay part to be covered by gravel to render it reasonably permanent for loaded teams."[235] Bouck expected this state-financed road (see fig. 3.6) to receive heavy traffic since it spanned the final seven-mile gap in the water route from the Atlantic to Lake Erie.

Though Lockport, from it's beginning, had been "a creation of The Erie Canal,"[236] with the arrival of canal boats in the autumn of 1824, it finally became a functioning canal port. This event, though long anticipated, must have appeared revolutionary, particularly to pioneers like the Comstocks who had lived through years of dreary isolation. Suddenly their lives were connected to a vast and dynamic world via a water route that reached to the Atlantic. The Lewiston *Sentinel* gushed that "we, of this frontier . . . have witnessed so far this season, more of the traveling mania, than in any one year within our remembrance." It also observed "a sensible revival of trade and business in general" encouraged by the "great perfection to which the *means* of transportation has been brought in this state," a taste of the frontier's future prospects when "the waters of Erie are passing through the Mountain Ridge."[237]

Even in its incomplete state, New York's canal was already affecting places farther west. A Dayton, Ohio, newspaper relayed news in 1824 from a "gentleman of this place who has recently returned from Detroit," a place that "looked remarkably dull" until ice left the river on May 8. Then, "steam and other vessels arrived bringing upwards of 800 settlers into Michigan Territory, and before he left there, 200 more landed making upwards of 1,000 in a week." The paper concluded: "such is the benefit of the New-York canal."[238]

The people of such western places followed the canal's progress like Americans of the 1960s followed the Apollo Missions. The canal seemed to embody a futuristic sense of progress that promised palpable and practical effects in their lives. The pivotal work at the Mountain Ridge was beginning to make Lockport a place of national prominence. Accordingly, the

1824 *Gazetteer of The State of New York* included a long description of the work in progress there—calling the Deep Cut a " 'Big Ditch' in good earnest," and cataloguing the town's meteoric rise. Lockport already contained twenty-four mechanic shops, five law offices, eight physicians, eight inns, four schools, and a population of 1,458 exclusive of laborers. Taking into account that this survey was based on earlier information, and adding in the laborers, gives a total population in 1824 of about 2,500 to 3,000. The *Gazetteer* also described the water power potential of Lockport "from which great expectations are formed" of an industrial future.[239] At the end of 1824 there was still much work to be done to usher in not only Lockport's future but that of countless other places—from New York to Chicago—that would feel the canal's effects.

1825

The state *Gazetteer* and local newspapers always estimated Lockport's population "exclusive of laborers." While workers were undoubtedly more difficult to count, this practice is another indication that, from the citizens' point of view, common laborers were not part of Lockport: they simply did not count. That the few workforce estimates we have always round off to the nearest five hundred (e.g., one thousand or fifteen hundred) only shows what crude guesses they were. There are some hints in the contracts suggesting that the total number of navvies at the Mountain Ridge may have risen to two thousand or more, although the paucity of evidence makes it impossible to draw a firm conclusion.[240] In addition, we have absolutely no estimates of the number of women, girls and young children who worked off site as laundresses, cooks, grocers, and so forth; even if outnumbered by males four-to-one, there would have been between three and five hundred of them. It is conceivable, therefore, that the laborers' community was larger than that of Lockport's citizens, perhaps even substantially so.

The commissioners had announced the completion of the three northerly rock sections at the end of 1824.[241] As contractors completed their sections, Bouck assigned them to move their crews where needed. The commissioner obviously found it useful to retain some of the flexibility he had become accustomed to under the innovative 1822 contracts. Bell and Richardson, for example, employed their laborers on the sections of six different contractors.[242] Crews led by Otis Hathaway, Darius Comstock, and Seymour Scovell also moved from place to place on the Deep Cut, presumably still feeding, housing, and paying their workers on a monthly basis. Imported crews usually performed tasks that, while often part of the original contracts, were separate from actual excavation. These included constructing drains and towpaths, pumping out water, and removing timber from canal banks. Bouck also issued dozens of minor contracts to construct stone culverts, erect walls, divert small streams, and complete various other tasks.[243] With the entire canal to the east completed, and so much attention on the work at the Mountain Ridge, the whole enterprise finally seemed to be functioning like one large construction company with Bouck as the active manager directing engineers, as well as contractors and their crews, to apply their efforts where needed.

Surviving records do not clearly indicate which sections were the last completed.[244] Crews finished excavating the section from the Tonawanta Creek to Black Rock on June 1, opening the canal's second western terminus. The line from Black Rock harbor to Buffalo was also progressing rapidly, and with the special inducement of barrels of whiskey, it too would be completed before the Deep Cut.[245]

For grand public events, June of 1825 would be the liveliest month the people of Lockport had ever seen. The canal that had released them from isolation only a few months earlier was now bringing within their compass elements of the vast world. For the first time, Lockport's citizens, if not its common laborers, were about to experience a palpable sense of participation in national life. General Lafayette was coming to Lockport. The venerable general's two-year tour of the country he had been so instrumental in liberating nearly fifty years before seemed to catalyze nationalistic sentiments that were at once retrospective and prospective. Though Americans were buffeted by rapid change and increasingly divided by growing class and sectional tensions, historian Andrew Burstein argues that they nevertheless agreed on one thing: "that homage should be paid to their Revolutionary origins."[246] Lafayette's triumphal tour was one of several occasions of nationalistic celebration that characterized a period some have called the Era of Good Feelings. The festivities marking the opening of the Erie Canal, only four months away, and the Jubilee of Independence in 1826 would stimulate similar outpourings of emotion and rhetoric. Popular nationalism was just beginning to emerge in 1825, and, as we shall see, the Erie Canal would play a catalytic role in its growth.[247]

Lafayette had crossed the ocean at age nineteen to volunteer with Washington's Revolutionary forces. Two years later he journeyed back to France to argue for naval assistance. After returning to Virginia, he convinced Washington that, by combining French and republican forces at Yorktown, they could trap General Cornwallis and his forces. The French Navy arrived just in time to seal the decisive victory. Well aware of these now legendary events, Americans positively gushed at the sight of the old hero. His presence also presented an opportunity to compare the Revolutionary generation with their own and to celebrate both the nation's founding ideals and its subsequent progress, represented most clearly by the Grand Canal he would be the first foreign dignitary to traverse. Having survived a second war with Britain and doubled the size of their territory, many Americans believed that the potential for national growth was unlimited, producing what literary scholar John Seelye calls "geopolitical euphoria."[248]

On the morning of June 7th a party of seventy to eighty horsemen rode northwest from Lockport to greet General Lafayette and his traveling companions, who included the Marquis's son, George Washington Lafayette, and his secretary, August Levasseur, who recorded and commented upon the entire tour. Among those greeting Lafayette was the president of the canal commissioners, Stephen Van Rensselaer, who had apparently made a special trip in order to accompany his old fellow soldier east on the Grand Canal. The general had arrived from the west on a Lake Erie steamboat and had visited Niagara Falls before proceeding to Lockport. At the very moment when Lafayette and his escort emerged from the forest at the western edge of Lockport, "hundreds of small blasts, charged with powder by the workmen engaged in quarrying the bed of the rock to form the canal, exploded at the same moment, and hurled fragments of rock into the air, which fell amidst the acclamations of the crowd."[249] Peter Way suggests that this distinctive greeting expressed workers' pride in their own skill and daring.[250] "No where," Levasseur records, "have I ever seen the activity and industry of man conquering nature so completely as in this growing village." Everywhere he heard "the sound of hatchet and hammer." Trees were cut, "fashioned under the hands of the carpenter, and raised on the same spot in the form of a house." Lockport's streets were "traced through the forest, and yet encumbered with trunks of trees and scattered branches," and "in the midst of these encroachments of civilization on savage nature, there is going on, with a rapidity that appears miraculous, that gigantic work, the grand canal, which, in tightening the bonds of the American Union, spreads comfort and abundance in the wilds through which it passes."[251]

The carriages stopped in front of "an arch of green branches" though which local dignitaries led the general to a platform where thousands witnessed a brief public ceremony. A dinner at the Washington House with Lockport's leading citizens followed. At its conclusion, the toasts began. Finally, Lafayette rose and spoke: "To Lockport and the county of Niagara—they contain the greatest wonders of art and nature, prodigies only to be surpassed by those of liberty and equal rights."[252] In the next chapter, I will examine in detail the meanings people saw in the conquest of the Mountain Ridge. The words and pictures through which they articulated those meanings constitute a kind of public dialogue on certain contemporary preoccupations. Here, it is sufficient to note that Lafayette and Levasseur, the most worldly and articulate of Lockport's early visitors, identified what would be one of the most central themes of that public dialogue—namely, the relation of civilization (or art) to nature. The complex association of Lockport, with its artificial wonders, and the nearby natural wonder, Niagara Falls, was one way of dramatizing this relation. Indeed, contemporaries often viewed the canal as a whole in these terms, and increasingly the destiny of the nation as well. The next day at Rochester, for example, the patriarch of that village stood with the general upon a platform atop the great aqueduct and commented: "the foaming contract beneath beats in vain." In response, Lafayette described the canal as "an admirable work, which genius, science and patriotism have united to construct, and which grand objects of nature, which threatened to impede, have been made only to adorn."[253]

After the toasts at Lockport, Edna Smith recalls, the general "received the ladies and children, who thronged in from the surrounding country, in the parlors of the hotel." At about 7 o'clock Lafayette and his companions made their way to the basin below the locks—still the western end of the canal—where a flotilla of packet boats waited, "one fitted up elegantly for himself and suite, each drawn by six horses gaily caparisoned" and "decorated with flags and banners, having bands of music on board and crowded with invited guests to escort him on his way." Onlookers crowded every available vantage point, knowing that canal crews had prepared another unique spectacle. Mrs. Smith describes the scene:

> A salute was improvised by drilling one hundred holes a few inches deep on the north side of the ravine on the edge of the bank. . . . In each was put a small charge of powder, and a train laid from one to the other, and just as the boats started, the music struck up, the thousands of spectators cheered, the train was fired and the hundred guns roared and reverberated from what was then the wooded and rocky shores of the glen and everything went off in a blaze of glory.

Amid all this faux artillery, apparently at least one real cannon saw action in each of the two salutes that marked Lafayette's arrival and departure. The Lockport *Observatory* announced "with regret" that two men "had their right arms so severely injured that it became necessary to have them amputated." The accidents had occurred "at different times, by the discharge of the cannon, while the unfortunate individuals where in the act of ramming down the cartridge."[254] With flying rocks and maiming, even the otherworldly hiatus provided by General Lafayette's visit bore an uncanny resemblance to another day in the ditch.

A different kind of public spectacle animated the people of Lockport in mid June. Preparations were underway, in Niagara Square in Buffalo, for a triple hanging. These preparations included the construction of bleachers to accommodate the expected throngs. People began arriving at Lockport on packets and then either walking or taking a stage around the Deep Cut. On June 17, the three Thayer brothers hung for the murder of a peddler named John Love.[255]

In early June newspapers announced preparations for Lockport's next public event. "The grand and sublime ceremony," the Rochester *Telegraph* declared, "of laying the *top stone* of the Locks, at the Mountain Ridge [Lockport] will be performed by the brethren of the *mystick tie*" on June 24. The occasion was to mark the completion of "the locks connecting the Erie and Genesee levels of the canal." The announcement invited "all Masonick brethren in the vicinity" to attend, and urged interested parties to register for accommodations and packets that "will leave on [the] 23rd." The following week the *Telegraph* added that the event "is expected to be numerously attended by masonick brethren from Canada as well as the neighboring villages," and that "Rev. F. H. Cuming of this village, is appointed to deliver an address."[256]

The choice of June 24 was dictated not merely by the progress of lock construction, but principally because it was the feast of St. John the Baptist, a key day for Masons in part because of its traditional association with the summer solstice. Culver, Maynard, and their crew apparently completed the locks themselves at least several weeks prior to June 24, but Bouck issued many small contracts to put the finishing touches on what many considered the Erie Canal's centerpiece. Workers were therefore busy with such tasks as installing lamps—the locks would soon be operating twenty-four hours a day—drilling holes for railings, installing and painting railings, and constructing a wharf at the foot of the locks.[257] Bouck had also ordered two marble tablets in May; presumably he either approved the inscriptions or penned them himself.[258]

The Lockport Five, as they would soon be called, were unlike other locks on the Erie, or for that matter, on any canal. James Geddes had first proposed combined or staircase locks in 1817, although he would have located them over a mile to the west. On the western side of the Niagara River, Welland Canal engineers would opt for a gradual approach, climbing the side of the escarpment in a winding sweep of locks interspersed with levels.[259] The choice, first by Geddes and then by Thomas, to follow the path of a prehistoric river through a gorge it had etched into the escarpment, essentially precluded a gradual approach. The so-called natural Basin was a box canyon that terminated at a seventy-five foot waterfall. Ascending this cliff gradually would have multiplied the required rock excavation many times. Although the Lockport combine was the only staircase on the original Erie, there were precedents in Europe; in addition to the French masterpieces on the Canal Du Briare and the Canal Du Midi, there were several more recent British examples.[260]

What distinguished the Lockport Five from all other combined locks was that they were actually the Lockport Ten. It was David Thomas's conception of double combined locks that allowed two-way traffic at a point that otherwise would have proved a disastrous bottleneck. Apparently, many engineers submitted proposals to implement Thomas's plan. When the commissioners chose Nathan Roberts's plan, he considered it "the most triumphant moment" of his career, though he "rose to be a master of his profession."[261] Assuming a vertical lift of no more than eight feet per lock, the Erie Canal standard, Geddes had proposed a flight of eight locks to accomplish the sixty-foot rise. Roberts's innovation was to increase the steepness of the staircase, and thereby decrease the amount of rock to be excavated, and the number of locks needed. The increased steepness also enhanced the visual impact of the scene and helped make Lockport a distinctively vertiginous town. Because boats could pass through five locks more quickly than through eight, increasing the steepness of the staircase also enhanced its functionality. Taken together, the innovations of Roberts and Thomas greatly speeded up the future passage of boats between the Genesee and Lake Erie levels of the canal.

Commissioner Bouck, probably in consultation with Chief Engineer Roberts, directed that three separate inscriptions be affixed to the locks. An oval marble tablet, set centrally at the foot of the locks, contained the following straightforward message:

"The Erie Canal"
362 Miles in length was commenced the
4th of July, 1817
And completed in the year 1825
At an expense of about $7,000,000,
And was constructed exclusively
By the Citizens of the State of NEW YORK[262]

The capstone that was to be set in place on June 24 had a cavity into which certain unknown items were placed. A brass plate covered the cavity and bore this inscription: "First stone laid July 9, 1823 by N. S. Roberts, Engineer and Samuel Horn, Master Workman, O. Culver and J. Maynard, Contractors."[263] It is unusual to find an artisan, even a master mason like Horn, memorialized by name. Perhaps the Masonic influence, so evident in the ceremony of June 24, had something to do with this. There were, of course, a horde of other workers involved who remain unmentioned. Constructing locks required a crew with a higher percentage of skilled artisans than crews typical of excavation work. A large number of common laborers did participate in the considerable excavation necessary to prepare the site for the locks, a job that took longer than actual lock construction.

The final inscription appeared on a second marble tablet placed at the top of the locks:

Let Posterity be excited to perpetuate
our
Free Institutions
and to make still greater efforts than
their ancestors, to
promote
Public Prosperity
by the recollection that these works of
Internal Improvements
were achieved by the
Spirit and Perseverance
of
REPUBLICAN FREE MEN[264]

Despite the lack of federal financial backing, middle-class New Yorkers obviously saw the Erie Canal as a national achievement. Canal scholars have often referred to these words partly because they present, in Carol Sheriff's view, "a concise statement of the canal sponsors' attitude toward progress."[265] This generation would fulfill the promise of the virtuous republican past by spreading prosperity and "free institutions" westward. Of course, the people who built the Erie Canal were a "more motley crew" than the inscription indicates.[266] "Very few of the thousands of men who worked on the Deep Cut or the combined locks," Sheriff observes, "would have qualified as republican citizens—in either their own or other minds."[267]

The Lockport chapter of the Ancient Order of Hibernians, 164 years later, finally erected a small monument recognizing canal laborers. They placed the monument, which was also approved by New York State legislature, at the north end of the Big Bridge. The inscription, along with local newspaper coverage of the event, had a retrospective quality, with Lockport's now successful Irish American community emphasizing how far they had come. It could have been a moment to recognize the contributions of all laborers who helped build the canal—whether black, Welsh, Irish, or other—but instead the inscription singled out only the Irish: "In memory of the many Irish Immigrant Laborers whose endurance in the construction of the Grand Erie Canal brought untold wealth to the area in which they settled."[268]

On June 24, 1825, a large number of masons, "together with the citizens and strangers formed a numerous assemblage."[269] The Reverend F. H. Cuming's address was, once again, both retrospective and prospective. He began by waking dead heroes, telling them to "throw off their load of earth" and witness the improvements and prosperity of the republic. If he could, he would bring them to "this romantic place" where "wilderness skirts our view" and observe "this magnificent work, around which is gathered this great multitude of freemen." Stunned by such sights, they would only be able to guess that the "village, locks, canal, yea, every thing we now see, were the effects of magic." Next, he would resurrect Washington and show him "this most splendid part of the whole mighty undertaking," and ask of him "whether his children had not well improved the rich legacy." If the nation's first leader "should look at nothing else save this one work," he would be "satisfied that his country had already done enough to immortalize itself." Though Cuming took "the whole line" in his view, he first articulated the notion that this place, and particularly these locks, somehow encapsulated the entire project. He was about to "let down into its place" the "last stone on the most prominent point" of the canal. "A few more days," he proclaimed, "and *here*, HERE the *waters meet*" (emphasis in original).

Although the Deep Cut—which was yet unfinished—was by far the more arduous of the two accomplishments by which the canal overcame the Mountain Ridge, Cuming mentions it only once. He likened the remaining excavation to "the removal of the smallest mole hill." Cuming therefore also helped to establish a tradition of discounting the Deep Cut in favor of the locks that were so much more serviceable, both visually and rhetorically, to national and middle-class ideals. Engineers Wright and Thomas, who had once underestimated the task of cutting through the Mountain Ridge, had learned that even a small portion of it should hardly be called a "mole hill." Excavating the Deep Cut was grunt work, performed largely by men who, as Sheriff puts it, "represented neither the virtues of the republican past nor the prosperity of the commercial future."[270] Beginning with this elaborate ceremony and Cuming's address, the Deep Cut would be slowly elided, while the locks—the "most striking and finished point on the whole line"—moved Cuming to reach into the future and bid "posterity never to forget the names of those who projected, superintended and executed it."[271]

Reporting on this event, Thurlow Weed, the editor of the Rochester *Telegraph*, was moved to abandon his usual journalistic perspective. At the ceremony, "the most perfect harmony prevailed." The combined locks were "a work which will remain for ages as a monument of American genius and patriotism." They were "unquestionably the most important works of the whole line of the canal; and as they are approached from the deep ravine below, present one of the most grand, and interesting views which can be imagined." He also noted that the canal was now "completed except for a few miles which will be completed in September."[272]

Another distinguished foreigner visited Lockport on August 20. Karl Bernhard, the Duke of Saxe-Weimar in the Netherlands, was a military officer who, as a Dutchman, was very familiar with artificial waterways. Like so many who would follow him to Lockport, the Duke paid homage to both the locks and the town. He seemed impressed that the locks were "arranged in two parallel rows." Because of this, "the navigation is greatly facilitated, and the whole work, hewn through and surrounded by large rocks, presents an imposing aspect." He found Lockport "an extremely interesting place." It was also "perfectly wild, yet this appearance will no doubt vanish in the course of four or five years, so that it will present as splendid an appearance as Canandaigua and Rochester." Unlike most subsequent visitors, the Duke also described the Deep Cut where work was still in progress. "On our arrival," he wrote, "the canal was still unfinished for about five miles," though "it was supposed" that all would be completed "before the close of the year." He described the cutting "through solid rock, generally about thirty feet deep, for a distance of more than three miles," a task "mostly effected by blasting." Bernhard also noticed that "several hundred Irishman were at work," residing in "log huts, built along the canal," and suffering "severely in consequence of the unhealthy climate, especially from fevers, which not unfrequently prove fatal."[273]

It is unclear whether Bouck and Roberts were still anxious about the prospects of removing that last "mole hill" before western New York's infamous winter descended. Certainly a great many others throughout the state were indeed anxious when the first of September passed with no news from the Mountain Ridge. Finally, on September 29, William Bouck sent this long awaited announcement:

> *To the Hon. Stephen van Rensselaer, President of the Board of Canal commissioners:*
>
> Sir—The unfinished parts of the Erie Canal will be completed and in a condition to admit the passage of boats, on Wednesday, the 26th day of October next.
>
> It would have been gratifying to have accomplished this result as early as the first of September, but embarrassments which I could not control, have delayed it.
>
> On this grand event, so auspicious to the character and wealth of the citizens of the state of New York, permit me to congratulate you.
>
> Wm. C. Bouck, *Canal Com.*
>
> Lockport, Sept 29, 1825[274]

The embarrassments to which Bouck refers began when water inundated the southernmost portion of the Deep Cut, forcing the suspension of work while Scovell and other contractors cut a drain through the unfinished sections to the brow of the mountain.[275]

One week earlier a group of prominent citizens had gathered at the Cottage to discuss "the propriety of celebrating the completion of The Erie

Canal." They formed a committee, named Darius Comstock chair, and issued an announcement titled "Meeting of the Waters." The committee resolved that "whereas this happy consummation of the great undertaking will be witnessed at Lockport, a point on the canal exhibiting the grandeur and immense labor of the work—where the waters which waft the boats from Lake Erie will commingle with those that bear them to the ocean," that the celebration should naturally take place in their village. Therefore, they resolved to invite the governor, the lieutenant governor, "the present and late canal Commissioners and Engineers, to attend the celebration at this place."[276] Perhaps to the dismay of the committee's members, the Erie Canal would not be opened with a single celebration at the site of its completion, such as the celebration that would take place in 1869 at Ogden, Utah, where the golden spike marked the completion of the transcontinental railroad. Instead, Governor Clinton and the canal commissioners planned a ten-day sequenced celebration that would culminate in New York City.[277]

The actual completion of the canal—and the "meeting of the waters"—occurred without ceremony on the evening of October 24. Orasmus Turner, editor of the Lockport *Observatory*, remembered that "on the evening of the 24th of October, the work was completed, the guard gates were raised, and the filling of the Lake Erie level commenced."[278] Although this event left no other historical trace, it could still be imagined in fiction. Walter Edmonds, a celebrated regionalist of the 1930s, in his novel *Erie Water* (1933), tells the story of the building of the Erie through the eyes of two laborers, an Irishman and a black, who follow the construction work west and happen to be in Lockport on the night of October 24, 1825.

> They walked down to the Deep Cut, and they saw that the canal was more than half full. Coming slowly for the new banks, the water had some small impression till the lock gates dammed it. Then it had risen quickly. It made a dark straight track along the towpath wall stretching back into the still blackness of the stone. But even in that blackness, it held reflections of the stars.[279]

4

Writing Lockport

A Dialogue on American Progress

> The Erie Canal . . . had an effect . . . all over the world, somewhat akin to our first Declaration of Independence. It was a practical snapping of the finger, on the part of a great people, against the obstacles which nature had thrown in the way of their advancement. . . .
>
> The difference between the great works of antiquity with which we can compare our own achievements in art, like the Locks at Lockport, are that these were erected as monuments to human pride or for religious uses, while ours are made to promote the happiness and welfare of the people who constructed them; they were the results of mere physical strength and endurance, while ours have been accomplished more by the mind than by the body; science and skillful enginery, have done the work, which only armies of laborers, whose lives were sacrificed in their work . . . accomplished these.
>
> —C. F. Briggs, *The United States Illustrated*

If the conquest of the Mountain Ridge were a contemporary event, it would be possible to observe it directly and to interview the people involved. John Locke once compared human understanding to "a Closet wholly shut from light, with only some little opening left, to let in external visible Resemblances, or Ideas of things without."[1] Humans, for Locke, have no direct access to "reality": all they can experience is the evidence presented through the windows of the senses. In attempting to grasp an event that took place nearly two hundred years ago, we are compelled, it would seem, to rely on a great deal less direct evidence. Information on the conquest of the Mountain Ridge survived largely because people with the means to create a historical trace considered it meaningful. Our evidence about the event has been filtered through the perspectives of those people and others who later preserved its memory, again because of the meanings they saw in it. It is only because of these meanings that we can see the event at all, and, in a sense, we can only see it through them. Yet we are not hopelessly in the dark, trapped in a closet, within a closet, within a closet. People create meanings in interaction, not in isolation. Although it was individuals who wrote Lockport, they did it through an intertextual process that was itself embedded within a wider context that is not entirely illegible to us.

James Geddes was the first to glimpse the potential significance of the Mountain Ridge site. Like Thomas and Wright later, he could instantly see the site's importance to the success of the canal project, and he knew that the canal itself was near the center of public discussion about internal improvements, America's destiny, and progress. It did not take an engineer to recognize these things. Two days after Lake Erie water first entered the Deep Cut, an elaborate celebration unleashed a torrent of rhetoric about what the people of New York State had just accomplished, and what it meant for the future of the state and the nation. Then, for the next quarter century, when the Erie Canal was the most important artery to the West, throngs of settlers, traders, and travelers traversed New York State. Tourists from both sides of the Atlantic streamed west to see the newly accessible falls of Niagara and, in the words of Lionel Wyld, "the marvel of the Lockport flight of locks and other feats of engineering which the Erie Canal had to offer."[2] Dozens of travelers recorded their impressions in journals, articles, and published books. The canal and its marvels—like the accomplishments at Lockport—became the focus of a public dialogue on the meaning of progress and its relation to America's future. In addition to travelers' writings and the speeches that accompanied the canal's opening celebration, newspaper editors and other local voices joined the dialogue. These voices represented the views of adventurous Europeans and middle-class Americans, especially those from the Northern states.

We cannot assume that these views of Lockport and the Erie Canal represent those of Americans as a whole. The dialogue—although it contained a considerable range of opinions—was a privileged forum with few points of entry for those with little social or economic standing and just about none for those at the very bottom of Northern society, the common laborers who had actually dug the great ditch.

The introductory passage by C. F. Briggs appeared in *The United States Illustrated* (1853),[3] a volume edited by Charles A. Dana at a time when the United States was vastly expanding its continental hegemony, and when the role of the Erie Canal in that expansion seemed evident. I will analyze Briggs's essay more thoroughly at the end of this chapter. Here I simply want to point out that Briggs, writing in 1853, caught up many of the themes that, as we shall see, had appeared in thirty years of writing Lockport. For him, the canal was a defining American achievement, comparable to the Declaration of Independence. A "great people" merely had to snap their fingers to make the continent open before them. The Lockport locks were a serviceable symbol for all of this. His words also reveal what had been written out of the story. The canal and the "Locks at Lockport" were made "to promote the happiness and welfare of the people who constructed them." They were works of "mind" requiring no "armies of laborers." This fantasy about the meaning of the canal, and especially of the achievements at Lockport, had been thirty years in the making, but it had been present from the beginning. In assessing the stories people have told about the conquest of the Mountain Ridge, a central challenge will be to attend to the absences and to ask, in turn, about the meaning of their erasure.

I have already discussed the first attempts to ascribe meaning to the struggle to overcome the Mountain Ridge; these included comments in the yearly commissioners' reports, newspaper accounts, the *Gazetteer of the State of New York* (1824), the speeches and plaques connected with the capstone ceremony, as well as the comments of foreign visitors such as William Hamilton Merritt, Karl Bernhard, and General Lafayette.

In the following section, I examine the rhetoric surrounding the canal's opening celebration. Then, in a second section, I assess the aftermath of the Mountain Ridge conquest, considering the immediate destinations of the engineers and laborers and the rippling impact of the event across the continent. I examine next the views of travelers and other writers, largely from the canal's heyday, roughly 1825–1855. Finally, I evaluate lithographs, engravings, and other graphic depictions of Lockport.

The Meeting of the Waters

On the 29th of September 1825, Canal Commissioner William Bouck, finally convinced that the Deep Cut could be finished before winter, issued his letter from Lockport announcing October 26 as the date of completion. DeWitt Clinton and a host of dignitaries would have to make their way to Buffalo to begin the ceremonial journey of the fleet—led by the packet *Seneca Chief*—that would celebrate the opening of the Erie Canal, culminating with a colossal public celebration in New York City. The commissioners rejected the pleas of the committee of Lockport citizens that had recommended that the principal celebration take place in their village, perhaps because, although the triumph of art seemed total there, the costs had been immense. Delays, injuries, deaths, and riots were not easily adaptable to the rhetoric of jubilation that gushed forward at every public occasion during this central event of the Era of Good Feelings.

One indication of the importance prominent New Yorkers saw in the completion of the canal was the publication of an unprecedented large-format volume, *Memoir at the Celebration of the Completion of the New York Canals*, by Cadwallader D. Colden. The author, a former congressman and mayor of

New York City, from an illustrious family, was also the editor of a state–commissioned volume that compiled all the commissioners' reports and other documents of the canal's construction period.[4]

Colden's *Memoir*, an essay on the history and significance of the canal, occupies only the first quarter of this 428–page volume. Also included is a "Narrative of the Festivities" by New York journalist William L. Stone, who served as a kind of official chronicler of the ceremonial fleet's journey from Buffalo to the Atlantic. The largest section of the volume describes, in minute detail, the elaborate celebrations in New York City; here, we can observe—through correspondence and original planning documents—the behind–the–scenes workings of the largest public celebration North America had ever seen. Most remarkable of all were the elaborate illustrations, the first lithographs published in the New World. As a work of art and an example of American progress, here was a book to match the canal itself.

Early in the *Memoir*, Colden articulated the Enlightenment notion that, to many Americans, indicated that their country was becoming the very center of progress, both material and moral: "To use and enjoy the reason and power with which man is endowed by his creator, he must have liberty and independence."[5] For Colden, the Erie Canal was the most impressive fruit of the Revolution to date. Continuing in this nationalistic mode, Colden then evoked Niagara Falls, a presence hovering in the background of the entire canal enterprise:

> There is as much difference between man, the subject of a despotic government, and the citizens of a free & representative republic, as there is between waters diverted to some artificial channel, and the deep current of Niagara, pouring through its natural course, irresistible, but by the hand of the Almighty.[6]

Here was a strangely twisted metaphor for American artifice: its accomplishments—the Constitution and the canal—were natural, inevitable developments, like the "irresistible" Niagara River rather than "some artificial channel." Hence, the anticipated fruit of this "natural" canal, the expansion of the continental hegemony of the United States, was also a natural development.

Colden next recalled the principal obstacles overcome: the eastern section where dozens of locks climbed from the Hudson Valley to Schenectady, the aqueducts at Little Falls and Rochester, the Irondequoit embankment, and the works at the Mountain Ridge:

> The deep cutting towards the western extremity of this section has cost more money and required more labor, than any other work on the canals. To pass the mountain ridge, there has been a necessity for excavating

> seven miles to an average depth of twenty–five feet, three miles of which is through hard rock. The combined locks, at the brow of the mountain, the commissioners describe as a work of the first magnitude on the line, and as one of the greatest of its kind in the world.[7]

This litany of marvels at Lockport, like those at other key sites, would be repeated endlessly in the coming decades, one indication of the complex intertextuality among the writings of travelers, journalists, politicians, and artists.

Like many of the writers who followed him, Colden at times stressed the naturalness—almost the inevitability—of the canal and, at other times, the human agency that created it. The Erie Canal, for him, was much more than an example of the creative energies republican freedoms could release: it was a device for uniting America's far–flung citizens, preventing the country—for the present as least—from dissolving into "many distinct governments."[8] That a product of their own handiwork could accomplish such heavy lifting inspired in the breasts of people like Colden "hopes of continued prosperity, and of a rich and powerful maturity," and allowed them to "delight in the promised sunshine of the future."[9]

Governor Clinton, along with delegations from New York and Rochester, arrived at Buffalo on October 25. The great celebration commenced at nine o'clock the next morning with an artillery salute and procession "consisting of the different mechanical professions, with appropriate badges, and Flags ornamented with paintings—officers, civil and military—citizens—professional gentlemen—committees—canal officers—strangers—orator and clergy—Governor, Lieut. Governor, &tc.," in short, an organized, hierarchical display that included all the ranks of society except the common laborers. At the canal basin, the governor and his party boarded the packet *Seneca Chief*, where Jesse Hawley of the Rochester delegation addressed the gathering. The Erie Canal "will constitute a lever of industry, population and wealth to our Republic," Hawley proclaimed, showing the world that "National pride" can be expressed through "the arts of peace in domestic improvements." Judge Forward, on behalf of the Buffalo delegation, briefly replied, emphasizing that the canal project had "invigorated the moral and physical energies" of New Yorkers. When the judge finished his remarks, a cannon was discharged, commencing "the grand state salute," and "the boat then started for the ATLANTIC OCEAN, amid the cheers of our citizens and a feu de joie from the rifle company."[10]

Cannon had been placed about five miles apart to relay the signal all the way to New York City. Here was another elaborate contrivance of art—like Colden's volume, or the canal itself—that pointed directly to the future: an aggressive attack on the past. Colden, in the midst of actually writing the *Memoir*, heard the signal that had traveled "five hundred and thirteen miles" in one hour and twenty minutes. "Who that has American blood in his veins," he exclaimed, "can hear this sound without emotion?" Samuel F. B. Morse, whose telegraph would signal another stage of American progress, also heard the grand salute as he toiled in his painting studio.[11] A "Celebration Ode" sung at Buffalo called the salute "th' red flash, the line along," that "Tells to the world, with echoing roar / Matter and space are triumph'd o'er." William Stone suggested that there was also a defensive reason for the grand salute, namely "to guard against the disappointment that might arise from any unforeseen accident, which might have retarded the work beyond the specified time." There had been so many delays at the Mountain Ridge—Bouck even apologized for "embarrassments" that had delayed work in his letter announcing the completion date—that the commissioners were still not entirely confident of completing the Deep Cut in time.

The ceremonial fleet seemed concocted to represent the continent's still exotic interior, the very names of the boats conjuring vastness, Native Americans, and Niagara Falls. Leading the way, the *Seneca Chief* recalled a nation that, until recently, controlled the Niagara Frontier; and reminded citizens at the time of the celebration that there were still many Seneca reservations, including one adjacent to Buffalo. The *Seneca Chief*'s key symbolic cargo consisted of two barrels of Lake Erie water destined to be mingled with the waters of the Atlantic at the culmination of the voyage. Other boats included the *Superior*, the *Commodore Perry*, and the *Buffalo*. A boat "more novel" than the rest was called *Noah's Ark*; on board were "a bear, two eagles, two fawns, with a variety of other animals, and birds, together with several fish—not forgetting two Indian boys, in the dress of their nation—*all products of the West*."[12] The *Ark*, a vessel "from the unbuilt city of Ararat, which was to arise on an island near the western termination of the Canal," was loaded with "all manner of living things, to be found in the forests that surround the Falls of Niagara."[13] Ararat was the dream of Mordecai Noah, a prominent New Yorker who wanted to create a homeland for the Jews on Grand Island, just above the falls.[14] For Noah, the North American interior was a place of possibility; in the decades to come, the entire canal region would become studded with utopian communities as well as social and religious experiments.

The packet *Niagara* from Black Rock left before the fleet's arrival and later joined the fleet at Lockport. Perhaps General Porter and Black Rock's other citizens were bitter that Buffalo seemed to be eclipsing their village. Later, the *Niagara* moved ahead and reached Albany three days before the rest of the fleet, as Ronald Shaw suggests, "to enable the General to escape the eulogies of Clinton at every stop."[15] One additional craft, the *Young Lion of the West*, also joined the flotilla at Lockport.[16] This vessel, fitted up by the citizens of Rochester, contained barrels of produce, flour, and products of Rochester's workshops, as well as "a collection of wolves, foxes, raccoons, and other living animals of the forest."[17] The "West" this fleet represented seemed rich, lush, and full of possibility but also primitive and crying out for civilization.

When the fleet left Buffalo, the festivities there continued. The procession proceeded to the courthouse where Buffalo attorney Sheldon Smith addressed the citizens. Smith touched all the familiar themes but emphasized the great difficulties that had just been overcome at the nearby Mountain Ridge. Straining to account for such accomplishments, he reached for a familiar biblical analogy that seemed to divinize American artifice:

> As the rock of Horeb, when smitten with the rod of Omnipotence, opened to water the wilderness to the east; so has the mountain rock been cleft asunder by the rod of enterprise to cheer and enrich the wilds of the west.

In a statement similar to Lafayette's, Smith proclaimed that Buffalo "stands amid the rarest works of nature, and the most sublime of art." Finally, he noted the canal's continental implications: only when "all those fertile regions, that extend west to the Pacific Ocean," have felt the touch of civilization "shall the utility of the Grand Canal be known and appreciated."[18]

At Lockport, a thirteen–gun salute at sunrise awakened the villagers. A procession commencing at nine o'clock proceeded from the Washington House, the town's finest hotel, to "the grand natural basin at the foot of the locks" where all awaited the cannon signal from Buffalo. Orasmus Turner, the editor of the Lockport *Observatory*, noted that "all were looking forward to a gala–day—a period of joy and hilarity."[19] The relay cannon at Lockport was a thirty–two pounder from Perry's Lake Erie fleet. Hearing the signal from Buffalo, the gunner, who had been a lieutenant in Napoleon's army, discharged the cannon. The canal commissioners and engineers, who had forgone the Buffalo celebration and planned to join the fleet at Lockport, all boarded the packet *William C. Bouck*, designated as the first to pass through the locks. Over two hundred ladies crowded upon the packet *Albany*; the remainder of the procession boarding other vessels.

At the sound of the signal cannon, a band—whose leader had been held prisoner by the Barbary Pirates—began to play, but was drowned out by artillery fire and explosions in the rocky walls of the basin.[20] Turner's retrospective account, published in 1842, relates that, as soon as the signal sounded,

> the lock gates were opened, and the fleet commenced ascending to the Lake Erie level. As it ascended the stupendous flight of locks, it's decks covered with a joyous multitude, it was greeted with the constant and rapid discharge of heavy artillery, thousands of rock blasts or explosions

> prepared for the occasion, and the shouts of spectators, that swarmed upon the canal and lock bridges, and upon the precipices around the locks and basin.[21]

Turner's newspaper account, published four days after the event, is only slightly less subdued:

> The large and elegant boats, and their decks covered with an admiring and joyful multitude, traveling majestically up this stupendous flight of locks, to ride for the first time on the bosom of the water of lake Erie—the roaring of artillery—the banks of the canal, locks, bridges, and other high precipices around the locks and basin, literally swarming with spectators,—all conspired to render the scene grand beyond description.[22]

When the little fleet reached the top of the locks, Judge John Birdsall delivered an address. Again Turner portrayed the scene he witnessed:

> Stepping upon an elevated platform upon the deck of one of the boats, in the stillness that had succeeded the earthquake sounds, and the shouts of human voices, he exclaimed: "The last barrier is past! We have now risen to the level of lake Erie, and have before us a perfect navigation open to its waters."

After the address, the little fleet started west "under a discharge of artillery, [and] the explosions of rocks."[23] They proceeded seven miles through the Deep Cut to Pendleton, where the canal met the Tonawanta Creek. Here they joined the *Seneca Chief* and its flotilla. Bouck, the engineers, and the entire Lockport contingent were no doubt keen to introduce to these honored guests, firsthand, the object of their long and exasperating labor: the Deep Cut. George Catlin's lithograph *Deep Cutting, Lockport* depicts the *Seneca Chief* gliding through the artificial canyon (fig. 4.1). The fleet entered Lockport "under a discharge of artillery."[24]

The numerous dignitaries convened at the Washington House to partake of "a well provided table." Then the toasts began:

> *The Meeting of the waters*—the noblest triumph of human enterprise and ingenuity
>
> *The Erie Canal*—a rich stream from *deep* to *deep* o'er which old Neptune might be proud to ride
>
> The people of the State of New York—whose will makes mountains give way—streams revert—and the troubled waters of the inland seas, with the heaving ocean, unite in peace

Figure 4.1. *Deep Cutting, Lockport*, George Catlin, lithograph, in Cadwallader D. Colden, *Memoir at the Celebration of the Completion of the New York Canals* (Albany: State of New York, 1825), opposite p. 299. Print Collection, Miriam and Ira D. Wallach Division of Art, Prints and Photographs, The New York Public Library, Astor, Lenox and Tilden Foundations. Used with permission.

> *The Mountain Ridge*—Here has been accomplished what a Xerxes could only threaten: the mountain has been leveled with the sea [emphases in original][25]

William L. Stone began his official "Narrative" by praising the people of New York for their victory "not only over the doubts and fears of the wary, but over the obstacles of nature—causing miles of massive rocks at the mountain ridge to yield . . . turning the tide of error as well as that of the Tonnewanta—piling up the waters of the mighty Niagara, as well as those of the beautiful Hudson—in short causing a navigable river to flow with gentle current down the steepy mount at Lockport—to leap the River of Genessee—to encircle the brow of Irondequoit."[26]

Lockport was "the spot where the waters were to meet when the last blow was struck," a place where

> nature had interposed her strongest barrier to the enterprise and the strength of man. But the massive granite of the "Mountain Ridge" was compelled to yield. The rocks have crumbled to pieces and been swept away, and the waters of Erie flow tranquilly in their place.

Stone next turned his attention to the "grand natural basin."

> This basin, connected with the stupendous succession of locks, and the chasm which has been cut through the mountain, is one of the most interesting places on the route, if not the World, and presents one of the most striking evidences of human power and enterprise which has hitherto been witnessed. A double set of locks, whose workmanship will vie with the most splendid monuments of antiquity, rise majestically, one after the other, to the height of sixty-three feet: the surplus water is conducted around them, and furnishes some of the finest mill-seats imaginable, a marble tablet modestly tells the story of their origin.[27]

Stone's narrative stressed the triumph of art over nature's "strongest barrier." He was also the first of many to compare the locks favorably to the "monuments of antiquity."[28]

Interspersed in the portion of the *Memoir* dedicated to Stone's "Narrative" are eight lithographs, four of which depict Lockport scenes. In fact, images of Lockport outnumber those of any other place in the volume, an emphasis that would persist during the coming canal era. Although there would be no single iconic image of the Erie Canal, representations of the Lockport locks perhaps come closest to such a status.[29] All of the Lockport lithographs were drawn by George Catlin, a New York artist who is remembered today principally for his later depictions of Native Americans. A separate section of the volume, "Views on the Canal," describes the eight lithographs in detail.

The first image, *Entrance to the Harbor, Lockport* (fig. 4.2), warrants an extensive commentary.

> After passing over a monotonous level of sixty-two miles without a lock, the eye of the traveler is suddenly arrested by a formidable terrace of rocks. . . . The canal enters this terrace of seventy or eighty rods [about a quarter mile], through a natural ravine, forming a convenient harbor for an hundred boats, or more. . . .
>
> Approaching Lockport from the East nothing of the village can be seen until the boat is just doubling this cape, when in an instant the whole scene opens to view, and the sound of the bugle announces its approach.
>
> The singularly romantic appearance of this place, with its striking contrast and sudden transition from the tedious monotony of the country below, must fill the mind of every traveler with peculiar delight as he approaches it from the East.

The second lithograph, *Lockport from Prospect Hill* (fig. 4.3) presents "a more general view of the village and harbor." Employing a conceit that

Figure 4.2. *Entrance to the Harbor, Lockport,* George Catlin, lithograph, in Cadwallader D. Colden, *Memoir at the Celebration of the Completion of the New York Canals* (Albany: State of New York, 1825), opposite p. 296. Print Collection, Miriam and Ira D. Wallach Division of Art, Prints and Photographs, The New York Public Library, Astor, Lenox and Tilden Foundations. Used with permission.

Figure 4.3. *Lockport from Prospect Hill,* George Catlin, lithograph, in Cadwallader D. Colden, *Memoir at the Celebration of the Completion of the New York Canals* (Albany: State of New York, 1825), opposite p. 297. Print Collection, Miriam and Ira D. Wallach Division of Art, Prints and Photographs, The New York Public Library, Astor, Lenox and Tilden Foundations. Used with permission.

would soon be trite, Stone suggested that "from the singular and appropriate form of this basin it would almost seem as if nature had formed it for the purpose to which it is now applied." The basin, at "its head" was of a "circular form," enclosed in "a rampart of rocks." It was "similar to a funnel." Indeed, since the Mountain Ridge was the last obstruction in the canal's path, the fertile lands of Ohio, Michigan, Indiana, and Illinois were all accessible at the top of these locks, and the produce of that vast territory would soon be squeezing through this funnel-shaped basin. The village of Lockport, Stone added, "is situated on both sides of the canal immediately above the locks; its foundation is on solid rock." Some of its buildings "are remarkably fine, being built of stone; the remainder are built of logs, and intended only for temporary use."[30]

The two remaining lithographs depict the Deep Cut. The first, *Process of Excavation, Lockport* (fig. 3.4), came in for considerable scrutiny in the last chapter. The final plate, *Deep Cutting, Lockport* (fig 4.1) is a "representation of the Canal on the upper level." The *Seneca Chief* itself is plying the passage through the Erie Canal's summit. Stone's text turned this passageway into a stunning fairyland:

> A ride through this chasm for three miles, on a canal-boat, between these formidable walls of solid rock, where nothing is to be seen above their summits, though in the midst of forest, is calculated to excite in the susceptible mind the most pleasing and singular sensations. . . . the whole wall is formed of geodiferous limestone rock, containing in its cavities the most beautiful clusters of crystallization, on which the passing stranger is continually catching his eye and reluctantly passing.[31]

From this, one could hardly guess that the gleaming channel, only a few days earlier, had been the site of toil and strife, let alone explosions, injury, death, and riot. The description of the *Process of Excavation* lithograph concludes with a similar erasure of the violent scene represented. After a long technical discussion of how the cranes had operated during construction, the reader is informed that they had since all been destroyed, leaving behind only "huge piles of rocks." The description then suggests that "as nothing of these are to be seen from the Canal-boat, while passing through the excavation, I would recommend to the curious and romantic traveler to spend a day along these banks, where he can easily amuse himself among the beautiful minerals with which these heaps abound."[32]

In the great wave of tourism that the Erie Canal itself would initiate, "curious and romantic" middle-class travelers would often assume a certain aesthetic distance from the realities before them. John Sears has examined this dark aspect of nineteenth-century tourism that, for example, turned the child laborers of Pennsylvania's Mauch Chunk mines—barefoot and blackened by coal dust—into picturesque objects of the tourists' gaze.[33] Ignoring, for the moment, Stone's text and placing Catlin's two lithographs of the Deep Cut side by side, one might be reminded of a medieval diptych: on the left, the smoke and confusion of perdition; on the right, the crystalline beauty and order of paradise. But in the American context, the ideal often seems to lie ahead, temporally or spatially. The channel may be beautiful, but it is the future it portends and promises—as well as the vast West that it makes accessible and binds to the nation—that is sublime. The diptych is perhaps more likely to remind contemporary Americans of the scene from Cecil B. DeMille's film, the *Ten Commandments*, where the passage out of captivity is opened through the Red Sea at the command of Moses. One moment the passage is thronged by Pharaoh's legions, and the next the waters have poured in to submerge the eminently expendable evildoers. The laborers in the ditch did not fare as badly as the Pharaoh's soldiers, but they—just as effectively and guiltlessly—have been submerged and forgotten, while the governor's boat glides effortlessly westward.[34] The immediate aestheticization of the Deep Cut elides the Erie's bloodiest and most frustrating task, hiding both the process and the cost of the great accomplishment.

It was already dark on the evening of October 26 when the ceremonial fleet prepared to leave "the rugged scenery of Lockport."[35] The village was "brilliantly illuminated, and the boats descended the locks and departed east" amid more artillery fire.[36] Because the work at the Mountain Ridge had only ended two days earlier, most of the laborers were presumably still on hand to participate in the "joy and hilarity" of the entertaining spectacle. Perhaps as Lockport's prominent citizens were drinking toasts with the governor's party at the Washington House, laborers were enjoying less pretentious libations at the countless groceries and taverns of the village. Workers were obviously involved in the preparations for the occasion, and certainly they executed the repeated rock blasts in the steep walls of the natural basin. There is even some evidence that workers had been driven to exhaustion right up to October 26. The *Observatory*, in the same issue that contained the account of the celebration, reported an "Unfortunate Occurrence." One Orrin Harrison, a workman employed at the locks, drowned.

> He was leaning against one of the balance beams, and from excessive fatigue fell asleep, and was precipitated into one of the locks, in about 8 feet of water. One of the culvert gates having been previously opened, his legs were drawn in through the gates, and before he could be extracted life became extinct.[37]

The standard Erie Canal locks had a rise of eight feet. It was Nathan Roberts's innovation to increase the steepness of the Lockport flight by increasing each lock to a twelve-foot rise. Although this reduced the number of chambers, and the speed of ascent and descent, is it also made the lock pit a deeper and more dangerous contraption.

Another tragedy occurred when the flotilla arrived at Weedsport, just west of Syracuse. "A twenty-four pounder was accidentally discharged, and Mr. Remington and Mr. Whitman, who were acting as gunners, were instantly killed."[38] Many proponents of the canal agreed with Thomas Jefferson that it would demonstrate "the superior wisdom of employing the resources of industry in works of improvement rather than of destruction."[39] Yet it seems that the implements of war—even when employed for peaceful purposes—will still claim victims. Even "works of improvement" carry a human cost.

When the fleet arrived at the Rochester aqueduct, local citizens greeted the dignitaries from the deck of a boat with this prearranged dialogue:

> *Question*:—Who comes there?
> *Answer*:—Your Brothers from the West, on the waters of the great Lakes.
> *Question*:—By what means have they been diverted so far from their natural course?
> *Answer*:—By the Channel of the Grand Erie Canal.[40]

These expressions at Rochester, together with Sheldon Smith's proclamation at Buffalo that "the mountain rock [had] been cleft asunder by the rod of enterprise to cheer and enrich the wilds of the west," show that, in 1825 at least, community leaders in the western part of the state understood the hydrological centrality of the Deep Cut. Indeed, they seemed awed by their own power: by conquering the Mountain Ridge, New Yorkers had turned the "waters of the great Lakes" toward the Atlantic.

At Rome, Clinton's flotilla encountered its most explicit protest. The decisions of the Black Rock contingent and that of Bouck and the other engineers to snub the opening ceremonies at Buffalo had been submerged under the prevalent "good feelings" of the occasion. The citizens of Rome, incensed that the canal was bypassing their village, filled a "*black barrel*" with water from the old canal of the Western Inland Lock Navigation Company and marched "with muffled drums," to the new canal, "into which they poured the contents of the black barrel."[41]

Stone's description of Little Falls mirrors those of countless visitors who would soon be plying the canal's waters. "Next to The Mountain Ridge," he wrote "the construction of the Erie Canal at Little Falls, was the most formidable labor executed." Like Lockport, here was a scene "exceedingly wild and picturesque."[42]

The celebration at Albany surpassed the grand ceremony in 1823 that had marked the completion of the canal's remarkable descent from Schenectady to the Hudson. As was perhaps fitting for the state's capital, the celebration also exceeded—in terms of pomp, ceremony, and expense—any displays of the western villages. It is sufficient to note that six hundred dined at two parallel tables, each 150 feet long.[43]

Orasmus Turner remembered the flotilla's journey as a "jubilee, such as has never, upon any other occasion, been witnessed in our country." At every town along the route, the fleet encountered music, artillery salutes, festooned bridges, illuminated buildings, and patriotic signs—a "protracted 4th of July." But, as Turner noted, "it was reserved for the Empire City of the Empire State, to add the grand finale."[44] A "gentleman who had witnessed the fleet upon the Thames" celebrating the final defeat of Napoleon reported that the "spectacle upon the waters of New York, far transcended that in the metropolis of England."[45]

The New York celebration took place on both water and land. The "Grand Aquatic Display" consisted of forty-six elaborately decorated vessels, including twenty-nine steamboats and the four packet boats from the West. Amid artillery blasts and the cheers of thousands lining the shores, the fleet proceeded through the Verrazano Narrows to Sandy Hook. In the "crowning ceremonial," Governor Clinton poured a keg of Lake Erie water into the Atlantic. Cadwallader Colden then presented the governor with a copy of the *Memoir*.[46]

The only significant departure from prearranged plans occurred when the fleet was returning. There were four British war sloops in the harbor for the celebration. Two of them fired a salute as the flotilla approached the Battery. "In consequence of this compliment," the flag ship gave a signal, and "the whole squadron passed around them in a circle." On deck, the British bands played "Yankee Doodle" while the American bands returned the courtesy with "God Save the King." Although this incident denoted "good feeling, fellow-ship, and union of sentiment," there were, once again, aggressive implications. Stone "could not but reflect on the facilities of defence which, by means of steam navigation, our city would possess in the event of hostilities with any maritime power."[47] Such a circular maneuver would have been impossible for a wind-driven fleet. Indeed, John Seelye notes that a similar circle of steamships would be employed at the 1861 Battle of Port Royal in South Carolina to reduce "two confederate forts at once." In spite of the nationalistic rhetoric surrounding the celebration, Seelye argues that because the canal would unite the North and the West at the expense of the South,

> it might be said that the slowly moving columns that churning paddles made into a circle at the mouth of the Hudson River—a vast engine turning in the outpouring current that had floated the *Seneca Chief* from

> Buffalo—was a fatal machine, a penultimate expression of changing forces that would in time disrupt the very Union whose preservation the canal was supposed to guarantee. [48]

To the celebrants, advances in steam navigation were part and parcel of the same progressive spirit that had inspired not only the canal itself, but also the confident belief that the future belonged to the United States. It is therefore fitting that the first lithograph prepared for the *Memoir*—and thus the first lithograph made in North America—was Imbert's *Plan of the Fleet* depicting the spontaneous circular display.[49]

A seven-thousand-strong "City Procession" marched from Greenwich Village to the Battery where the "Aquatic Party" had anchored. "This close approach of the boats to the Battery," the *Memoir* informs us, was intended to give each party "a view of the other," becoming, Seelye suggests, in one transformational moment both actors and audience."[50] Later, the occupants of the fleet, "preceded by the aborigines from Lake Erie," fell in at the rear of the procession—adding eight hundred to its number—and all proceeded to City Hall.[51] Like the "Aquatic Display," the parade through New York was regulated and hierarchical. The order of marching groups seems designed to reflect the entire social spectrum with the glaring exception of common laborers. The first group consisted of "foresters with axes to cut down the trees and clear the earth for cultivation," followed by "tillers of the ground." Next came the societies of tailors, millers, bakers, brewers, and other artisans; then merchants, teachers, students of Columbia College, doctors, clergy, lawyers, military officers, state legislators, and finally, the governor, canal commissioners, and engineers. The evening concluded with a grand ball and elaborate fireworks.[52]

The significance of the accomplishments at the Mountain Ridge were recognized not only in the words of Colden and Stone, and the lithographs of Catlin, but also at the New York celebration itself. The float, or "Large Car" that accompanied the two hundred tinplate workers and coppersmiths featured

> in the center part of the car the Five Double Locks at Lockport represented in copper; twenty-four tin stars on each side of the Locks representing the states of the Union; the boat *Lady Clinton* ascending and the boat *Lady of the Lake* descending through the locks—three other boats lay in the basin. The locks were filled with water from a reservoir under the canal by means of a forcing pump.

This remarkable display, which elevated the Lockport locks to a kind of national centerpiece, was also intended to educate New Yorkers about the hydraulics of the artificial river that promised so much to their city. This was a functioning model, and the smiths who created it marched behind "at their own request, as it afforded them a better opportunity of exhibiting the process of ascending and descending in the Locks." The front of the float held another image of "the Five Double Locks at Lockport, made of tin." At the rear of the float, a banner proclaimed: "Grand Canal-Locks at Lockport—Completion of The Erie Canal, October 26, 1825."[53] The symbolic articulateness of the Lockport locks was already eclipsing the more important achievement of the Deep Cut.

The Lockport locks also appear on the elaborate invitation presented to those favored citizens who would "witness the ceremonies, and the immixtion of the mild waters of Lake Erie with the briny floods of the Atlantic, at Sandy Hook." The steel engraving, by Asher B. Durand, shows female figures representing "Liberty and Justice" (fig. 4.4). The city of New York is on "the left hand of Liberty" at the "Southern extremity of The State; and on the right hand of Justice are seen some of the wonderful series of Locks, at Lockport, near the Western terminus of the Canal." Between the

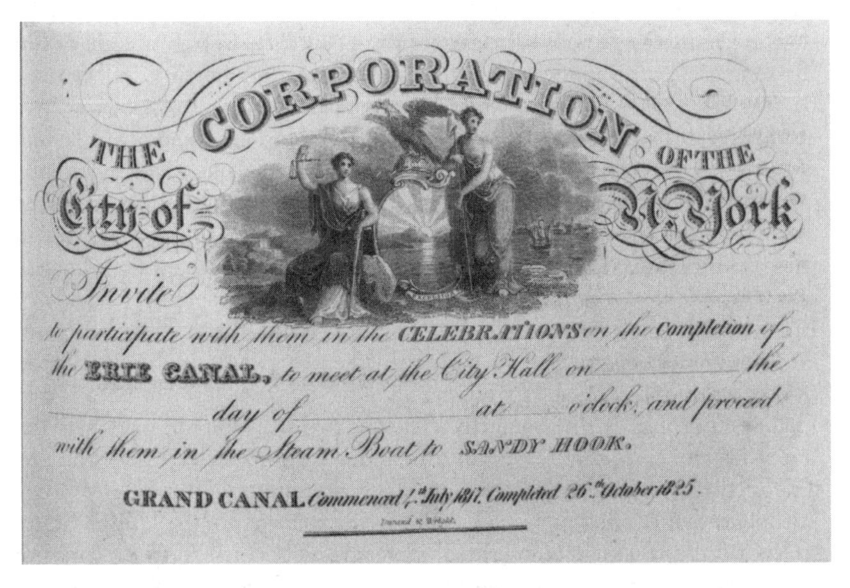
THE CORPORATION OF THE City of N. York

Invite

to participate with them in the CELEBRATIONS on the Completion of the ERIE CANAL, to meet at the City Hall on the day of at o'clock, and proceed with them in the Steam Boat to SANDY HOOK.

GRAND CANAL Commenced 4th July 1817. Completed 26th October 1825.

Figure 4.4. Engraving of Liberty and Justice, with Lockport Locks and New York City, depicted on the "Invitation Card" for the ceremony of "the immixtion of the mild waters of Lake Erie with the briny floods of the Atlantic," by Asher B. Durand, in Cadwallader D. Colden, *Memoir at the Celebration of the Completion of the New York Canals* (Albany: State of New York, 1825), opposite p. 344. Print Collection, Miriam and Ira D. Wallach Division of Art, Prints and Photographs, The New York Public Library, Astor, Lenox and Tilden Foundations. Used with permission.

two allegorical figures was a shield exhibiting "a rising sun," and above it "a full fledged young Eagle on a terrestrial sphere, and expanding his pennons for flight." Although the ascending eagle was an established convention, the symbolism could not be more clear: the waters of Lake Erie, 570 feet above those of the Atlantic, pour through the Mountain Ridge to fertilize the prospects not only of New York City but of the entire nation whose dawn has arrived: the eagle is about to take flight. Although it was forgotten by the time New York Mayor Wagner visited Lockport with John F. Kennedy in 1960, the fates of New York and Lockport were often recognized as intimately linked during the antebellum period.[54]

A recurrent theme throughout the festivities was the relationship between the canal and the continent. That a product of their own handiwork had brought the West into direct intercourse with the coastal metropolis inspired in many New Yorkers a kind of geographical rapture. "In one year more," Stone exclaimed, the Erie Canal "will carry our trade to the Falls of St. Mary, and will eventually give us access to the most remote shore of Lake Superior!"[55] For Colden, an imaginative vision of the West was almost necessary to appreciate the canal's significance. "When we consider that the immense regions surrounding the Lakes," he wrote, "are all within the temperate zone" and "populating with astonishing rapidity, . . . we must be impressed with ideas of the importance of the great works we have accomplished." Carried away by this reverie, Colden imagined a vast system of "internal navigation from the Atlantic to the Pacific" and eventually "a northwest passage to India, for which Hudson was searching when he discovered the river which bears his name." Should the reader find error in these speculations, Colden suggested, "they may be corrected by reference to a map of the United States, from a view of which they were made."[56] For Colden, maps were necessary to understand the accomplishment. Accordingly the *Memoir* opens with a thirteen-by-twenty-inch foldout map of the continent, followed by an equally large map of New York State that includes a profile of the canal. In addition, two lithographic maps—of the state and the continent—were also included.[57] John Seelye traces the origins of these sentiments to the immediate aftermath of the War of 1812. Although that conflict effectively halted the expansion of the United States to the north, at least temporarily, it nevertheless secured the boundaries of an immense realm, eliciting "an intense feeling of geopolitical euphoria."[58] Other scholars argue that the idea of the geographical unboundedness of the American realm has deeper historical roots.[59] Whatever its origin, geopolitical euphoria not only helped to inspire the canal but also was given a decisive fillip by the canal's completion. Indeed, the canal commissioners, in their 1824 report, had expressed a similar euphoria in describing the Niagara Frontier's prospects:

> When our interior seas shall have a population on their borders equal to that on the borders of the Mediterranean—when our whole territory, between the Atlantic and the Pacific shall be filled with enterprising, prosperous, free, and happy inhabitants, there will be found no spot in the interior of this continent, presenting more motives to industry, more business, or more wealth, than the shores of the Niagara.[60]

A great deal of rhetoric at the celebration pointed to the virtue of employing energy and intelligence to improvements rather than to warfare. Lieutenant Governor James Tallmadge, for example, speaking at the Albany ceremonies, disparaged "the conquering hero" whose banners were stained "with the tears of the widow and the orphan." In contrast, the canal, he believed, inspired a peaceful agenda: "May our glories ever be numbered by the benefits conferred, rather than by the injuries inflicted upon our fellow-mortals."[61] Yet, one only needs to remember that the shores of "our interior seas" were already occupied to recognize the aggressive undertones of this geopolitical euphoria: the next generation would name it "manifest destiny." Laurence Hauptman argues that the Erie Canal was part of an imperial thrust that dispossessed the Iroquois.[62] To the extent that the canal helped to secure American dominance of the continent, it was a device that doomed the native inhabitants of the trans-Appalachian West and threatened Mexicans, Russians, and any others with a territorial stake in that vast region. This was, as Seelye suggests, part of a deep American tradition of "conflating Exodus with the imperial impulse."[63]

Although euphoric sentiment suffused the rhetoric of the celebrations, the metaphors employed in that rhetoric indicate a fundamental contradiction: was the great accomplishment a marriage or a war? The popular New York poet and printer Samuel Woodworth actually used both metaphors in his specially commissioned "Ode for the Canal Celebration." The Typographical Society's float featured two printing presses—recently manufactured in the city—that cranked out copies of the ode during the entire procession. Four boys, "costumed as heralds and Mercuries, distributed over 8,000 copies to the assembled multitudes."[64]

Woodworth begins the ode by stating the martial analogy directly: " 'Tis done!—Proud Art o'er Nature has prevailed!" This was the same "Art" that had "prompted Columbus to seek a new world," and "lighted the path of the pilgrim band." It had inspired "Columbia's immortal sages" to create a republic and enabled Franklin "To arrest the forked shafts of Jove, / And play with his bolts of thunder." This art was the enemy of the "darkness" that once "veil'd each nation," and presumably still held sway in North America's interior. Art, in short, represented what people would soon refer to simply as "progress," a word that would continue to denote both technological and moral improvement. Those who fought on the side of art in its battle against nature's obstacles deserve to be remembered like conquering warriors:

> Then hail to the Art which unshackles the soul,
> And fires it with love of glory,
> And causes the victor who reaches the goal,
> To live in deathless story.

The marriage metaphor was implicit in the "crowning ceremonial" at Sandy Hook, "The Marriage of the Waters." Dr. Samuel L. Mitchill, perhaps the country's most prominent living scientist, presided over these ceremonies, as he had those at Albany in 1823. At that earlier celebration, he had introduced the idea that "the Lord of the seas wedded the Lady of the Lakes." After Governor Clinton had poured out the keg of Lake Erie water, Mitchill delivered an address emphasizing the nuptials between the "Monarch of the boundless Main" and the lady who arrives "rich in friends and produce" from the distant lakes. "Never was there an indication of so much wealth and fortune brought by a wife to her husband." It was, therefore, "no wonder that the Ruler of the Deep should have become enamoured of our incomparable Belle."[65]

In the central stanzas of the ode, Woodworth takes the marriage metaphor in a more explicitly erotic—and revealing—direction:

> 'Tis done!—the monarch of the briny tide
> Whose giant arm encircles earth,
> To virgin Erie is allied,
> A bright-eyed nymph of mountain birth.
>
> To-day, the *Sire of Ocean* takes
> A sylvan maiden to his arms,
> The goddess of the chrystal lakes,
> In all her native charms!
>
> She comes! attended by a sparkling train;
> The *Naiads* of the West her nuptials grace:
> She meets the sceptred father of the main,
> And in his heaving bosom hides her virgin face.[66]

That the construction of the canal could be figured as both a war and a wedding—a conquest and a voluntary union—indicates a deep confusion about how to understand the entire project. The fact that art vanquishes nature proves that the combatants are unequal. Similarly, the "Wedding of the Waters" is no marriage of equals. During this period, it is important to remember, art and nature are themselves gendered categories. That the continent's interior should be represented as feminine is hardly a novel notion.[67] The "goddess of the chrystal lakes" is a needy, weak, infantilized figure who eagerly "hides her virgin face" in her spouse's "heaving bosom." She is a fresh young "maiden" from the forest and lakes, "bright-eyed" and laden with "native charms"—in short, a fantasy of the male gaze.

If, for a moment, we take this conceit seriously, it seems to imply that the whole canal enterprise, from conception, to construction, to interpretation, was suffused with eroticism. One indication is the frequent use of double entendre words like "intercourse" and "consummation" in expressions of the marriage metaphor. Nine days before the canal's completion, for example, the Rochester *Telegraph* noted: "The wedding of the waters of Lake Erie, with those of the Hudson, is to be solemnised on the twenty-sixth instant, and we are happy to observe that marriage feasts are making ready in every part of the state." The *Telegraph* then added: "Our brightest, highest hopes, are all consummated."[68] Consummation, of course, can mean both a completion and "making a marital union complete by sexual intercourse."[69]

The symbolic moment of consummation was the "Wedding of the Waters" at Sandy Hook where Clinton poured the Lake Erie water into the Atlantic. At the moment of consummation, it was appropriate that fertilizing fluids be exchanged. A keg of ocean water returned on the *Seneca Chief* to Buffalo where the canal celebration finally concluded—over a month after its commencement—with a small ceremony during which the keg was poured into Lake Erie. This reciprocal exchange, like the return of the cannon signal on October 26, was sufficiently anticlimactic to underscore, once again, that this was hardly a marriage of equals.

While the symbolic consummation happened in New York harbor, the actual completion of canal construction occurred in the Deep Cut. Darius Comstock and his committee of Lockport citizens, in the letter to the commissioners requesting that the principal celebration should take place at their village because the "consummation of the great undertaking will be witnessed at Lockport, a point on the canal exhibiting the grandeur and immense labor of the work—where the waters which waft the boats from Lake Erie will commingle with those that bear them to the ocean." Referring to the intense pressure to finish that last section of the canal, Orasmus Turner wrote that it was "prosecuted with a vigor that public anxiety and expectations demanded, as the great work approached nearer and nearer to a consummation."[70] Here, what Stone referred to as "nature's strongest barrier" seemed to repel the courtship of art. It was as if the goddess was saying "no," but her suitor believed she really meant "yes." In the terms of the erotic metaphor, was the Mountain Ridge the goddess's maidenhead? Marriages are consummated by a penetration, but abduction and rape accomplish the same deflowering without consent. Despite the rhetoric contrasting the peaceful benefits of internal improvements with the horrors of war and conquest, constructing the Erie Canal was, in a sense, an act of violence.

In a surfeit of metaphor, canal celebrants introduced another kind of symbolic union, that of Neptune and Pan, a neoclassical motif with Masonic roots.[71] This appears on the celebration's medal and on the official badge where Pan, in place of the goddess, is accompanied by a Native rower. Is this Indian also being united with, or married to, the Atlantic? Or is he part of the losing side in the battle just completed between art and nature? The ambiguity about whether the canal represents a conquest or a marriage creates a tension, particularly evident in Woodworth's poem. In both a wedding and a war there are two parties. Presumably, the goddess represents the same nature that art, with considerable resistance, has conquered. The juxtaposition of these metaphors, at the very least, invites an interpretation of a union that rests less on mutual attraction than on violent coercion. Despite these ambiguities, no one at the canal celebration doubted that the continent's interior had indeed been united with New York and the Atlantic, nor that this was a prodigious connection.

Aftermath

Because there had been no preexisting urban community at Lockport, Bouck and the canal engineers and contractors were not outsiders to the community of citizens that congealed during the long construction period. Indeed, there had been but a few scattered houses when Geddes arrived in 1816, and Thomas had only finalized the route, determining the location of Lockport, late in 1820. Before the canal's completion, the construction tasks were the only reason for 2,500 people to be there. The class tensions of the construction period, as I have argued, further united the middle-class citizens whether or not they worked on the canal. The canal's completion provided Lockport commercial and manufacturing opportunities that would assure its post-construction existence. In chapter 5, I will consider the complex ways that the canal's success affected the social, economic, and cultural life of Lockport.

Another result of the Erie Canal's success was the spread of canal mania throughout the continent, an enthusiasm that was particularly intense in Pennsylvania, Ohio, Michigan, Indiana, Illinois, and the Canadas—areas already feeling the canal's impact. This created a great demand for engineers, contractors, and laborers with construction experience; in 1825 there was no greater concentration of such expertise than at Lockport where so much effort had been expended during the final phase of construction. As many took advantage of these opportunities, the distinction between canal builders and other citizens became apparent. I will begin by examining the dispersion of engineers and workers whose experience was in demand. They played a key role in furthering Lockport's reputation during the canal era. In a separate subsection, I will consider how that reputation led to the creation of new "Lockports" in several regions of North America.

Dispersion

Because Lockport by 1825 had become a thoroughfare—a place connected to a network of other places from the Atlantic coast to Chicago—there would be a large and continuous turnover of migrants, as well as a great deal of back-and-forth travel following lines of commercial, ethnic, religious, and family connection. Separating the construction people from the other citizens was not always tidy. David Thomas, for example, after being forced out of Lockport by enemies of Clinton, found that his unique skills were in demand. The Welland Canal between Lake Erie and Lake Ontario was just beginning construction in 1824, and Thomas served as chief engineer during the first year of construction. He had several other positions but turned down an offer to take charge of the Pennsylvania canals because of an illness in the family. Then in 1843 Thomas returned to Lockport and married Edna Smith at the Friends Meeting House. Dr. Isaac Smith had died the previous year. Thomas had known the Smiths since their arrival in Lockport in 1821. They had attended the same log meetinghouse, the village's first place of worship. The remarkable Aunt Edna married, first, the man who named Lockport, and then the man who located it. David and Edna lived in Lockport until 1854, when they moved to Cayuga County. He devoted the last thirty years of his life to studies and practical experiments in botany, horticulture, and pomology. He was the first president of the North American Fruit Growers Association and wrote extensively in practical periodicals. An outspoken abolitionist, he also wrote a number of antislavery articles.[72]

Commissioner Bouck, building upon his well-known success at Lockport, entered politics and eventually was elected governor of New York.

Engineer Alfred Barrett worked on the Welland Canal and later became chief engineer of the Erie Canal. He spent the last part of his life working on the canals of Lower Canada. His son, Alfred W. Barrett Jr., was born in Lockport, educated in Montreal, and returned to become assistant engineer on the Western Division of the Erie Canal.

David Stanhope Bates began as a surveyor and later joined the Erie Canal as an engineer. He worked on the Irondequoit embankment, the

Genesee aqueduct, and the Lockport locks. He left Lockport to become the principal engineer of the canals of Ohio. Like many other former Erie Canal engineers, Bates was also a pioneer in railway construction.

Nathan Roberts, who designed the Lockport locks and toiled in their vicinity from 1822 to 1825, became chief engineer of the western section of the Pennsylvania Canal. He also worked on the Chesapeake and Ohio Canal and built the bridge over the Potomac at Harper's Ferry.

The Erie Canal's most experienced engineers, James Geddes and Benjamin Wright were more advanced in years but still remained active after 1825. Geddes worked on canals in Maine, Maryland, Pennsylvania, and Ohio. Wright, the chief engineer of the New York canals from 1817 to 1828, worked as an engineer or consultant on canals in Connecticut, Rhode Island, Massachusetts, Delaware, Maryland, Illinois, and the Canadas.[73]

It would be difficult to overestimate the role of the Erie Canal in the development of engineering in the United States. Canal Commissioner Stephen Van Rensselaer endowed the Rensselaer School at Troy in 1824, the first American institution of higher education devoted to science and engineering.[74]

Engineers were near the top of the canal workforce pyramid. Equally in demand were skilled contractors and experienced men with teams and wagons. Manufacturing in Upper Canada was so little developed when Welland Canal construction began that even picks, shovels, wheelbarrows and horse-drawn carts and wagons had to be imported from the United States. In addition, many contractors "who had learned on-the-job on the Erie" crossed the Niagara River with "their horses, oxen and equipment to work on the Welland." One of these was Oliver Phelps, who had secured a Mountain Ridge contract under mysterious circumstances in 1823. He began in Upper Canada as a subcontractor on the Welland Canal's own Deep Cut and later took over the work. Experience at the Mountain Ridge was particularly pertinent because the Welland was traversing a route so similar in topography and geological structure. Phelps is remembered principally for inventing an earth-moving machine that employed animal power and pulleys to remove debris. Phelps placed a series of these machines at intervals along the Deep Cut—a less costly and elaborate version of the giant cranes that lined the Erie's Deep Cut. Clearly, his intimate acquaintance with the Mountain Ridge cut was crucial to his success west of the Niagara River.[75] "To no one," wrote William Hamilton Merritt, "was that part of Canada more indebted for its foundation, clearing, growth and prosperity . . . than to the energy, wisdom, foresight, and industry" of Oliver Phelps, "the Master builder of the Welland Canal, and the Father of St. Catherines."[76]

Common laborers, the broad base of the workforce pyramid, were also in demand as canal construction escalated across the continent. Because so few early Lockport newspapers have survived, we have only a sketchy record of advertisements for workers. One year after completion, an announcement in the Lockport *Observatory* called for "500 laborers" for the Welland Canal.[77] Since it was so close, the Welland construction would have been particularly attractive to laborers who had to use their own resources to reach the job site. In the coming decades, as the supply of canal laborers began to exceed demand, the plight of navvies would become increasingly desperate. Canal companies would call for many more workers than they needed in order to force down wages, even below the levels needed to reproduce the workforce. The result was an increase in riots, uprisings and militia interventions.[78] The bloody culmination of Erie Canal construction at Lockport was a harbinger of what was to come for North American navvies. Lockport historian Clarence O. Lewis argues that more than half of the Mountain Ridge laborers stayed to work in the nascent industries of the village.[79] Such a conclusion is speculation since, unlike engineers, navvies have left only the faintest of historical trails. Although these engineers, contractors, and workers were leaving, many more people were arriving at Lockport. One of these was Jesse Hawley, the man whose early writing had stirred such interest in the Erie project. Before too long he was writing again from Lockport—this time urging the enlargement of the canal.[80]

Other Lockports

In the year of the canal's completion, N. B. Holmes published a pamphlet claiming that the commissioners had made a serious error by allowing the canal to descend and rise again between Utica and Rochester, rather than following Gouverneur Morris's plan of a continuous descent from Lake Erie to the Hudson.[81] Although such a scheme would allow Lake Erie water to fill the entire canal, Wright and Geddes had estimated that it would be highly impractical and costly. From the favorable comments he makes about General Porter, it is obvious that Holmes was a bitter enemy of Clinton. Holmes's idea was to create a "Second Lockport," a second double tier of combined locks. His profile shows three levels between Buffalo and Utica; the first Lockport connects the Lake Erie and Rochester levels; the second Lockport would connect the Rochester and Utica levels.[82] In a footnote, he adds: "The place for the locks is called by the temporary name of Second Lockport, until another is determined on, by the commissioners or the inhabitants." Interestingly, this pamphlet shows that Lockport, even before the canal's completion, was not only a well-known place, but also a *kind* of place. Moreover, since Lockport was obviously a recent creation of art, it could be replicated. Indeed, during the quarter century that followed the Erie's completion, at least twenty-one new Lockports would appear in North

America. There were six Lockports in Pennsylvania, five in Ohio, two in Indiana and Michigan, and one each in Kentucky, Illinois, Kansas, Louisiana, Tennessee, and Manitoba.[83] Most of these Lockports were on canals or other watercourses, often at key locking points.

A traveler heading west in the 1840s on the Erie Canal and passing through the first Lockport would encounter other Lockports whether traveling to Pittsburgh via Erie on the Erie Branch, or to the Ohio River via Cleveland on the Ohio and Erie, or to Cincinnati via Toledo on the Miami and Erie, or to the Ohio River via Toledo on the Wabash and Erie, or to St. Louis via Chicago on the Illinois and Michigan.[84]

Lockport, Louisiana, founded in 1836, in a terrain and climate that could hardly be more different from Niagara County, New York, is also the Lockport most distant from the original. It is located at the key point on the Barataria and Lafourche Canal, a project promoted by wealthy sugar producers who wanted to open to commercial cultivation the low-lying areas southwest of New Orleans. The canal, built entirely with slave labor, had two locks: one at the Mississippi opposite New Orleans and the second at Lockport where the canal joined Bayou Lafourche, which connected to the Gulf of Mexico. "Undoubtedly," comments historian Thomas A. Becnel, Lockport, Louisiana "was named after the more famous town on the Erie Canal in New York."[85]

The most significant—and interesting—of the new Lockports sprouted southwest of Chicago on the Illinois and Michigan Canal. The I and M, constructed between 1836 and 1848, was arguably the most historically important canal—apart from the Erie—in North America. Although there were once four separate canals connecting the Great Lakes to the Ohio and Mississippi rivers, today there is only one: the I and M's successor, the Illinois Waterway. Chicago emerged as the metropolis of the interior at the point where the continent's two major river systems were most easily joined. In addition to Lockport, several other New York names appear along the one-hundred-mile canal, including Seneca, Morris, and Utica.

Unlike its namesake in Louisiana, Lockport, Illinois, had direct and ongoing contact with the first Lockport, although the two settlements were over a thousand miles apart via water navigation. The most important link between the Lockports of New York and Illinois was forged by William Gooding and his extended family. Gooding moved from western New York to Will County, Illinois, in 1833. He worked as an engineer on the Indiana canals between 1834 and 1835, was then appointed chief engineer on the I and M in 1836, and remained in its service until 1871.

The first Goodings to arrive in Lockport, New York, were John and Marry Anne, who came with four sons and a daughter in 1823. They are listed as founding members of Lockport's First Presbyterian Church where

John later became a deacon. John's brother, Stephen, who also arrived in 1823, opened a store and employed his brother's son Stephen F. Gooding in 1832. The younger Stephen became a surveyor and later worked as an engineer for railroad lines and eventually on the enlargement of the Lockport locks. He later moved to Lockport, Illinois, where his son Stewart was born. Deacon John Gooding was the principal force behind the Pioneer Line of stages that shunned all Sunday travel. The *Souvenir History of Niagara County* comments that "Deacon John Gooding, patron of Pioneer Hill and after whom Gooding Street is named, removed to Lockport, Illinois and died there in 1840."[86]

The Goodings had come to Lockport from the village of Henrietta, southeast of Rochester. Less than twenty miles from Henrietta was William Gooding's hometown of Bristol, New York. William was most likely John's brother or nephew and he must have had intimate knowledge of the works at Lockport to which his relatives had moved, some of whom would later follow him to Illinois. Like John Gooding's son Stephen, William probably had experience as a contractor before migrating west.[87]

Though many of the Gooding clan removed to Illinois, others remained as prominent citizens of the first Lockport. In 1902, when four generations of New York Gooding men posed for a photograph, three of them were employed in the newspaper business.[88] Interactions between the Goodings of Lockport, New York, and those of Lockport, Illinois, were clearly continuous and mutually beneficial.

William Gooding was the key figure of the I and M. His salary of $3,500, higher than any other public employee in Illinois, was controversial. On his estate in Lockport Township, he built a magnificent Greek Revival house with Doric columns that Abraham Lincoln reportedly visited. The surrounding district later became known as Gooding's Grove. A silhouette statue of William Gooding at Lockport honors his contributions to the development of Illinois.[89]

The example of the Goodings is only one of many connections and parallels between the two Lockports that seem at once uncanny and logical. Lockport, Illinois, is about twenty miles from Lake Michigan but thirty-two miles from Chicago via the I and M. Lockport, New York, is about twenty miles from Lake Erie but thirty miles from Buffalo via the Erie. Each is the first locking down point on its respective canal. At both sites, the work had been arduous, resulting in riots, deaths, and military interventions. The work at the Mountain Ridge had produced the most contentious labor relations on the Erie Canal. By the time of the I and M's construction, the conditions of the navvies had become much more difficult. There were four major disturbances, one in which ten died. At Lockport in 1840, local citizens organized a posse after workers attempted to burn the canal office. Lockport was the headquarters of the I and M, so Gooding was directing the entire work from that office.[90] Construction crews encountered dolomite limestone at both Lockports; later commercial quarries opened and the rock was shipped via canal. There are so many limestone buildings in both cities that their architecture bears a family resemblance. Both Lockports developed an upper and a lower town. Both became industrial after the construction period by taking advantage of the drop in elevation into the "Grand Natural Basin" of Lockport, New York; and into the "Hydraulic Basin" of Lockport, Illinois. Industry in both cities began with flour and paper mills. The presence of industry, in each case, led to environmental problems. Today each city is struggling to remediate serious toxic waste sites.[91] Doppelgangers in light and shadow.

Perhaps the most significant parallel between the two Lockports is that each was a key site on a canal that was arguably the most important cause for the development of a metropolis. When the I and M opened in April of 1848, a cargo of sugar from New Orleans passed through Chicago and arrived at Buffalo. Like the Erie before it, the I and M was rearranging the geography of the continent. At the canal's northern terminus, extensive lumber docks and stockyards developed. Swift and Armour—a former I and M contractor—built slaughterhouses and meatpacking facilities, employing many former navvies under conditions that likely made them wonder whether they were still in the ditch. From a vast region of rich soil, livestock and grain began to pour through the I and M. "Almost single-handedly," the Illinois and Michigan Canal had "spurred the economic development of northern Illinois."[92] All of this had been made possible by the success of the Erie Canal. The I and M required the Erie just as the Lockport of Illinois required—and drew inspiration from—the Lockport of New York. When Bouck and the fifteen-hundred-strong workforce were struggling to overcome the Mountain Ridge, the future of New York City hung in the balance, and when Gooding, from his headquarters in Lockport, labored to keep the I and M progressing, the fate of Chicago was at stake. Two Lockports. Two canals. Two metropolises.

During the canal era, the names Lockport and Clinton, in the view of most North Americans, were intimately linked. In American toponymy, the name Clinton is nearly as ubiquitous as Jefferson and Franklin. We find Clinton counties in Pennsylvania, Ohio, Michigan, Indiana, Illinois, Kentucky, Missouri, and Iowa—a pattern that resembles the proliferation of Lockports. Throughout the United States and Canada, there are Clinton lakes, rivers, towns, villages, and streets. The replication of Lockports and Clintons is a kind of naming that must be distinguished from the borrowing of European place-names so common in eighteenth- and nineteenth-century North America. An appropriate parallel would be the proliferation of Manchesters.

The villagers at Niagara Falls decided to name their settlement Manchester because of the idea that the English city had come to represent: industrial power and prosperity. A century and a half later, the people of Portland, Oregon, designated their high-tech corridor the Silicon Forest—and the people of Sioux Falls, South Dakota, named theirs the Silicon Prairie, because they were hoping to replicate the futuristic prosperity that the Silicon Valley connoted in the 1980s and 1990s. The names Lockport and Clinton had similar resonance in the second quarter of the nineteenth century. If we had to choose one word to describe the idea they represented, that word—for most middle-class North Americans at least—was progress.

Progress was more than an idea: it was something people could palpably feel, particularly at places like Lockport. Travelers passing through the town, as well as journalists and other writers who commented on its significance, expressed a range of reactions to the place and to the notions of progress it seemed to embody. I now turn to this literature.

Travelers at Lockport

Less than two years after the canal's completion, Mrs. Anne Royall, a wealthy young woman from Washington, DC, approached Lockport from the west. She had just visited Niagara Falls, of which she had written: "it robs the beholder of senses; struck dumb with amazement, there seems to be nothing else in the world." The Deep Cut, a product of human art, evoked a similar response:

> It was growing dark when we entered this excavation, and the passage so deep from the surface of the earth it was terrific, and not long since caused a lady to faint.
>
> It staggers belief; and nothing but reality; the huge machines yet standing on the margin of the canal, with which the stone was removed out of it; the mountains of stone, heaped up on each side; and the certainty that you are actually sailing through a mountain, could wring the truth from us, that this stupendous work was performed by man. At the end of the excavation is Lockport.

And Lockport had a second marvel at its heart:

> If astonishment was not already wrought up to the highest pitch at seeing this deep cut, through a solid rock of three miles, it would certainly be so at the sight of those locks. By these locks the canal ascends what is called the mountain-ridge from the long Genesee level. Thus out-braving nature. . . .

At this point in her narrative, Mrs. Royall inserted a detailed description of how the locks actually worked. Then she offered an interpretation:

> These double locks are the greatest piece of masonry, perhaps, in the United States. Taking the beauty, symmetry, and the style in which it is executed into view, it is not exceeded by the capitol of the United States. The locks are ten in all; a row of five opposite to five others, and empty from one to the other. There are three rows of steps from the first lock to the lowest. Here, the passengers mostly get out and run down the steps. These steps, with the neatness of the whole architecture; the great depth to which the boats descend, looking as if they were sinking down to the bottomless pit, strikes the traveler with nearly the same awe and admiration he feels at the grandeur of the great falls.

Next, she turned her attention to the town itself:

> Lockport is a considerable village, of 1500 inhabitants, and although of recent date, is growing fast into notice. The length of the canal and the time it was constructing, with the amount of its cost, is inscribed in large letters on an arch in the center of the locks. It was nearly dark when we passed Lockport, by which means I was deprived of the pleasure of seeing more of this wonder of the western hemisphere.[93]

Anne Royall's response to Lockport in 1827 touched many of the themes that would persist in travelers' accounts throughout the canal's antebellum heyday: the sublimity of the Deep Cut and the locks, the fascination with the new hydraulic technology, the awe at human accomplishment "out-braving" nature, the sense that Lockport with its plaque describing the entire canal somehow encapsulated the Erie's achievement, and the comparison with Niagara Falls. The only major themes missing are an explicit reference to nationalism and the canal's role in securing America's continental destiny. Her upbeat, romantic interpretation of Lockport, however, was far from universal. Other writers, while often touching upon these same themes, evaluated their meaning and significance quite differently.

Because the progress of the canal's construction, and the unprecedented celebration, had been followed throughout the country and beyond, most visitors arrived with knowledge and expectations. In the decades that followed, travel and other writing about the canal and Lockport became increasingly intertextual, with many visitors explicitly mentioning the various earlier descriptions they had read. Hence, writings on the canal—and in particular on Lockport—constitute a kind of dialogue. As we shall see, the dialogue focused on a set of interconnected themes that cluster around notions of progress.

The discussion that follows examines the meanings that antebellum Americans and foreign visitors associated with Lockport. It is based, first, on fifty travelers' accounts, most of which are contained in published books. Of these, thirty-five were written by non-Americans, with British travelers representing the largest single group. In addition, numerous guidebooks, contemporary histories, gazetteers, periodical articles, and newspaper accounts also commented on Lockport. In a separate section, I consider about ten paintings, engravings, and other graphic images of Lockport beyond the Catlin prints and other images associated with the celebration; the visual interpretations they present deserve special attention.

The opening of the Erie Canal ignited North America's first wave of pleasure travel, or tourism. American travel writers were part of the new commercial and industrial middle class, a group that, for the first time, had the resources and leisure to undertake extensive journeys.[94] Foreign travelers were also people of means, but some of them were members of the aristocracy rather than the new bourgeoisie. Of those who wrote about Lockport, women represented a larger proportion among American (27 percent) than among foreign (9 percent) travelers. Despite their common economic status, these writers crafted a remarkable range of responses.

Travel writing theory has evolved a great deal in recent years. A key insight is Jonathan Culler's suggestion that complaints about the crassness of other travelers and nostalgia for a more authentic era have always been at the heart of the travel or tourist experience.[95] Travel writers sought to establish an identity, but since most writers hoped to publish and profit from their observations, they also had to appeal to the perspectives and longings of their readers. For both of these reasons, they endeavored to offer something distinctive. Examining the contexts from which travelers have emerged—their aesthetic, gender, class, and national commitments—must be tempered by an understanding that, to some extent, travelers were released from ordinary commitments and conventions. Writing about European travelers in North America, William Stowe argues that they traveled "as a way to shed familiar constraints of their culture and place, to act out experimental roles, and try out tentative personalities, measure themselves against foreign customs and values, and seek aesthetic or spiritual experience."[96] Travel thus opened the possibility of wrestling with the distinctiveness of places visited and crafting individual responses, although clearly travelers could never entirely transcend their contexts and commitments. Women writers, in particular, often created ambivalent, poly-vocal accounts. Linda Revie identifies a "feminine aesthetic" that "does not fall back on one type of description at the expense of another," but rather "pushes the conventions to their limits."[97] Revie's own approach is therefore to choose a few representative travelers and to contextualize them densely, paying particular attention to their subjectivity.

Because of my interest in the negotiated meanings of a place and the ideas it connotes, rather than the subjectivities of individual visitors, my approach will be different. While remaining sensitive to travelers' subjectivities, I will not test the reader's patience with dense contextualization of every visitor.

The quarter century that followed the Erie Canal's completion was a time of enormous flux. To many, the canal itself seemed the epicenter of change. Alexis de Tocqueville had no doubt that the Erie Canal and other improvements in transportation were key elements of progress. "I make no claim to have discovered that they contributed to the prosperity of a people," he wrote; "it is a truth universally felt and recognized."[98] While often disagreeing on the value of progress, few questioned that the Erie Canal embodied it in symbol and in substance. By connecting New York and the seaboard to an immense continental hinterland, it had spread the commercial economy first to the counties along its path and then to that vast region to the west. In addition to enabling commercial agriculture in regions that previously had relied primarily on subsistence activities, it also helped to spur industry. The industrial revolution now spread from New England and other coastal areas to cities along the canal like Utica, Rochester, and Lockport. As we have already seen, these industrial and commercial transformations had enormous social implications. At Lockport and other places along the canal, travelers sought to understand these changes and grapple with their meaning.[99]

The canal's opening also coincided with the first serious attempts to define the cultural distinctiveness of the United States. Hudson River writers such as Washington Irving, William Cullen Bryant, and James Fenimore Cooper, along with painters such as Thomas Cole and Frederic Church, soon followed by the writers of the New England Renaissance, all emerged during the canal's heyday to address a national and even an international audience. Irving's story of Rip Van Winkle resonated partly because so many Americans felt that they had suddenly awoken to a strange new world.[100] Of course, there was hardly unanimity of opinion about the virtues of this new world. Elite writers and artists often assumed the role of Jeremiahs, pointing out the dangers and paradoxes of progress. The question of American identity, for these elites and indeed for most European writers, almost always focused on the young nation's distinctiveness vis-à-vis Europe. Asserting its connection to North American nature could help establish this difference, yet, for a society of settlers moving into newly conquered land, there was a need to define another difference, and here the imperial rhetoric of civilizational superiority was mobilized. Hence, nationalist discourse was marked by paradox, tension, and instability.[101]

Foreign travelers, particularly those from Great Britain, were also wrestling with the costs and benefits of industrial, commercial, technological, and social change. At sites like Lockport, many explicitly contested the opinions expressed in the published works of their own countrymen. Foreign visitors often had one additional agenda: assessing the merits of the democratic experiment and its relation to other kinds of change. Their analyses were not always as penetrating as that of Tocqueville, but, again, they offered a surprising array of insights and commentary.

The canal also helped to give structure to an emerging North American Grand Tour, often followed by European travelers and Americans from the eastern seaboard.[102] The itinerary usually included the Hudson Valley, Saratoga, the Erie Canal, and always Niagara Falls, which—as one British visitor put it—was "the goal and object of western travel."[103] Guidebooks soon appeared, covering practical issues—accommodations, route maps, timetables, and estimated costs—as well as the virtues of the natural and artificial sights along the route, often including descriptive quotations and poetry from the writings of previous visitors.[104] Many travel writers strayed from suggested routes, perhaps hoping to offer their readers distinctive insights. This was an age of growing individualism. Travel writers were keen to project a unique voice. Their comments on the various sights were often reflexive, emphasizing their own emotional and intellectual responses. In an age of change, many hoped that the journey might effect a change within themselves. Nathaniel Hawthorne, and many others, referred to the journey to Niagara as a pilgrimage, and like traditional pilgrims, hoped for some kind of personal transformation.[105]

Carrying a guidebook, like reading the accounts of earlier travelers, could increase the intertextual quality of a travel writer's comments. Spafford's *Pocket Guide* of 1824 was more lyrical than later guidebooks. Spafford waxed poetic about the combination of nature and art at Little Falls, called the Rochester Aqueduct the "grandest single feature," and after a long description of the famous locks, proclaimed Lockport "the backbone of the Erie Canal." He also described the "ravine through which the canal rises the Mountain Ridge" as a "singular and interesting entity."[106] Gordon's 1836 description simply asserted that the double locks were "of excellent workmanship, with stone steps at the center, and at either side guarded with iron railings."[107] Holley's *Picturesque Tourist* (1844) stated that Lockport "has sprung up and become one of our largest inland villages."[108] By the 1840s, visitors could also read about Lockport's growth in books and magazines. An 1845 article in the *National Magazine* described Lockport's past, its present importance for industry and continental transportation, and concluded that "when the extent, resources, and rapidly increasing population of the mighty West, are carried forward to the future, the mind becomes bewildered in their contemplation."[109]

Travel writers who visited Lockport crafted a diverse set of responses to what they experienced there. They substantially agreed, nonetheless, in the identification of certain important contemporary issues that made the place worth visiting. The Erie Canal at Lockport was about progress, America, and continental hegemony in relation to nature, Niagara Falls, and obstacles to U.S. hegemony. My analysis of this process, which I call writing Lockport, will be structured thematically according to a series of interrelated connections that many travel writers made. The first was that the works at Lockport—the Deep Cut and especially the locks—represented or encapsulated the Erie Canal as a whole. It was, therefore, an appropriate site at which to sum up and assess the entire project. Next, travelers often saw the rapid growth of Lockport from a wilderness site to an urban place as a demonstration of the canal's benefits. Third, the experience of traveling through Lockport, for some, had an aesthetic dimension that they attributed to the juxtaposition in the landscape of the artificial and the natural. Then, many visitors, regardless of whether they commented on aesthetics, took the opportunity at Lockport to speak in broad terms about the relation of nature to art. Most associated Lockport—and hence the canal—with art or progress, often explicitly American progress. The clearest way to demonstrate the significance of the locks and Deep Cut, the canal, and art, was to contrast them with nature, and particularly with the nearby falls of Niagara. Mention of Niagara, for many, then flowed naturally into awareness of continental issues. Many travelers made specific reference to the canal's role in U.S. continental hegemony. This, in turn, led to many comments on the relationship between the United States and the Canadas. Finally, some foreign travelers explicitly addressed what the canal and the accomplishments at Lockport said about the nature of America. The fact that visitors employed these thematic connections does not mean that they agreed on fundamental issues such as the value of progress. In the discussion that follows, I will endeavor to address these connections separately without losing sight of the fact that they are complexly interrelated.

Lockport and the Canal

Sibyl Tatum, an educated young woman in her late teens, traveled in 1830 with her parents from the east coast to Ohio via the Erie Canal. Concerning Lockport, she wrote:

> Lock Port appears to be improving, some handsome buildings. The locks are a great curiosity as well as beautiful, there are five durable locks one immediately after the other in which the water rises 60 feet. It still is an astonishment to me what the energy of man will lead them to do. We are passing on the canal where it was blown through solid rock for miles. The Erie Canal was commenced 4th of July 1817 and completed 1825, 362 miles and expenses 7,000,000 of dollars.[110]

Sibyl Tatum is one of many travelers who, after expressing admiration for the human ingenuity that created the locks and Deep Cut, launched into an assessment or summary of the entire canal. A number of other travelers chose to discuss the whole canal project at Lockport as well.[111]

An illustrative example is that of Basil Hall, an aristocratic Scottish naval officer, who floated through Lockport in 1827. After a brief description of the busy town, he turned to the already famous locks:

> Lockport is celebrated over the United States as the site of a double set of canal locks, admirably executed, side by side, five in each, one for boats going up, the other for those coming down the canal. The original level of the rocky table-land about Lockport is somewhat, though not much, higher than the surface of Lake Erie, from which it is distant, by the line of the canal, about thirty miles. In order to obtain the advantage of having such an inexhaustible reservoir as Lake Erie for a feeder to the canal, it became necessary to cut down the top of the ridge on which Lockport stands, to bring the canal level somewhat below that of the lake. For this purpose, a magnificent excavation called the Deep Cutting, several miles in length, with an average depth of twenty-five feet, was made through a compact, horizontal limestone stratum—a work of great expense and labour, and highly creditable to all parties concerned.
>
> The Erie canal is 363 miles in length, 40 feet wide at the surface, 28 at the bottom and four feet deep. . . .

After making this abrupt transition from Lockport to the Erie Canal, Hall took six paragraphs to describe the canal's history and economic benefits, especially to the New Yorkers along its path. Next he warned of a kind of canal bubble. The Erie's success had given birth to a canal mania, including projects without "any good purpose." Hall nonetheless recognized that certain areas were about to boom, like the region between the Ohio River and the Great Lakes. "By means of canals stretching from the very center of that fertile region," he wrote, "they can now send their produce to Lake Erie; from whence it may enter the Grand Canal at Buffalo, and so find its way to Lockport, Rochester and Albany; and from thence its course to the sea at New York down the Hudson." Finally he warned of competition for the trade and allegiance of the continent's heart from both the Mississippi and St. Lawrence alternatives to New York's artificial river.[112] Unlike some tourists, Hall clearly understood the significance of the locks and Deep Cut: they made it possible to fill the canal to the east with the waters of the upper Great Lakes. He also saw the continental implications of the canal's success. He decided to pause and assess the whole canal project at Lockport not merely because he was following a convention, but rather because he grasped the significance of what had been accomplished there.

Ten other travel writers explicitly noted that, at Lockport, the canal gained its summit level, the level of Lake Erie and of the continent's interior.[113] Lockport was a place where the canal's continental importance seemed particularly visible. The Mountain Ridge was the canal's final obstacle—the last barrier to the continent's interior—and the place of the "Meeting of the Waters." These considerations helped to convince contemporary, and indeed future, writers that, if they had to choose one place to represent the Erie Canal, Lockport was better than most.

I do not mean to suggest that Lockport was the only place that served as a symbol of the Erie Canal or the only place at which travelers decided to offer summary assessments, but rather that it was perhaps the most prominent among many canal sites. Moreover, the difficulty of overcoming the final obstacle seemed to recapitulate previous challenges. Historian Ronald Shaw points out that the Lockport locks have "almost invariably been chosen by authors from that day to this to illustrate any work on the Erie Canal."[114] It helped that Lockport was a picturesque spot where the contrast between nature and art—the natural ravine and the stunning locks—seemed to encapsulate a central issue of the dialogue on the meaning of the canal.

Writers' opinions about the value of progress or of art vis-à-vis nature did not necessary depend on gender, nationality, or the decade in which they wrote. Hence we find a spectrum of opinions among women writers; some are enthusiastic about the canal and all it seems to represent; others are appalled. The same is true for other categories, such as British visitors. This shows how general the debate was: it penetrated everywhere. Accordingly, in the discussion that follows, I will not separate the written accounts by decade or by the gender or nationality of their authors. In the cases where these differences seem to matter, I will highlight them.

The "Immaculate Conception" of Lockport

In his 1951 article, "The Erie Canal: Mother of Cities," Rochester historian Blake McKelvey pointed out that the Erie Canal "gave birth to but two cities—Syracuse and Lockport." But even Syracuse had drawn its early residents from numerous surrounding towns and the people who clustered about its salt works. "But it was at Lockport," McKelvey writes, "that Mother Erie achieved immaculate conception."[115] Before the canal's arrival, there was simply nothing there, nothing even in the vicinity, nor any real prospects of change. Nathaniel Hawthorne's suggestion that the Erie Canal, by uniting "two worlds, till then inaccessible to each other" had "conferred inestimable value on spots which nature seemed to have thrown carelessly into the great body of the earth, without foreseeing that they could ever attain importance," seems particularly applicable to the case of Lockport. "Surely," Hawthorne observed,

> the water of this canal must be the most fertilizing of all fluids; for it causes towns—with their masses of brick and stone, their churches and theatres, their business and hubbub, their luxury and refinement, their gay dames and polished citizens—to spring up, till, in time, the wondrous stream may flow between two continuous lines of buildings, through one thronged street, from Albany to Buffalo.[116]

Over and over again, travelers and supporters of internal improvements pointed to the growth of towns as proof of the Erie Canal's "fertilizing" power. The unexpected growth of Rochester was the classic example. "No town in America, perhaps none in the world," McKelvey points out, "had ever before mushroomed so rapidly."[117] European travelers found such places particularly novel. The outskirts of Rochester, Mrs. Basil Hall commented, "looks as if a box of houses had been sent from New York, the lid opened, and the houses tumbled down in the midst of blackened stumps."[118] James Boardman, a London businessman who traveled through Rochester in 1829, was so taken by the juxtaposition of "the state of nature" and "refinement and polish," that he fell into a "prospective reverie, excited by the strangeness of the scene."[119]

Although Lockport never grew as large as Rochester, its growth nonetheless impressed visitors because it had started from scratch, and wilderness continued to surround it for many years. Henry Tudor, a British barrister who believed the United States had been unjustly slandered by British travel writers, remarked that Lockport's history was "concise enough."

> Five years ago, or thereabouts, it existed not; a wide waste of wilderness occupying its site—and now there are between 300 and 400 houses, of which it owes, entirely, the existence to the presence of the Erie Canal that runs through it. One thing, however, may be safely predicted, that in ten or twenty years from this time it will have become a handsome, wealthy, and important town. Though in its infancy, it possesses, notwithstanding, the finest work on the whole line of the Erie canal, in the splendid locks whence it derives its designation. This gigantic work consists of ten locks, of fine hewn stone, formed in a double range of five in each, placed in juxtaposition, and which have been constructed for the purpose of surmounting the rocky ridge dividing the two levels on each side of it.[120]

The Scot traveler Adam Fergusson noted that Lockport "has been rapidly transformed from the wilderness into a thriving village of perhaps 2,000."

British army lieutenant E. T. Coke began his 1832 account with a similar comment: "we entered the village of Lockport, which, like Rochester, or most places on that line of communication, has sprung up in almost a day."[121] To emphasize this mushroom-like growth many travelers contrasted the present population with that of an earlier date, or, like the Long Island businessman Moses Cleveland, the number of houses. "In 1821 as I am informed there were but two houses in the village," but by 1831, the year of Cleveland's visit, there were "nearly 400."[122] One Scot visitor wrote:

> Lockport is another of those mushroom villages which the Erie Canal has raised into existence. In 1821 there were only two houses at this place, now there are several hundreds. It does appear strange that a village of this size should not contain an inhabitant above twenty years of age born in this country. The inhabitants are strangers from many parts of the world, who have been led by various causes to make this place their adopted home.[123]

Other visitors were impressed with the general bustle and energy they found at Lockport. Theodore Dwight, a New Englander who arrived at the village in the early 1830s, noted, like so many others, that "within a few years, the spot has been changed from a wilderness into a village of several hundred shops." It was not the mere size that struck him, however:

> This is one of the interesting places on the canal. Here is the noblest display of locks, two ranges, made of fine hewn stone, being constructed against the brow of the Mountain Ridge where the foaming of the waste water, the noise of mills, and the bustle of occupation excites many lively feelings.[124]

Dwight's "lively feelings" recall the words of James Boardman, who was so amazed at Rochester that he experienced a "prospective reverie, excited by the strangeness of the scene." Indeed, when Boardman reached Lockport, he commented that "the scene, from the number of boats passing through the locks, is extremely animated."[125] In 1837, Connecticut minister James-Hanmer Francis described Lockport simply as "a place of much business."[126]

Nathaniel Hawthorne described the experience of entering and leaving a typical canal town in similar language:

> Passing on, we glide now into the unquiet heart of an inland city . . . and find ourselves amid piles of brick, crowded docks and quays, rich warehouses and a busy population. We feel the eager and hurrying spirit of the place, like a stream and eddy whirling us along with it. Through the thickest of the tumult goes the canal, flowing between lofty rows of buildings and arched bridges of hewn stone. Onward, also, go we,

> till the hum and bustle of struggling enterprise die away behind us, and we are threading an avenue of ancient woods again.[127]

Many travelers shared Hawthorne's unease with the bustle of places like Lockport. Basil Hall, who visited the village less than two years after the canal's completion, described the scene:

> On the 28th of June, we proceeded to Lockport, a straggling, busy, wooden village, with the Erie Canal cutting it in two, and hundreds of pigs, stage-coaches and wagons, occupying the crowded streets; while a curious mixture of listlessness and bustle characterized the appearance of the inhabitants.[128]

Another British traveler, Edward Abdy, found Lockport more substantial in 1833, but hardly worthy of sustained attention:

> It was ten o'clock before we reached Lockport, where I found a clean and comfortable bed, after a tedious journey, the fatigues of which were scarcely indemnified by the novelty of the adventure.
>
> I left the rising town of Lockport without viewing its wonders, beyond what could be seen through the windows of the coach, as it passed by many substantial well-built houses: the recent birth of which was attested by the stumps of trees about them.[129]

Many visitors commented on the substantial built environment of the canal towns, and at Lockport they also emphasized the "solid masonry" of the locks.[130] William Lyon Mackenzie, who would later be among the leaders of the Canadian Rebellion of 1837, first visited and described Lockport in 1825. Here, I want to consider his return visit seven years later, when he described a more substantial urban landscape.

> Lockport thrives amazingly; there are now two towns, the upper above the locks and the lower below them. Many of the buildings show signs of great wealth, and the appearance of the place, taken as a whole, evinces a sense of security, comfort and industry.[131]

Hawthorne was equally impressed with the solidity of Rochester. Though it had

> sprung up like a mushroom, . . . no presage of decay could be drawn from its hasty growth. Its edifices are of dusky brick, and of stone that will not be grayer in a hundred years than now. . . . The most ancient town in Massachusetts appears quite like an affair of yesterday, compared to Rochester.[132]

The extent to which this perceived solidity is attributable to American viewers as opposed to American builders is debatable, but neither wanted to believe that this canal and these "solid" Yankee towns bore any resemblance to boom-and-bust mining encampments. For both groups, perhaps, solidity served as a bastion against the tensions of an era of earth-shaking transformations and demographic mobility—tensions that the canal not only embodied but had helped to instigate.

The very name, "Lockport," composed of two preexisting words, indicated change in the canal era. "Lock" itself was a high-tech word, comparable to "digital," or "silicon," in our times. "Port" was a pretentious appellation for an inland town. Heading west from Rochester, the canal traveler passed through Spencerport, Brockport, Newport (later Albion), Middleport, Gasport, and Lockport. Most of western New York's population at this time had roots in New England where we find Bridgeport, Rockport, Kennebunkport, and dozens of other "ports," but these places were at or very near the sea. Brockport had originally been called Brockway, after its founder, but the "way" was changed to "port" presumably to indicate a place connected to the wider world. As Carol Sheriff has suggested, such a designation helped to attract settlers to a new town because it "promised to link them to markets as both producers and consumers."[133] Horatio Spafford complained that "excepting Lockport, all of these *ports*" seemed ill named.[134]

Lockport Sublime

The experiences of some travelers at Lockport had an aesthetic dimension. As we have already seen, Mrs. Anne Royall felt that the Deep Cut "staggers belief," and at the locks she noted "the great depth to which the boats descend, looking as if they were sinking down to the bottomless pit, strikes the traveler with nearly the same awe and admiration he feels at the grandeur of the great falls." She was personally moved in a way that she, and many others, compared to the experience of visiting Niagara Falls. Carl David Arfwedson, a Swedish author and playwright, on an extensive tour of North America in the early 1830s, arrived at Lockport at 3 a.m.

> Darkness prevented me from examining attentively the five locks built here: I was obliged to content myself with walking up a few granite steps, made on the sides of the canal, and, with the assistance of a stick, trying to grope my way as well as I could. When at length I had reached the upper lock, about sixty feet above the basin, I found myself, all at once, in the middle of Lockport, a village entirely built on and surrounded by rocks. It was a singular sight to look down from this point on the double row of locks, built close to each other, dimly

> lighted with lamps, and in the dark appearing as so many flights of steps. Each lock is twelve feet wide: the stone-work is executed with much care and taste. To obviate the possibility of the detention of the canal-boats at this place, two sets of locks are built, by which arrangement one boat is able to ascend, whilst another descends. This happened at the time I was examining their excellent construction. A boat, laden with produce from Ohio, was lowered to the right, with the same rapidity as ours was raised on the left. The lanterns on deck were the only mark by which I could perceive whether the craft rose or fell, the noise of the rushing water entirely drowning the voices of the steersmen. The effect of the glimmering light between the black stone walls was like magic. No traveler should visit Lockport without witnessing such a scene.[135]

Although Arfwedson also praised the Deep Cut as "a most gigantic and wonderful piece of work," most visitors who used words like "beautiful," "romantic," and "sublime" focused, like the Swedish traveler, on the famous locks.

Adam Fergusson, a Scot who plied the canal in 1831, partly to gather information for his countrymen who might want to immigrate to Upper Canada, also emphasized the locks over the Deep Cut, although for him their juxtaposition enhanced the effect. He began by noting that "the canal near this place has been cut through solid calcareous rock, for a distance of five miles."

> At the termination of this stupendous excavation are placed five double locks, of handsome and solid masonry. The descent, I believe, is about 70 feet, and the scene is altogether picturesque and interesting. After having been for above an hour immersed by the rock on either side, you find yourself suddenly emerge, and approaching steadily to an abyss of somewhat threatening aspect. Presently, however, a halt takes place, and the beautiful mechanism of the hydrostatic ladder transports you in safety to the plains below. You have in fact descended the falls of Niagara, for it is the same ridge which intersects Canada, and you are now upon the level of Ontario.[136]

Fergusson's evocation of an "abyss of somewhat threatening aspect" tamed by the "beautiful mechanism of the hydrostatic ladder" by means of which the traveler has "in fact descended the Falls of Niagara" would seem a perfect example of what David Nye has identified as the "American technological sublime." At places like Niagara Falls, pilgrims anticipated an experience of the natural sublime, a concept articulated by eighteenth-century aesthetic theorists. Although overwhelmed by the apparent infinitude of a wild natural phenomenon—like the high Alps, a storm at sea, or a colossal cataract—the viewer may nonetheless be exhilarated. According to Edmund Burke and Immanuel Kant, such experiences awaken in the viewer a sense of his or her own infinite worth—a recognition of an inner human dimension of imagination, reason, or spiritual depth. In applying the word "sublime" to works of human artifice, Nye expands sublimity to anything that produces astonishment. He argues that both the natural and the technological sublime became important unifying ideas in the young United States, partly because they did not refer to ethnicity or the past, subjects that could only be divisive. "While voters might disagree on the issues of the day," Nye writes, "they could agree on the uplifting sublimity of Niagara Falls, the Natural Bridge, or the Erie Canal at Lockport."[137] He identifies the Erie Canal, and particularly the Lockport locks, as the first technological marvel to be celebrated in a nationalistic manner. They would hardly be the last. He notes that both the Erie Canal and the Baltimore and Ohio Railroad were launched on July 4. He identifies a number of subsequent technological feats that served to unite otherwise contentious Americans, including the Brooklyn and Golden Gate bridges, the Empire State Building, the atomic bomb, and the space shuttle launches. In his more recent writings, Nye has taken a more nuanced approach to the role of technology in the United States, suggesting several possible narratives.[138] While it is indeed reasonable that what appears as sublime can evolve, and also that human accomplishments can serve as unifying symbols in a nation with no stake in the past, it is not clear exactly how this connects to the natural sublime. Facing Niagara Falls, it is the individual who feels his or her own inner dimensions, but technological feats are necessarily collective accomplishments.[139] Yet, as in the case of the natural sublime, technological wonders gain their significance by the natural or wild forces they confront: chasms conquered by bridges, outer space conquered by spacecraft. The Deep Cut and Lockport locks conquered the Mountain Ridge, but as Fergusson's account suggests, by the "beautiful mechanism of the hydrostatic ladder," you have "in fact descended the Falls of Niagara." If there were technological sublimity at Lockport, it could be appreciated particularly in contradistinction to Niagara Falls. It is worth noting that the awe of technological achievement was hardly confined to Americans, as the example of Fergusson underscores.

The aesthetic dimension of a visit to Lockport clearly involved the contrast between the artificial and the natural. In the early 1830s, Maximilian, the Dutch prince of Wied, approached Lockport from the west. The canal, he wrote,

> is cut through a stratum of gruwacke, which rises from four to fifteen feet above the water; but the depth of the ravine soon increases, and the bridges are thrown, at a great height, over the canal. At Lockport,

> an extensive place, situated on the eminence, the canal is conducted, by means of five sluices, down a slope of at least sixty feet. The prospect from the eminence is very beautiful. The canal descends between two hills, connected at a considerable elevation by a bridge, under which the boats pass.[140]

The contrast between the bridges and locks, on the one hand, and the vertiginous landscape of Lockport on the other, clearly impressed the prince and rendered the village "beautiful." The fact that the canal descended "between two hills," helped to frame the technological accomplishment as well as the visitor's immediate experience.

When canal travelers passed through Lockport, it was, literally, a moving experience. They often described a sequence of steps. Entering from either direction presented a series of dramatic contrasts—approaching the locks after miles of travel though the artificial canyon of the Deep Cut, or suddenly turning into the natural basin after sixty miles at the same level. Bridges, buildings, spires, entered and faded from view. Almost everyone disembarked and walked up or down the steps separating the two flights of locks; the whole vicinity was crowded with groceries offering alcohol, food, and souvenirs. The changing perspectives afforded by this new mode of travel introduced Americans to the exhilaration of movement, an obsession that many still consider particularly American.[141]

Approaching Lockport in 1836, Caroline Gilman, a young woman from South Carolina, felt "far from home, to think of our vicinity to these great waters." The Great Lakes-St. Lawrence Valley was at the margin of her sense of national consciousness, but Gilman was no isolated Southerner. She had read several Erie Canal travel accounts as well as the works of Emerson and other transcendentalists.[142] Lockport reminded her of a famous city in Scotland: "I would suppose this place to be situated something like Edinburgh; it has its upper and lower town, and the natural wildness that I have heard attributed to that great city. . . . Here the great Erie Canal has defied nature and used it like a toy; lock rises upon lock, and miles are cut in the solid stone.[143] The stark juxtaposition of "natural wildness" and defiant human technology was important for those who found Lockport "romantic," "wonderful," or "beautiful." Captain Oldmixon, a retired British naval officer, recalling his earlier journey on the canal, remembered "the scenery through the forests and beautifully wild and romantic spots, particularly near the little falls, or the Mohawk, and at Lockport."[144]

Caroline Westerley, the seventeen-year-old daughter of an early and wealthy settler in southeast Ohio, wrote a series of letters to her younger sister back home during her travels to the coast with her parents. She noted that "the first locks which occur on the canal after leaving Lake Erie are at

Lockport. A descent of sixty feet, down what is called the *mountain ridge*, is here passed by means of five locks." Then, after a long and detailed explanation of exactly how the marvelous locks work, she mentioned that she had read "Eaton's Survey of the Canal" but was disappointed in her search for the mineral specimens it promised because "so many travelers resort here, that every thing of this kind near the locks is picked up." Finally, she offered this assessment of the place:

> Lockport is really a most wild and picturesque spot. The work of art, great as it appears in the excavations I have described, is by no means more striking than the aspect of nature at this place. The village is divided into an upper and lower part; the descent from the one to the other is over the *mountain-ridge*, composed of naked and almost perpendicular rocks. Occasionally a tasteful white cottage, with its green window blinds, its courtyard and shrubbery, appears from a point of view, evidently chosen for its romantic situation on some commanding cliff, rather than its proximity to other dwellings.[145]

Like Caroline Gilman and Prince Maximilian, Westerley was fascinated by the combination of art and nature, technology and wilderness. The place was "wild and picturesque"; the canal's marvels, she observed are "by no means more striking than the aspect of nature at this place." Trying to describe the Erie Canal to South Americans, Ishmael—Melville's narrator in *Moby Dick*—declares: "Through three hundred and sixty miles, gentlemen, through the entire breadth of the state of New York; through numerous populous cities and most thriving villages; through long, dismal, uninhabited swamps, and affluent, cultivated fields, unrivalled for fertility; by billiard-room and barroom; through the holy-of-holies of great forests; on Roman arches over Indian rivers; through sun and shade; by happy hearts or broken. . . ."[146] Neither the "Roman arches" nor the "Indian rivers" were nearly as impressive alone as they were in combination. The canal at Lockport followed the path of an ancient river; the locks descended into the ravine that the river had carved. The striking contrast between what humans and what natural forces had created enhanced Lockport's interest for many visitors.

The counterpart to these versions of the Lockport sublime were the accounts of several visitors who were unimpressed with the marvels of Lockport and revolted by the combination of the wild and artificial they found there. I will examine two of these antisublime accounts here. They both involve difficult weather; perhaps the balance between the natural and the artificial seemed obliterated by this meteorological wildness.

As part of an extensive tour of North America between 1833 and 1835, Tyrone Power, a successful British barrister, took a canal packet from the western terminus. Almost immediately, the situation deteriorated:

> All this day the air stood absolutely still. At our places of halt we were joined by men who had left the stages in consequence of those vehicles not being able to travel. Our pace was considerably reduced. . . . At Lockport we found business nearly at a stand-still; the thermometer was at 110 degrees of Fahrenheit. We passed several horses dead upon the banks of the canal, and were compelled to leave one or two of our own in a dying state. Here more persons joined than we could well accommodate and I found positively that all movement by the stage route was at an end, forty horses having fallen on the line the day previous.

Although Power obviously traveled through the Deep Cut and saw the famous locks, he did not even mention them. He considered stopping at Lockport, but "to attempt abiding in any of the places along the canal," he concluded, "would be an exchange for the worse." But things did get worse. Unable to abide the crowded, stifling cabin at night, Power filled his hat with cigars to ward off mosquitoes and spent the night on deck where, he mentioned, "dozing a little," but "never foregoing my cigar for a minute."[147] His descriptive subtitle for the "Erie Canal" chapter was "Packet-boat.—Heat.—Cedar Swamp, Long Swamp, and Musquito Swamp.—Utica." It reads more like a journey through the lower rings of Dante's Inferno than through New York State.

Willis Gaylord Clark, an American traveler from Philadelphia, came to Lockport with expectations. "Lockport is famous for its deep cut," he wrote; "representations of this great achievement I have seen in print, and had supposed that it was a marvel of the first order." Unfortunately, Clark and his companions had traveled by stage from Lewiston through a heavy rain, and as they approached Lockport on the wretched corduroy road, they encountered "mud, without end or bottom, alluvial pudding thickened and gurgled on every side." He wrote that, after entering the town,

> we looked earnestly for the Deep Cut. We continued to gaze until we reached the hotel, when we sallied forth in the rain, with a friend or two, in rabid quest of the wonder. The first view we obtained was from the village bridge. Never was there a more complete disappointment. The line of the canal, to the west, appears very like its usual long and snake-like length; and I put it to the reader, if one very often looks upon a more *common* thing than a canal, after you have traveled across, and alongside, and around it, for some two or three hundred miles? This, then, was the Deep Cut! Oh, minimum of marvels!

Clark then concluded that, since "it was a rainy day," and "everything was gloomy and dismal," he resolved "to give the Deep Cut a *dead* cut."[148] Many of Niagara's visitors arrived with similar expectations, and some found the reality a disappointment, at least initially. "I had come thither haunted with a vision of foam and fury and dizzy cliffs, and an ocean tumbling down out of the sky," Nathaniel Hawthorne wrote of his 1834 visit to Niagara, but after struggling "to adapt these false conceptions to the reality, a wretched sense of disappointment weighed me down."[149] Clark admitted that many had told him "that Lockport is possessed of delightful haunts" and that "the neighborhood around is like a paradise in summer," but these were not his sentiments. Boarding the stage again, he observed "a wide domain of mud, and meadows filled with stumps, and ancient logs, reeking with the rain. Everything looked remorselessly unprepossessing." Finally, he "solemnly" confided to a companion "that if Lockport were built of ducats, and the abdomen of every little hill in its neighborhood pregnant with precious stones and jewels, I would not there reside."[150]

From a twenty-first-century perspective, it is hard to see a canal as anything more than "a common thing." But even in the 1830s, the canal usually only inspired awe in those with a framework of ideas and associations that illuminated its significance. Power and Clark mentioned no connection with Niagara Falls, America's continental destiny, innovative technology, or progress. Their reactions were visceral and immediate.

Art and Nature

Perhaps the master trope for Lockport and the Erie Canal as a whole, for both their admirers and detractors, was the antithesis between nature and art. Nathaniel Hawthorne, who was very suspicious about the ultimate value of the progress his countrymen so enthusiastically embraced, could appreciate the thrill of new experiences the canal made possible. One of these new possibilities was a journey to Niagara Falls. One night, as he traveled toward Niagara on a packet boat, unable to sleep, he went on deck:

> A lantern was burning at each end of the boat, and one of the crew was stationed at the bows, keeping watch, as mariners do on the ocean. Though the rain had ceased, the sky was all one cloud, and the darkness so intense, that there seemed to be no world, except the little space on which our lanterns glimmered. Yet, it was an impressive scene.[151]

Unwittingly, Hawthorne penned a paean to the new experiences that a work of artifice had opened. The contemplative tone of this passage would be echoed in Antoine de Saint-Exupéry's description of flying across the Sahara at night, navigating the blackness solely by his glowing instrument panel, or numerous depictions of a large sedan cruising through the American night, illuminating a small world with its headlights. The thrill of movement would be a recurring theme, especially in the United States.[152]

Yet Hawthorne felt uneasy about the canal's bold penetration into the realm of ancient trees—a "great ditch" through a "forest avenue." In one particularly desolate spot, he imagined that "the wild Nature of America had been driven to this desert-place by the encroachments of civilized man. And even here, where the savage queen was throned on the ruins of her empire, did we penetrate, a vulgar and worldly throng, intruding on her latest solitude." At odds with the world around him, Hawthorne journeyed "through a gloomy land and among a dull race of money-getting drudges."[153]

Finally reaching Niagara Falls, he encountered more of the "vulgar throng":

> Next appeared two traders of Michigan, who declared, that, upon the whole, the sight was worth looking at; there certainly was an immense water-power here; but that, after all, they would go twice as far to see the noble stone-works of Lockport, where the Grand Canal is locked down a descent of sixty feet.[154]

For Hawthorne, the fact that some Americans were more enthused about the "noble stone-works of Lockport" than the great falls proved their vulgarity. Hawthorne's comparison also indicates, once again, how closely Lockport and Niagara were linked during this period.

For many, Lockport clearly functioned as a symbol of art, a futuristic emblem of the amazing things being accomplished by human ingenuity. This was clear in the rhetoric surrounding the canal's opening celebration, and we have already seen a number of travelers' accounts that equate Lockport with art. Caroline Gilman wrote that "here, the great Erie Canal has defied nature and used it like a toy." Anne Royall was surprised that "this stupendous work was performed by man" who she characterized as "out-braving nature." William Leete Stone said that "the work here is a splendid monument of the ingenuity and enterprise of man, in surmounting obstacles seemingly insurmountable." Sibyl Tatum, the young woman from Ohio, wrote: "It is still an astonishment to me what the energy of man will do."[155]

Henry Tudor, an English barrister who had read the works of many earlier travelers, arrived at Lockport in 1831. Traveling by stage from Rochester, he had been struck by "a picture of man, just emerging from a state of savage life into the arts of civilization. All was wildness, rudeness, and disorder." Of Lockport, he suggested, "a tolerably correct idea may be formed, by considering it the counterpart of the country through which I have just been leading you. Indeed, its foundations are but just laid; and it looks, at present, like the element of order struggling with, and rising out of chaos." He described the locks as "the finest work on the whole line of the Erie canal," but was even more impressed by the Deep Cut. The "grand excavation" had required "enormous" labor:

> It was here that the greatest obstacles of the whole 363 miles had to be encountered; and it certainly strikes the beholder with astonishment, to perceive what vast difficulties can be overcome by the pigmy arms of little mortal man, aided by science and directed by superior skill. In many places, the explosive power of gunpowder alone could have torn asunder the massive and deeprooted rocks, which seemed to defy the power of all except the great Being who created them.[156]

Tudor's description shows, again, that works of art "strike the beholder with astonishment" when contrasted with a wild natural background. Hence, because of the primitive country surrounding it, Lockport can seem like "order struggling with and rising out of chaos," and the "pygmy arms of little mortal man" impress because they defy the "massive and deeprooted rocks" of the Mountain Ridge.

Stairway to Niagara

Lockport, from its birth, was joined with Niagara. The entire canal project drew inspiration from the prospect of connecting the coast to Niagara Falls, and the falls became "the goal and object of western travel"[157] only because the canal made it accessible. In the nineteenth century, no one knew that Lockport had been the site of a prehistoric pair of waterfalls, a twin to Niagara, eighteen miles eastward on the same escarpment. Still, there were enough parallels for a great many visitors to notice. Twentieth-century scholars have made an explicit visual connection. Ronald Shaw suggested that the Lockport locks had "something of the appearance of a cataract" partly because they "stood at the head of a natural basin, flanked by steep banks a hundred feet high on either side." Moreover, "the herringbone arms of the lock gates at each step emphasized the suggestion of a waterfall."[158] John Seelye calls them "a veritable Niagara of double locks."[159] Historian Marvin A. Rapp notes that "to some, this flight of locks gives the appearance of a cataract standing at the head of a natural basin."[160] Hudson River painters helped to frame an aesthetic that is still with us. In his "Essay on American Scenery" (1836), Thomas Cole singled out water as the "most expressive feature" of American landscapes and the waterfall as "the voice of the landscape" that "presents to the mind the beautiful, but apparently incongruous idea, of fixedness and motion."[161] The waterfall belonged in the center of the picture. The Lockport locks, the reverse waterfall that stood in the center of so many pictures, with its plaque proclaiming "the spirit and perseverance of Republican Freemen," was a sort of voice for New York's artificial river.

We have already seen how visitors, again and again, reminded their readers that at Lockport one ascended the same Mountain Ridge that Niagara tumbled over. Adam Fergusson even proclaimed: "You have in fact descended the falls of Niagara." Indeed, every drop of water that descends the escarpment at Lockport diminishes Niagara Falls by the same amount. Dr. Thomas Nichols, a Connecticut physician who was often critical of American policies, made a similar point about the Lockport locks. He described his 1837 journey from Albany to Buffalo as sublime:

> We never forget it. A thousand landscapes fill the gallery of our memory. We have passed over dizzy viaducts, and through miles of deep-cut ravines, where hundreds of lives were lost in making them; we have ascended steep hills, through a succession of locks. At Lockport, we are gently lifted up the very precipice over which Niagara rolls, fifty miles away, and are on the level of Lake Erie, whose waters have floated us up, up, up, to their own level. But to keep that level there are miles of a deep, dark cutting through the hard blue limestone, where tons of gunpowder were burnt . . . [162]

Nichols recognized that having been "lifted up the very precipice over which Niagara rolls," meant that one had achieved the level of Niagara's brink, and therefore that of the upper Great Lakes. Many other travelers noted that, at Lockport, one reached the canal's summit and the level of Lake Erie and the Niagara River above the falls.[163] Thomas Woodcock, a Manchester engraver who spent sixteen years in the United States, made the connection between the top of the locks and the top of the falls even more explicitly. On his way to Niagara in 1836, he visited Lockport and wrote: "there are five double locks of excellent workmanship which elevate us 60 feet. we [*sic*] are therefore 560 above the Hudson and have attained the same elevation as the 'falls' from which place we are only about 12 Miles."[164] The Lockport locks were, literally, a stairway to Niagara.

Another reason to compare, and contrast, Lockport and Niagara Falls was the tradition begun by Lafayette who, in 1825, had toasted "Lockport and the county of Niagara—they contain the greatest wonders of art and nature, prodigies only to be surpassed by those of liberty and equal rights."[165] Stewart Scott, a Lower Canadian law student, made a similar comparison in 1826: "as Niagara Falls are the greatest *natural* wonder," he observed, "so Lockport, its Locks and the portion of the Canal adjacent, are considered to be the greatest *artificial* curiosity in this part of America" (emphasis in the original).[166]

Reaching the summit level of the canal had continental implications. Indeed, for many Niagara Falls represented a gateway to the interior. William Leete Stone, who had been the official chronicler of the ceremonial voyage that opened the canal, returned four years later and described Niagara as "the grand outlet of the great inland seas of the still greater West!" Noticing the relative youth of the Niagara Frontier's population, already accumulating near the canal's western terminus, he concluded: "the young & vigorous swarm forth to make their fortunes by subduing and cultivating the wilds of the West. The tide rolls on, wave succeeding wave, like the heaving ocean. When it will be checked, or what barrier to form the boundary of the West, time alone will determine."[167] Notice the metaphorical association of American possession of the continent with the inevitability of natural hydrological processes.

Margaret Fuller, the Massachusetts transcendentalist and close friend of Emerson and Thoreau, came to Niagara in 1843 as the first stop on her tour of "The Western Eden." Thomas Cole had called the Niagara Frontier "the mighty portals of the golden west," and, for Fuller, it opened the way to something limitless: "Kinmont says, that limits are sacred; that the Greeks were right to worship a god of limits. I say, that what is limitless is alone divine, that there was neither wall nor road in Eden."[168] Getting to Niagara, for many Americans, had continental implications. Recall that Colden had compared the citizens of a republic to "the deep current of Niagara, pouring through its natural course, irresistible, but by the hand of the Almighty."[169] Paradoxically, some Americans saw their national project as at once the vanguard of progress and the embodiment of natural, and even providential, forces. Elizabeth McKinsey argues that, in the antebellum period, Niagara Falls became an icon of American promise and vitality.[170] For many antebellum citizens of the United States, Lockport, Niagara, and their country's continental aspirations were closely associated. This association was further enhanced by the fact that many of those traveling to Niagara left the canal at Lockport and proceeded by stage, and after 1837, by a short wooden strap railroad.[171]

Canada and the Canal

The westward trending Erie Canal pointed directly to Niagara Falls. It also pointed directly to Upper Canada. When the canal opened in 1825, peace between the United States and British North America was only ten years old. At the start of the War of 1812, Henry Clay had bragged that "the militia of Kentucky alone are competent to place Montreal and Upper Canada at your feet"; and John C. Calhoun claimed that in four weeks, "the whole of Upper Canada, and a part of Lower Canada will be in our possession."[172] The

Niagara Frontier had been the war's most contested and continuous front. Three times U.S. forces had tried to cross the Niagara River and three times they had been repulsed. When Americans traveled to Niagara in the decades that followed, they encountered a contested border, yet few of them had experience with any international border. No one used the term "manifest destiny" in 1825, but the concept of a continental United States was well established. From this perspective, the Canadas were in the same category as the rest of the West. The "geopolitical euphoria" that, as Seelye notes, was already present in 1812 continued to wax. Such a vision had helped to inspire the Erie Canal project. To the extent that the canal helped to secure American dominance of the continental interior and doom its native inhabitants, it also, potentially at least, threatened the Canadas.

Thomas Nichols, the Connecticut physician who had described Lockport in glowing terms as the place where "we are gently lifted up the very precipice over which Niagara rolls," proceeded immediately to Niagara Falls. He had arrived at the Niagara Frontier in 1837, the year of Canada's rebellion. He had no doubt that, as he put it, "Americans have never given up the desire nor abandoned the design of separating these loyal provinces from the government of Great Britain, and adding them to their confederation." He recalled that Americans had marched on Canada in 1775, and that in 1812 "our old dreams of conquest and annexation were revived." In the year of his arrival, a force gathered on Navy Island, just upstream from the falls, with the intention of invading Canada. "Nine-tenths of the insurgent forces," he wrote, "were Americans." He noted that most of his countrymen "looked upon the annexation of the British Provinces of North America as a most desirable event, and one certain, sooner or later, to be accomplished." Nichols concluded that "England still hinders American progress by keeping her grasp upon large neighboring territories."[173] Canadian historian Ramsay Cook once remarked that "the very existence of Canada, most nineteenth-century Canadians realized, was an anti-American fact."[174]

The Erie Canal pointed directly west, but it terminated at the north-south line of the Niagara River. George Wurts, a young journalist from New Jersey, was one of many who found the fact that the United States did not control both banks of that river disturbing. Wurts, who would later become a prominent editor and New Jersey secretary of state, composed this poem at Niagara Falls in 1851:

> How gloriously her waters roll
> Between her banks of green
> What different feelings swell the tone
> While gazing at the scene
> There spreads Victoria's realm away

'Tis sad yon lovely shore
Should tremble 'neath a monarch's sway
And fear the Lion's roar

Now turn we where the eagle flies
The empire of the west
Our country 'neath yon spreading skies
Lies beautiful and blest
'Tis sad to think a monarch's rod
Should rule one foot of western shore
God hasten on the glorious day
When far from North to South away
Columbia's flag shall soar[175]

Many American travelers seemed almost surprised to encounter a North American place over which their flag did not soar.[176] The border at Niagara, like Lockport, became a site of intertextual dialogue about these geopolitical issues. American travelers, while often assuming the naturalness of continental expansion, nevertheless expressed a variety of reactions to the border. At times, some even described Canada as an inviting—if probably temporary—alternative to the Unites States. For them, the Erie Canal was a passage beyond America. William Leete Stone, for example, explains that he had such an "utter loathing" of President Jackson and his "scurvy" administration,

> that it was a relief to me to get beyond his jurisdiction. I seemed to breathe a purer air; and although I love my own country best, and its institutions, yet I regretted that my circumstances were such as to compel me to return within the United States, until the people shall have returned to their sense, and this disgraceful state of things terminated.[177]

Although we have little written evidence, it is safe to assume that the large numbers of escaped slaves, many of whom also left the canal at Lockport, shared Stone's relief that an alternative to the United States existed. It is also noteworthy that, at times, local residents on both sides of the border appreciated the regional economic advantages that resulted from improvements such as the Erie and Welland canals that connected them to alternative markets and routes.[178]

For some British travelers, the very existence of the United States was an anti-British fact. Basil Hall, whose accounts of his 1827 travels were widely read and quoted, crossed into Canada at Niagara Falls, and observed:

"I have seldom been conscious of any transition from one country to another more striking than this." Noticing the rising power of the United States, Hall proposed the construction of a "powerful fortress" twelve miles from the falls in the center of the Niagara Peninsula "to show the Canadians, as well as their neighbors, that we are in earnest in our determination to maintain the integrity of the colonies."[179] In a similar vein, William Howard Russell noted "that the Americans in all their wars with the mother-country have sought to strike swift hard blows in Canada, and that hitherto, with every advantage and after considerable successes, they have been driven, weather-beaten back, and bootless home."[180]

The Scotsman Alexander Mackay saw great significance in the border that the Niagara River represented, but he expressed little partisanship toward either side. "It is the dividing line," he wrote, "between the different jurisdictions of Canada and New York, where the two systems stand confronting each other, which are now battling for supremacy throughout the world. There can be little question as to which of them is ultimately to prevail, whether for good of for evil, in the New World."[181] Even some British visitors seemed to agree with the dominant American perception that the border was ephemeral, destined eventually to be obliterated. Indeed, one Canadian historian has suggested that Americans' confidence in this "must be reckoned as one of the major factors that account for the survival of Canada in America."[182]

In 1838 Henry A. S. Dearborn, a progressive New England proponent of internal improvements, proposed to place a giant memorial at Buffalo where the canal entered Lake Erie. He envisioned "a colossal bronze statue of the illustrious DE WITT CLINTON, a hundred feet high,—holding aloft in one hand, a flambeau, as a beacon light, to designate, in the night, the entrance, and pointing with the other, in the direction of the route of the Erie Canal."[183] Even if, instead of Clinton, Dearborn had suggested Liberty holding her beacon aloft, constructing such a monument would have been a remarkably aggressive gesture during this period because the proposed statue faced northwest. To the Canadians who would have been able to see it clearly from the other bank of the Niagara River, it would hardly have been a comforting sight.

America's Stairway

Because so many writers connected Lockport—with its locks and Deep Cut—to the canal as a whole, and to art, progress, the republican experiment and its continental implications, it is important to remember that—although I am presenting them separately—they were not separate issues from the perspectives of antebellum writers. In fact, it was their connection that illuminated their significance and rendered Lockport an appropriate site at which to ponder and articulate their meanings. For foreign visitors, in particular, Lockport seemed to invite an assessment of America itself.

Reverend Isaac Fidler, who traveled from England to America in 1832, was delayed at Lockport because of frost and flood damage to the canal. "The American Canal," he complained, "like most of their works not executed by Englishmen altogether, is not so substantial as might be wished, and requires repairing continually."[184] Some British travelers projected a colonial disdain that reduced both Americans and Canadians to incompetent infants.[185] Those British visitors who, unlike Reverend Fidler, could see the economic and political value of the Erie Canal, were more generous. After a careful analysis, British lawyer and businessman John Galt—in his 1836 volume on the Canadas addressed to "Emigrants and Capitalists"—concluded that

> the eminent success of the New York system of internal navigation, has had the most beneficial influence, not only on her own happy destinies, but also upon those of the whole nation. It has more than accomplished the predictions of its most sanguine advocates. As one of the bonds of union, between the western and Atlantic states, (and it was the precursor and cause of many such bonds,) it is cherished by every patriot. As the means of communication between the Hudson, the Mississippi, the St. Lawrence, the Great Lakes of the North and the West, it is the greatest inland trade in the world.

By creating this canal, New York "has demonstrated to the people, the greatness of their power." Galt began the section titled "The Erie Canal" in this way:

> On a marble tablet, which is placed on the front of the upper lock of the Erie Canal, where it crosses the ridge between Rochester and Lockport, is the following inscription, which may be deemed authentic—"The Erie Canal, 363 miles in length, was commenced 4th July, 1817, and completed in the year 1825, at the expense of about 7,000,000 dollars, which was contributed exclusively by the state of New York.[186]

The Erie Canal was "cherished by every patriot" because it "demonstrated to the people, the greatness of their power." And this accomplishment was commemorated in stone "on the front of the upper lock on the Erie Canal." Of course, the other plaque was more specifically nationalistic, with its reference to the "Spirit and Perseverance of Republican Freemen." Galt clearly perceived the national and nationalistic importance of the canal. To the extent that visitors saw Lockport as representing the entire canal, it therefore also represented America, its energies, and its prospects.

Frederick Gerstaecker was another traveler who saw a national meaning in Lockport's landscape. He was a German hunting enthusiast who styled his 1837 excursion to North America a "safari." After a trying packet journey from Albany, he offered this brief description:

> In the morning we arrived at Lockport, where the canal has a fall of sixty feet, with two sets of locks, one for boats ascending, and the other for those descending: a noble work for so young a country. Here I left the boat for the purpose of seeing the Falls of Niagara.[187]

William Lyon Mackenzie first visited Lockport on December 1, 1825, about five weeks after the canal's opening. In less than a decade, he would become an influential newspaper editor, assemblyman, and the first mayor of Toronto. He was also a reformer who admired the American Constitution, its suffrage, and the separation of church and state, while denouncing the institution of slavery. He holds a key place in the history of Canadian reform, but when the reform movement morphed into the Rebellion of 1837, he was judged disloyal by many Canadians.[188] Mackenzie's reform sentiments were already visible when he wrote this description of Lockport just thirty-seven days after the canal's opening:

> I seated myself on a large gray stone, on the high ground above the canal basin, on the morning of the 1st of December, and surveyed the scene around me—the canal—the locks—stone and frame houses—log-buildings—handsome farms—warehouses—grist-mills—waterfalls—barbers' shops—bustle and activity—waggons, with ox-teams and horse-teams—hotels—thousands of tree stumps, and the people burning and destroying them—carding machines—tanneries—cloth works—tinplate factories—taverns—churches. What a change in four short years from a state of wilderness! Kings build pyramids; but it was reserved for a popular government to produce a scene like this.[189]

All the elements of this description—the stumps, the buildings, the hum of industry and commerce—when compared with the wild landscape of four years previous, created a picture of rapid change that people like Mackenzie at least would no doubt label progress. Seven years later, Mrs. Frances Trollope would describe the same set of landscape elements, but would evaluate their meaning quite differently.

Mrs. Trollope's cantankerous view of the American scene developed during an extended stay of over three years. Financial ruin had driven her from England to Cincinnati where more catastrophes followed, leaving her so destitute that she decided to write and publish the travel book that finally restored her position and family honor. In *Domestic Manners of the Americans* (1832), she praised the abundance and natural beauty of the United States, but of the "population generally," she admitted, "I do not like them. I do not like their principles, I do not like their manners, I do not like their opinions."[190] Near the end of her long travels through the continent, she approached Lockport in 1832 from the east via the corduroy road where she remarked, "we were most painfully jumbled and jolted over logs and through bogs, till every joint was nearly dislocated." Next, she turned her attention to the growing village:

> Lockport is, beyond all comparison, the strangest looking place I ever beheld. As fast as half a dozen trees were cut down, a *factory* was raised up; stumps still contest the ground with pillars, and porticos are seen to struggle with rocks. It looks as if the demon of machinery, having invaded the peaceful realms of nature, had fixed on Lockport as the battleground on which they should strive for mastery. The fiend insists that the streams should go one way, though the gentle mother had ever led their dancing steps another; nay, the very rocks must fall before him, and take what form he wills. The battle is lost and won. Nature is fairly routed and driven from the field, and the rattling, crackling, hissing, spitting demon has taken possession of Lockport forever.
>
> We slept there, dismally enough. I never felt more out of humor at what Americans call improvement; it is, in truth, as it now stands, a most hideous place, and gladly did I leave it behind me [emphasis in original]. [191]

Trollope's diatribe against "what Americans call improvement" employed the trope of a battle that pits art against nature, a device we first saw in Woodsworth's ode. While she concluded that "the demons of machinery" have driven nature "from the field," she left no doubt that her sympathies lay with the vanquished. The next day, she traveled to Niagara Falls where nature still held sway and where, accordingly, she was overwhelmed with "wonder, terror and delight." Although Trollope self-consciously projected an outsider's perspective, the American "improvements" that Lockport seemed to embody in her account assume the status of a dangerous obsession, echoing a well-known passage from *Moby Dick*:

> [T]he harpooners wildly gesticulated with their huge pronged forks and dippers; as the wind howled on, and the sea leaped, and the ship groaned and dived, and yet steadfastly shot her red hell further and further into the blackness of the sea and the night, and scornfully champed the white bone in her mouth, and viciously spat round her on all sides; then the rushing *Pequod*, freighted with savages, and laden with fire, and burning a corpse, and plunging into that blackness of darkness, seemed the material counterpart of her monomaniac commander's soul.[192]

Because it was written by a citizen of the United States, Melville's implicit critique—like Thomas Cole's cautionary series, *The Course of Empire*—might be categorized as a jeremiad against what his countrymen "called improvement." British writers, too, were concerned about the rapid changes their own country was experiencing, but in the United States, the relationship between progress and nature was at the very center of the discourse of nationalism.[193]

Picturing Lockport

Apart from travelers' accounts and guidebooks, Lockport came to public notice primarily through paintings, engravings, and other graphic images, some of which illustrated magazine and newspaper articles. I have already discussed many of the most important images, including the four lithographs by George Catlin included in Colden's *Memoir*, as well as the other images and representations of Lockport that appeared at the canal celebration in New York. Here I consider ten additional images that received various degrees of promulgation. The presence of so many images testifies to Lockport's prominence as an icon of American technology and progress.

Two conclusions regarding these images are immediately apparent. The first is that the Deep Cut disappears from view after Catlin's two prints. This parallels the prominence of the locks over the Deep Cut in written accounts, but is even more pronounced. The second is that eight of the ten images show Lockport from the east, with the locks at the center surrounded by the developing urban landscape at the top of the Mountain Ridge. The other two depict Lockport from the west, showing the top of the locks and emphasizing the natural basin rather than the urban landscape.

In Catlin's two lithographs of Lockport from the east (figs. 4.2 and 4.3), the nascent village is indicated only by a few low buildings. This tends to emphasize the natural basin or ravine, although Catlin did establish the iconic version of Lockport centered around the locks. The primitive watercolor depiction by Mary Keys (1832), gives the town, boats, and especially the locks more prominence (fig. 4.5).[194] W. Wilson's 1836 drawing (fig. 4.6) has been widely reproduced.[195] It depicts a solid, prosperous Yankee town, centered on the canal and locks. The buildings depicted are accurate enough to have been sketched and enumerated by local historians.[196] The *Northeastern View of the Locks of Lockport, N. Y.* (fig. 4.7), an 1843 engraving that accompanied a descriptive article on Lockport in the magazine *Rural Repository*,[197] is a slightly closer view that follows the iconic pattern of Wilson's drawing. The article described the marvels of Lockport and its industrial and commercial prospects. It also mentioned that "its buildings,

Figure 4.5. *Lockport on the Erie Canal New York*, watercolor by Mary Keys, 1832, Lockport, NY. The Munson-Williams-Proctor Arts Institute, Museum of Art, Utica, NY. Used with permission.

Figure 4.6. *View of the Upper Village of Lockport, Niagara Co. N.Y.*, drawn by W. Wilson, published by Bufford Litho, 1836, negative number 26273. Collection of The New-York Historical Society. Used with permission.

Northeastern view of the locks at Lockport.

Figure 4.7. *Northeastern View of the Locks of Lockport, N.Y.*, the *Rural Repository* 20, no. 10, Saturday, December 30, 1843, Hudson, NY, 1.

public and private, are mostly built of the excellent stone which is here quarried." Like the Wilson drawing, the accompanying engraving showed a much more substantial built environment than the "wooden village" Basil Hall described in 1827.

We can identify the individual structures adjacent to the locks with specificity because, when the canal's enlargement began in 1839, the chief engineer in charge of expanding and rebuilding the locks, Thomas Evershed, created a large-scale map that labeled these structures. He also produced a remarkable drawing of the locks under construction (fig. 4.8).[198] Although Evershed's depiction is a construction drawing, it too follows the iconic pattern with the locks surrounded by the busy town. We find this same view again in an official volume, *Maps and Profiles of New-York State Canals* (1859). The engraving of Lockport (fig. 4.9) adorns a detailed drawing of a typical canal lock.[199] There is a similar view (fig. 4.10) in an 1881 *Harper's* article, "Water Routes from the Great Northwest,"[200] that discussed the competition between the Erie Canal and the St. Lawrence. The iconic image is repeated over and over again, in lantern slides and stereo cards in the 1890s, on countless modern photographs and postcards, and finally on the seal of the City of Lockport.

The two depictions of Lockport from the west are from 1840 and 1841. They express an evolving pastoral perspective on the relation between art

Figure 4.8. *View Showing the Progress of the Work on the Lock Section, Taken Sept. 1st, 1839*, drawn by T. Evershed, construction drawing by Thomas Evershed in "Working Plans of Enlarged Lockport Locks." Courtesy of the New York State Archives.

and nature that naturalized even the canal itself. As we shall see in the next chapter, this perspective informed many initiatives including the urban parks movement and had a significant effect on Lockport. The first image is an engraving (fig. 4.11) that illustrated James Silk Buckingham's 1841 book on his North American travels. Buckingham also depicted the scene in words:

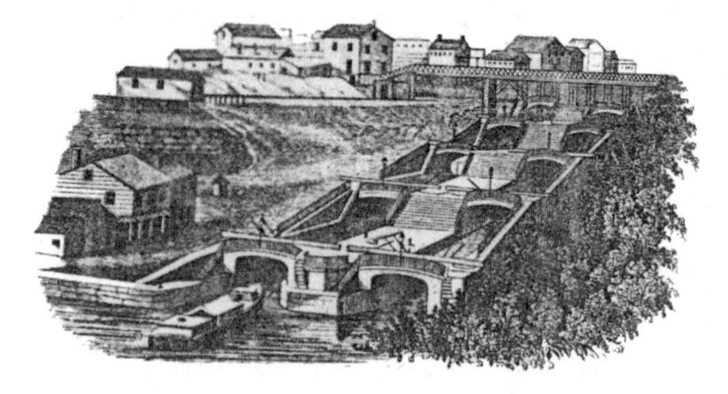

Figure 4.9. *Lockport Locks*, engraving, in Map and Profiles of New-York State canals designed under direction of Van Rensselaer Richmond, State Engineer and Surveyor, to accompany his report for 1858 (Albany: C. Van Benthuysen, printer, 1858). Courtesy of the New York State Library.

Figure 4.10. *Lockport Locks in 1881*, F. G. Mather, "Water Routes from the Great Northwest," *Harper's* 63 (1881): 414–35, 429.

"Our pleasure was greatly increased by arriving just at sunset at the Lockport station, where the boat was lifted up, by five successive locks. . . . The masonry is of the most solid and excellent workmanship, and everything about it is well calculated for durability." After specifying how much Lockport had grown

Figure 4.11. *Lockport from the West*, in James Silk Buckingham, *America, Historical, Statistic, and Descriptive*, 2 Vols. (New York: Harper and Brothers, 1841), 136–7. Courtesy of the New York State Library.

in the preceding twenty years, Buckingham explained: "such is the rapid growth of settlement along the track of the canal."[201] Yet the effect of his picture—despite the bridge, the locks, the buildings, the human figures—is to deemphasize the struggle that created this place and made the canal possible by diverting Lake Erie through this mountaintop town.

The other view from the east was one of the most widely viewed of all images of Lockport. The original was by William Henry Bartlett (fig. 4.12) and appeared in N. P. Willis's widely disseminated illustrated volume *American Scenery* (1840),[202] a compendium of romantic American scenes that included Trenton and Niagara falls. According to Seelye, Bartlett emphasizes "not the accomplishments of engineering, but the towering cliffs and forests of the natural scene." While Catlin "celebrates technology," Seelye argues, Bartlett "obscures it."[203] Although written and illustrated by an Englishman, *American Scenery* was a kind of coffee-table book that achieved great success on both sides of the Atlantic and helped to define for tourists what was worth looking at in North America. Bartlett stressed the natural aspects of Lockport, but he would hardly have included this scene if it were not already well known because of what humans had accomplished there. The view from the east that we see in the Catlin lithographs and read in descriptions like Mackenzie's underscored human achievement—what "a popular government" can produce—but Bartlett's view emphasized that nature carved this ravine and left only the slightest artificial tweaking to

Figure 4.12. *Lockport, Erie Canal*, drawing by William Henry Bartlett, engraved by W. Tombleson, in N. P. Willis, ed., *American Scenery*, 2 vols. (London: 1840), I, 160, negative number 57937. Collection of The New-York Historical Society. Used with permission.

divert the continent's waters, and trade, toward the east. Emphasizing the naturalness of the canal not only resonated with waxing romanticist sentiments, it also rendered the expansion of U.S. hegemony a natural thing. In the text accompanying Bartlett's picture, N. P. Willis did not entirely neglect the Deep Cut. "The canal boat glides through this flinty bed," he wrote, "with jagged precipices on each side; and the whole route has very much the effect of passing through an immense cavern."[204] The traveler, for Willis, experiences even the Deep Cut as a natural spectacle.

In the first chapter I discussed the mural by Raphael Beck, *Opening of the Erie Canal, 1825* (fig. 1.4), that the Lockport Exchange Trust Company commissioned for the centennial of the canal's opening and unveiled at the dedication of its new building in 1928.[205] As a retrospective view, the mural emphasized Lockport's patriotic connotations. At the top of these locks, travelers reached the level of the upper Great Lakes, the level of Chicago. At Lockport, perhaps for the first time, citizens of the U.S. were awed by their own power. The mural, with its many flags placed below and above the locks, depicts a country knit together by a work of human artifice. Although it depicts laborers, their part in the artifice is overshadowed by the prominence it gives to middle-class figures.

By the time of the Erie Canal centennial, the accomplishment at the Mountain Ridge had begun to fade from national—if not entirely from local—consciousness, but in the months preceding the completion of the Deep Cut, and especially during the canal's opening celebration, the process of writing Lockport had begun with a torrent of words. I conclude this chapter by returning to C. F. Briggs's 1853 essay. It provides a kind of bookend on writing Lockport during the canal's heyday. The essay, simply titled "Lockport," appeared in *The United States Illustrated*, edited by Charles A. Dana. The lavishly illustrated volume, a paean to national virtue and manifest destiny, so diligently submerged social tension that it betrayed a kind of panic.

"We are never so proud of our birth-right," Briggs wrote, "as when we look upon any of the great works of 'internal improvement' which have been so boldly conceived and so vigorously carried on to completion." Although the canal "gave a new impulse to commerce," its influence "upon the minds of people in this country," he wrote, "we regard as of infinitely greater importance." Briggs saw the canal as an important moment in the development of national feeling and confidence in the United States. Turning then specifically to Lockport, he continued:

> This deep cutting through the solid rock, is quite as great a wonder as the excavated temples at Elephanta, or the hewn-out dwellings of the ancient inhabitants of Petrea, who built their houses not upon, but in a rock, as though they were to live forever. The difference between

> the great works of antiquity with which we can compare our own achievements in art, like the Locks at Lockport, are that those were erected as monuments of human pride or for religious uses, while ours are made to promote the happiness and welfare of the people who constructed them; they were the results of mere physical strength and endurance, while ours have been accomplished more by the mind than the body; science and skillful enginery, have done the work; which only armies of laborers, whose lives were sacrificed in their work, and succeeding ages, accomplished these. . . . Our workmen are like "Singing Masons," who build "roofs of gold" for their own enjoyment, and not for their task-masters.

Briggs swept up the conquest of the Mountain Ridge, and the rest of the canal's construction, into a narrative of national progress. The cries of the laborers who toiled and died in the Deep Cut, their potential testimonies against such dishonesty, are absolutely mute. The site of Lockport prompted Briggs to launch into a justification for racialized empire.

> The Anglo-Saxons are born conquerors. . . . They have a patent from Heaven, to improve the earth, and most magnificently are they fulfilling their mission. To level mountains, excavate rivers, to bridge impassable chasms, to exterminate the useless tribes that encumber the earth without improving it, to unite together the utmost parts of the globe . . .[206]

In addition to whatever geoeconomic and geopolitical impetus the Erie Canal gave to U.S. imperial expansion in North America, it is clear that, for Briggs, the canal's more important effect was "upon the minds of the people." Within his narrative, the canal, and specifically Lockport, could serve rhetorically to demonstrate Anglo-Saxon superiority and justify expansion. For the "useless tribes" of the continent's interior, the Lockport locks were a dark instrument. Ancient empires had built the "hewn-out dwellings" of Petra "as though they were to live forever." The Americans built a stairway to empire as though their realm could not be limited in space. Recently a number of scholars have begun to notice that nationalistic rhetoric in the United States has often relied not only on comparisons to Europe, but also to the Middle East.[207]

Accompanying Briggs's essay is an engraving of Lockport (fig. 4.13), "having the appearance rather of a scene in some Italian city, than a view in a young town which has but lately grown up in the wilderness."[208] It is a copy of Bartlett's view, taken one step further into the denial of the struggle that gave Lockport birth. Italian palaces have replaced the picturesque frontier buildings of the earlier view. This is a Lockport of pure fantasy to match the story of "singing" laborers building "for their own enjoyment, and not for their task-masters."

Figure 4.13. *Erie Canal at Lockport, NY*, engraved by Hermann J. Meyer, following p. 162, in *The United States Illustrated; In Views of City and Country. With Descriptive and Historical Articles*, ed. Charles A. Dana (New York: H. J. Meyer, 1853), negative number 31201. Collection of The New-York Historical Society. Used with permission.

5

Losing Lockport

Afterlife of the Mountain Ridge Conquest

> In the novel's final and most chilling lines, the naming of a number of obscure American towns charts his shrinking stature: Batavia, Lockport, Hornell . . . ; he is "almost certainly in that section of the country, in one town or another."
>
> —Catherine B. Burroughs, "Of 'Sheer Being' "

F. Scott Fitzgerald's *Tender Is the Night* (1933) concludes with the character Dick Diver dissolving into the landscape of western New York. His "shrinking stature," as Catherine B. Burroughs argues, is signaled by the "naming of a number of obscure American towns."[1] Like Dick Diver, Lockport itself slowly faded into obscurity in the century following the canal's opening. It may seem curious that a formerly prominent place could be forgotten so entirely, but technological achievements often have an ephemeral purchase on human attention. How could the building of a canal inspire any awe after railroads, airplanes, and space shuttles? Should the technological feats of our age be similarly trumped in the next century, who will remember Cape Canaveral or Los Alamos? More intriguing is the question of historical significance. Given the Erie Canal's actual role in shaping economic and political power on the continent, why don't Americans celebrate it like the battles of Lexington and Concord? During the canal's heyday, Tocqueville reminds us, the importance of the Erie Canal and other improvements was "a truth universally felt and recognized." One was "convinced, not by argument, but by the evidence of all the senses."[2] Little by little, it seems, the Erie Canal became impalpable.

The installation of Raphael Beck's mural, just five years before the publication of Fitzgerald's book, indicates that the conquest of the Mountain Ridge still reverberated in local consciousness. Moreover, Beck was able to recapitulate a nationalistic narrative of the event—one that had been widely articulated in the antebellum era. Lockport's name, its landscape, and a typically American sense of local boosterism, made it impossible for local people to completely forget their city's genesis. Yet, the level of local consciousness about the actual significance of that genesis is also strangely shallow. Growing up in Lockport, it was my impression that most Lockportians—including myself—were unsure whether their city was most noteworthy because of the canal or the Big Bridge that spanned it. The 399-foot bridge in the very center of the city, we were repeatedly told, was the widest in the world, but what was there to say about the grimy waterway that now seemed as useless as the city's trolley tracks, which here and there could be seen poking through the asphalt?

I became seriously interested in investigating Lockport when confronted with the shocking juxtaposition of a hidden waterfall and a toxic dump. What is the relation of this "local" issue to the story of the stairway to empire? In this final chapter, I return to local concerns to address the significance of the conquest of the Mountain Ridge and the completion of the Erie Canal to the place where the waters met. Four sections address separate aspects of the canal's importance and resonance in Lockport. In the first section, I use Lockport as a kind of case study to consider the social, economic, and religious transformations that came in the wake of the canal. Next, I examine the ways in which Lockport's genesis affected its industrial development. In the third section, I explore the ways in which a pastoral impulse worked to naturalize both the canal and the town in the mid-nineteenth century. Finally I reconstruct the shadowy story of the Gulf, Lockport's uncanny doppelganger.

A New World

For those who had lived and worked in Lockport during the construction period, the decade that followed the canal's opening must have seemed startling. The town was now directly connected to the nation's metropolis and even the world beyond. A kaleidoscopic array of things, people, and ideas now entered Lockport. The canal was a conduit of new opportunities and possibilities for some, and continuous wrenching change for all.

Lockport in 1825 was little more than a large construction camp, buoyed by state funds. The market connection the canal provided gave the town a chance to survive as a commercial center serving its immediate hinterland. That connection also opened new possibilities for manufacturing, and it was this that largely sustained Lockport's growth during the antebellum heyday of the Erie Canal. The hydrological considerations that had convinced Geddes to route the canal through the Mountain Ridge meant that an artificial river of water had to flow from Lake Erie through Lockport to keep the canal navigable all the way to the Seneca River, over one hundred miles to the east. With Lake Erie as its millpond, and a sixty-foot drop in elevation, Lockport's industrial potential was clear. Entrepreneurs quickly built a race channel to divert water above the locks. Within ten years, the channel fed four flour mills, nine saw mills, three textile mills, as well as six boot and shoe manufacturers, four furniture factories, a furnace and plough factory, and a foundry. The town also contained several cooperages, carriage makers, harness and saddle makers, hat makers, as well as a printer, a bookbinder, two distilleries, a brewery, and a population of 6,092. Like Rochester and Utica, Lockport had become a diversified manufacturing center, doubling its population again by 1850.[3]

The labor requirements of manufacturing meant that Lockport's class structure continued to resemble that of its construction era. In the next section I will examine some specific ways in which Lockport's industrial development grew out of its canal construction experience. Here it is sufficient to emphasize that the new social divisions emerging in the industrial centers along the canal's path were something both Lockport's laborers and its "citizens" had already experienced. Indeed, the presence of industry, along with diverse kinds of canal work, allowed a significant portion of the construction workforce to remain.[4] Moreover, several additional construction projects brought large teams of laborers to Lockport. The largest of these was canal enlargement, an on-again, off-again process stretching from 1838 to 1858 at Lockport; at times it employed over eight hundred laborers. The construction of the Lockport and Niagara Falls Railroad between 1835 and 1837 brought in three hundred Canadian, Scotch, and Irish workmen. Finally, the Rochester, Lockport and Niagara Falls Railroad—part of the New York Central system after 1853—employed a large workforce in the two years before its opening in 1852.[5]

The presence of canallers, factory hands, and common laborers meant that class tensions persisted in Lockport. They were manifested in a number of ways. As wage laborers, workers' unpaid time was their own. In their own residential districts, and especially in the entertainment establishments clustered near the locks, there was space for an oppositional culture to develop. A central feature of this culture was internal dissention. Although recent Irish immigrants made up the single largest part of this group, they were joined by Welsh, Germans, African Americans, and others. There were fractures along lines of gender, race, religion, and ethnicity. What many workers had in common was resentment against the transatlantic commercial system that forced them to dance to the tune of those willing to buy their bodies and their time. Yet this same system gave them some freedom to develop patterns of life outside of direct control. To varying degrees, working-class males drank, whored, and brawled among members of their own class. Occasionally some directed their resentment at those they perceived as benefiting from the system they saw as eroding their communities, their families, and their sense their dignity. Laborers strove to use their wits, skills, and social capital to forge a living in a world that gave them new liberties, but also severed them from traditional sources of security. For many workers—both male and female—the market revolution and its accompanying commercial and industrial dynamics did not, on the whole, seem liberating.[6]

By 1845 Lockport had twenty taverns and fifty-three groceries—seventy-three drinking establishments in a village of 9,328. At the beginning of the enlargement process in 1838, a detailed engineer's plan superimposing the proposed new locks on the existing locks (fig. 5.1) showed the adjacent structures, many of which would have to be removed. Most of the buildings densely clustered in the vicinity of the locks were labeled "Groceries." By comparing this detailed plan with contemporary images of Lockport such as Wilson's *View of the Upper Village of Lockport* (1836; fig. 4.6), we can easily see that each of the nine clusters shown in the plan (unlike the tenth one labeled "Grocery") consisted of several distinct establishments. Hence we can estimate conservatively, that there were at least twenty-five drinking places, all within fifty feet of the locks.[7] This raucous scene must have appalled many middle-class villagers and genteel travelers.

Reverend Isaac Fidler, an Anglican cleric who came to the United States in 1832 to explore the possibility of immigration, commented on an aspect of Lockport's working-class culture about which we have very little

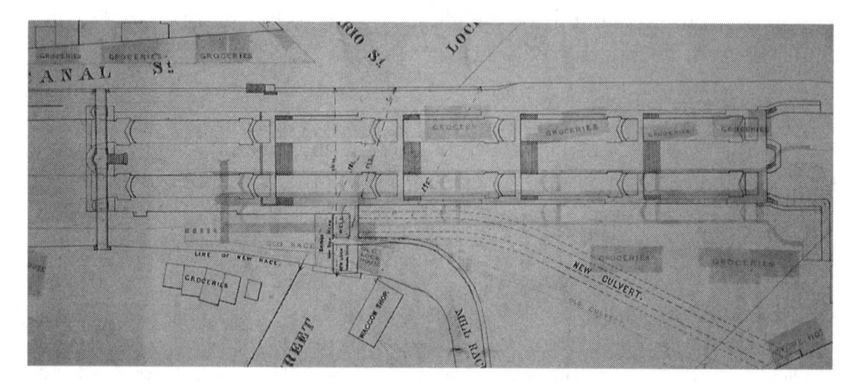

Figure 5.1. Groceries near the Lockport locks, 1836. There were more than twenty-five drinking establishments within fifty feet of the Lockport locks; Thomas Evershed, *Plan Showing the Relative Position of the Old and New Locks, Race and Culvert, Lockport*, Engineering Plans for the Enlargement of the Erie Canal, 1836–1837. Courtesy of the New York State Archives.

direct evidence. Frost and flooding had damaged the canal the previous winter, and he noted that "many men were employed" repairing it. Fidler decided to take the coach for Lewiston and Niagara Falls.

"On Leaving Lockport," he noticed that "two young girls were admitted into the coach, rather singular in their dress and manners. They were also more free in conversation, and with less reserve, than any American women I had seen before." At the first inn, the girls got out to warm themselves. "My companion," he continued,

> whose curiosity and suspicions were more acute than mine, expressed his sentiments to be that our female fellow travelers were not of good reputation; and stepped into the inn after them to make inquiries. I felt much at a loss to distinguish any particular criterion from which my amusing friend could have drawn prognostications so unfavourable to their character. Nothing escaped them, which could have excited in me such surmise or conclusion. He soon returned with the information that they were what he expected, and that the neighborhood abounded with similar characters. This was the only instance, in all my rambles through America, in which female behavior or language met my observation, betraying a departure from strict reserve; and the departure in this instance was of such a nature that, as to awaken no suspicions in a stranger's mind.[8]

Reverend Fidler's inability to see what was before his eyes reveals how alien the very possibility of prostitution was to him. Indeed, the rough culture emerging among concentrations of wage workers at places like Lockport was, for people like Fidler, an almost incomprehensible thing. To the middle-class villagers of Lockport—a place that "abounded with similar characters"—this culture was certainly more familiar, although not necessarily less alien. Many middle-class men had daily interactions with workers, but they certainly must have cautioned their wives and children about the moral dangers of places like the Lockport locks.

The Rochester *Observer* concluded in 1831 that there was so much "prostitution, gambling, and all species of vice" on the canal that it should be labeled the "Big Ditch of Iniquity."[9] According to the *New York Evangelist*, in 1835 over fifteen hundred "taverns, grog shops and houses of ill fame" lined the banks of the Erie Canal.[10] At the same time, the Yankee villagers of places like Lockport also understood that their prosperity, and the prospects of their nation, had been immeasurably enhanced by this same ditch. It was at once a conduit of progress and of disorder. Carol Sheriff labels this dissonance a "paradox of progress."[11]

At times, the tensions between the communities and the conduit became extreme. Because the Lockport locks were a bottleneck that slowed canal traffic, boat crews often fought over which vessel was to enter the locks first. In 1836, rival packet lines running between Buffalo and Rochester aroused the ire of journalists in both cities. Lockport's *Niagara Courier* was even more incensed:

> We feel it is our duty as publishing journalists, however unpleasant it may be, to put the traveling community on their guard, by cautioning them to avoid taking passage in either of the lines of packet boats plying between the cities of Buffalo and Rochester. The belligerent attitude these boats have assumed towards each other in attacking the respective crews with bludgeons and missiles whenever coming into contact, renders it totally unsafe for passengers to travel in them . . . the participants in the outrageous acts of violence which have disgraced their conduct and thrown the whole community in the vicinity of their broils into excitement, should be punished to the utmost vigor of the law. . . . Is it to be tolerated that the lives and property of persons traveling the canal are to be jeopardized by these continued riots?[12]

Middle-class people often viewed the boisterous world developing among wage laborers as an element of disorder entering their communities via the canal. In fact, the characteristics of free labor were but one part of the fundamental transformation to a commercial, market world—a transformation in which the Erie Canal was playing a key role. The market literally entered Lockport by the waters of the artificial channel. The new order, like the canal, was a human creation, although in time both would acquire the aura of the natural. Wage labor instrumentalized each worker's contribution to the new order: they worked because of the necessity to eat and to survive. The antebellum period did provide class mobility for a few, but it also created enormous income disparity. Among many members of the middle class, an almost ideological belief in individual responsibility took hold: since prosperity was the nearly inevitable reward of effort, poverty was moral failure. When middle-class males participated in the public realm of the new capitalist order, they were compelled to view their fellow citizens as limits on their own freedom and prosperity. Competition in the public commercial world seemed to demand selfishness.[13]

In addition to the challenges of adjusting to these changes, at Lockport there was another development made possible by the canal and the market revolution: manufacturing and other industrial ventures that required a concentrated workforce. In industrial towns, the new social dynamics were accelerated. One result of manufacturing and transportation improvements was the undermining of traditional home production, a process that had enormous effects upon women. In the new geography of everyday life, males ventured into the competitive struggle with their fellow men while women often remained at home, severed from their traditional economic roles. At the same time, many began to idealize the home as the last bastion of unselfish affection and moral purity.

We can see some of these social dynamics at work in the daily reports of Thomas Leonard (1815–1879), a Lockport carpenter who kept a diary from 1842 to 1849. When he wrote the early entries, Leonard was living in the community of Richfield, a few miles south of Lockport on the canal, where he was engaged in building a small schoolhouse. His first wife had died in 1839. His work and his social aspirations increasingly drew him into the orbit of Lockport where he eventually remarried and purchased a home. In late November 1842, he took a job constructing a gallows for an execution to take place in the Niagara County jail in Lockport. He described the structure and the workings of the gallows in great detail. In a dry run, he tried "putting 120 pounds in place of the criminal . . . several times," and found it to "work well."[14] On November 30, a long entry described what is still Lockport's only execution:

> The prisoner is David Douglas, an Irishman, 34 years old, without learning. He is to be executed for the murder of Henry Cunningham on the 29th of April last. The murder was committed when they both were intoxicated and grew out of some trifling circumstance. He spent most of the forenoon in singing and praying and one might judge from his present appearance that he would die a humble supplicant for mercy. But other circumstances seem to prove too clearly that he dies an unrepenting, hardened sinner not realizing the situation. . . . A little before two o'clock the prisoner is brought to the gallows . . . makes a few remarks; says he always lived in friendship with Mr. Cunningham, for whose murder he is about to die and never had any intention of killing him, etc. The rope is adjusted about his neck by Mr. Tucker—the Sheriff remarks "you have but a moment to live"; the murderer is suspended in air and, without a motion, except a slight heaving of the breast, in 7 ½ minutes life is extinct. The body remains on the gallows about 40 minutes, it is then taken down, put in a coffin and taken to the Cold Springs and buried. There were about 50 persons present at the execution.[15]

Leonard maintained a remarkable emotional distance from this event. As the diary continues, we see that Leonard was at great pains to distance himself from the rowdy society of laborers. The story of Douglas and Cunningham also reveals how violence within that society was usually focused internally. Leonard made no mention of the fact that the sentence had been appealed because the jury had been allowed to visit a tavern while taking a break from deliberations, or that his body was clandestinely removed from its grave by a doctor who wanted to train his students by dissecting the cadaver.[16]

As an artisan, Leonard was compelled to work with—and when in Lockport to board with—laborers. "I have been disgusted with the conversation of the hands with whom I work," he wrote in the summer of 1843, "and to see the minds they were possessed of and the subjects about which they got so much excited." He hoped that he "may not be contaminated by their example." He was particularly upset with their "vile practice" of using tobacco. "It is really enough to make one blush and hang his head to think he is obliged to be ranked in the same order of animals with those specimens of humanity so often met with here who degrade themselves by this filthy practice." Leonard was a very pious man from the countryside. In Lockport, confronted with "those specimens of humanity," he was worried about his own status. They next day was the Sabbath, but after hearing several inspiring sermons, he returned to his "boarding place again to be annoyed and disgusted by low vulgarity and filthy conversation," and at dinner, he felt "hurt and surprised at the train of conversation." He then wondered, "Am I too much of an 'old, stiff, Presbyterian' on the Sabbath? too rigid and strict?" Finally he addressed his future readers: "perhaps some will say where is the necessity of recording such things here? I would answer I can but notice them and as I have no friend to whom 'to open myself' I must either keep them hid in my own breast or record them here which in a measure, relieves my mind."[17]

Thomas Leonard deeply aspired to be accepted into Lockport's elite society. He regularly attended just about every possible service at the First Presbyterian Church. After describing one Sabbath sermon, he summed up his predicament:

> In meeting with people living here, let the occasion be what it may, I cannot but notice the respect there is paid to *caste*, as if a man's rank in society was determined by dollars and cents, or the fineness of the cloth he wears. This I am the more pained to see, among Christians. I would be glad to associate with them, as with brethren, but I seldom find an individual who seems willing to accept *my* acquaintance. I can go to meeting and enjoy a sermon and then go home but I go as among strangers, meet as among strangers and go home as among strangers and feel that I am alone, surrounded in crowds. I would that I was situated differently in life. I would that my long-cherished hope of friends, society and happiness in this world might be realized, for 'I have formed full many a sanguinary scheme of earthly happiness.' . . . I would be glad if some kind friend, would observe my conduct and report to me the impression it is likely to have on others [emphasis in original].[18]

Unlike the rural life he had known, in town Leonard found himself alone amid crowds. The new class tensions are almost palpable in the earnest longings he confined to his journal. Although he saw the hypocrisy of basing social status on economic success, there was nothing he desired more than status in this new order. He wished he had a friend to observe his conduct and report "the impression it is likely to have on others." Later we find him reading *Olcott's Young Man's Guide* to learn "the course a young man ought to pursue and the objects he ought to have in view." Yet he continued to differentiate himself from those he considered below his status, especially his fellow boarders. "They are of such a caste that I do not feel as if I ought to be very familiar with them." They are only interested in "the gratification of the animal." His fear of contamination made him unwilling to accept *their* acquaintance. "My standing aloof so much is probably looked upon by them as a singularity," he wrote; "be it so."[19]

Thomas Leonard may not have understood the social tensions of the new order, but he deeply felt their contradictory tugs. The presence of bawdy and boisterous laborers, and the new competitiveness of commercial and social life, made his striving for middle-class respectability an anxiety-ridden quest. Although he spent a semester at the Lewiston Academy in a teacher training curriculum, he continued with his more lucrative carpentry work. Eventually Thomas Leonard was accepted into the First Presbyterian Church where he taught Sunday School and finally became a church elder.[20] One thing was certain, middle-class respectability for men like Thomas Leonard in antebellum Lockport required not only a disciplined continence, but a public display of faith. As his private diary shows, Leonard internalized and embraced these requirements, but the fear of falling and the demands of rising give his writings a tense edge.

We can see the internal tensions Leonard felt only because we have access to his recorded thoughts and feelings. In Lockport, the tensions of the new order were also revealed much more openly. Alcohol consumption, as we have seen, was prodigious at Lockport during the canal construction period. But it reached an all-time peak nationally in 1830 when annual per capita consumption for those over fourteen reached 7.1 gallons of pure alcohol. That breaks down to an average of 3.3 shots (of 1.5 oz.) a day of 100-proof alcohol.[21] The increase was not simply among laborers. Charles Sellers points out that the number of young men from successful families who succumbed to alcohol—and also opium—addiction shows that "the great American whiskey binge was fed primarily by the anxiety of self-making men."[22] The temperance crusades that had already begun in 1830 were, on one level at least, a desperate attempt by the middle class to save itself. The self-discipline that the Northern middle class first imposed on itself and then, with some success, on the working class, eventually reduced consumption by 75 percent, but left no doubt about the direction of cultural hegemony.

Another indication of tension in antebellum Lockport was public disorder. In addition to the more or less continuous scuffles involving canallers, public demonstrations and other gatherings often turned violent. An interesting example is the so-called McLeod Excitement. During the Rebellion of 1837, or what New Yorkers referred to as the Patriot War, Canadian rebels and a larger group of American volunteers set up a base on Navy Island in the Niagara River with the intention of invading Canada. The steamboat *Caroline* was ferrying men and supplies to the island. The British conducted a midnight expedition, took possession of the *Caroline*, set it afire, and sent it adrift toward Niagara Falls. Two Americans died in the attack. Three years later, Alexander McLeod, while visiting Lewiston, bragged about participating in the raid. He was immediately arrested and taken to Lockport to await trial. Support for the Patriot War had been quite vehement in Lockport. The Lockport and Niagara Falls Railroad had come into service just in time to carry supplies from the canal to Fort Schlosser where the Caroline was docked. A daily newspaper was published at Lockport for the sole purpose of reporting on the war's progress; as the news arrived via the railroad, it was immediately reported in a column headed "By This Evening's Car." Command of the Patriot forces was offered to Major Benajah Mallory of Lockport, but he refused the offer.[23] The rebellion's leader, William Lyon MacKenzie, had visited Lockport in 1831; impressed with its growth, he had commented: "Kings build pyramids; but it is reserved for a popular government to produce a scene like this."[24] After the failure of the rebellion, one of it's leaders, Benjamin Lett, fled across the border and set up his headquarters in Lockport. Although the *Caroline* incident is hardly remembered today, one can easily imagine a counterfactual history in which popular sentiment for annexing Canada continued to swell beyond the border regions, adding Canada to the story of manifest destiny. Indeed, British authorities were worried enough about another war with the United States that a secret military document from 1825 outlined "the vulnerable points of America." "Within 30 miles of Buffalo, the mouth of the Canal on Lake Erie; at a place called Lockport, there are 5 double locks, close to each other, required to rise 60 feet in a space of 130 yards. These Locks are so close to the British Frontier, that there could not be much difficulty in sending over a sufficient detachment to destroy them."[25]

After a long hearing at the Courthouse in Lockport, McLeod was held for trial, but allowed to give bail. Eventually, two local men agreed to become his bondsmen. According to the 1878 *Illustrated History of Niagara County*, the idea of releasing McLeod,

> was highly displeasing to the citizens generally, and before he was set at liberty, a public meeting was held in Lockport, attended by a large assemblage, to express the prevailing indignation toward the bondsmen. Speeches were made, bells were rung and cannon fired. A cannon was

> placed in front of the courthouse, and bands of music marched back and forth playing the "Rogue's March" with vigor.[26]

One of the bondsmen was intimidated and withdrew his offer, and McLeod could not be released. The largest flour mill in Lockport burned during this excitement, and some blamed it on McLeod's friends.[27] Meanwhile the British government was demanding his release and many were concerned that he would be "taken out and lynched by those enraged by the memory of his murderous assault on the 'Caroline.' "[28] Because of this situation, McLeod was removed to Utica for trial. Canadians were also enflamed by this affair. During the trial, "a party of royal dragoons crossed the Vermont line, made a prisoner of an American citizen, and carried him off to Canada," though he was soon released.[29] The Caroline affair is remembered in international law, as the first attempt to argue for a justified preemptive strike on a sovereign nation. McLeod was eventually acquitted because of lack of evidence.[30]

Meanwhile, the Canadian refugee Benjamin Lett, no doubt with encouragement and aid from his Lockport colleagues, planted a bomb that destroyed the Brock Monument, Canada's memorial to the fallen hero of the Battle of Queenston Heights. Lett was engaged in a number of other "desperate acts." After crossing the Niagara River, he went to the house of a British officer, and "calling him to the door, shot him." Next he blew up a lock on the Welland Canal.[31]

At Lockport, the stresses brought on by the market revolution were augmented by the international tensions of an unsettled border zone. On a number of occasions which I will examine, violent public demonstrations erupted in Lockport once again. It is important to note that there was hardly unanimity on the meaning of the border in Lockport. There were a great many abolitionists in the village who were glad for an alternative America. Thomas Leonard looked across the lower Niagara River and saw

> the flag at Fort George is in full sight, as is Lake Ontario. The monument is in ruins, having been blown up three or four years since by some individual to revenge the wrongs done by the Provinsial [*sic*] Government to the so-styled Patriots. How base must be the mind that would seek redress, or could find pleasure in destroying this becoming mark of respect and veneration for a brave commander who fell in the service of his country.[32]

Replicating the canal construction era in yet another way, Lockport continued to be a dangerous place, especially for laborers. In addition to the intentional violence of the sort that Cunningham suffered, there was a constant stream of accidents. Thomas Leonard reports on July 4, 1842, that "two individuals lost their lives today in Lockport by the accidental discharge of a cannon while in the act of loading it." Accidents involving public celebrations, cannon, and no doubt alcohol, were already a familiar pattern in the village. Before the construction of the Big Bridge, the canyon created by the Deep Cut was a "chasm" that "cut the town in two."[33] In 1842, a local shopkeeper wrote: "The yawning gulf already boasts / The death of two or three, / And should it yawn a twelve month more, / May be the death of thee."[34] Three years later, Thomas Leonard was hired to build a railing after one man fell to his death. Leonard was not convinced it had been an accident; "there is a mystery," he wrote, "about [how] he came to his end."[35] Perhaps the railing was less than adequate, for a few years later "when a considerable number of people were leaning against the rail at the west end and it gave way precipitating about a dozen of them into the canal. Several were drowned."[36]

When cholera arrived in 1831, canal boats from the east were quarantined after the epidemic killed many including the former sheriff, a druggist and his wife. Yet undoubtedly, most of the victims were among the poor; when it returned in 1852, one observer noted that it was particularly "attacking black cooks and workmen."[37] Work, too, could be dangerous; there were deaths in mills, and a large number caused by blasting for canal enlargement. Early in 1839, the *Lockport Journal* reported: "Another workman was hurt while blasting on the section just above the Locks. On Friday last Mr. House was killed at the same place." On July 18, 1839, "a canal boat containing a thousand bushels of wheat was struck by a flying rock and sunk." And on March 25, 1840, "Mr. Crawford, overseer of Section 3 was instantly killed . . . by the premature explosion of a blast."[38] In many ways, conditions had worsened for common laborers since the days of the building of the original canal; an increasingly common, though largely ineffective, response was the work stoppage; laborers working on enlargement in the vicinity of Lockport struck in 1849 and again in 1850.[39] Workers were much more susceptible to all of these dangers, but since, as many of these examples show, middle-class people were not immune, accidents and epidemics added a distinct layer of tension to the texture of everyday life for everyone in antebellum Lockport.

The presence of large numbers of workers, several scholars have suggested, may have sparked something beyond class tension. In his classic study of the "Burned-Over District"—the region, adjacent to the canal, that experiences repeated spiritual fires—Whitney Cross noticed a difference between manufacturing and commercial canal towns. Lockport "came in time to serve in a smaller way the same kind of urban purpose as did Rochester, Utica, and Oswego, in contrast to the more purely commercial nature of Buffalo and Syracuse." When he mapped the "concentration points of all the enthusiasms within the Burned-over District"—including not only revivals, but also antimasonry, antislavery, temperance, perfectionism, Millerism and spiritualism—he found that the manufacturing towns were the primary hotbeds of enthusiasm. Cross explained this by arguing that manufacturing towns had constant contact with their rural hinterlands, and that the enthusiasms were products of the interaction between traditional Yankee agrarian piety and the new commercial world brought by the canal.[40]

In his study of Rochester, Paul Johnson offers a different interpretation of the relationship between manufacturing and enthusiastic religion. Canal towns like Utica, Rochester and Lockport joined other seaboard centers, especially in New England, to form the first wave of North America's industrial revolution. Large water-powered mills replaced household and small-shop production in places that were accessible to the growing market. Before the arrival of the canal, households in the vicinity of Rochester had produced their own cloth, and artisans housed their apprentices and workers under their own roofs. Free wage labor in larger and larger factories released workers from this direct control and paternalistic care. Workers began to form their own neighborhoods and cultural patterns. Revivals gained much of their force by helping "to solve the problems of labor discipline and social control in a new manufacturing city." Using church and civic records, Johnson reconstructs the pattern of who embraced conversion when. His data shows that "revivals and their related social movements were disproportionately strong among master workmen, manufacturers, and journeymen craftsmen," whereas "there were relatively few merchants and clerks among the converts, and even fewer day laborers and transport workers."[41] Through economic and social pressure, Johnson argues, masters and factory owners eventually convinced many of their employees to embrace religion and eschew alcohol.

> Revivals provided entrepreneurs with a means of imposing new standards of work discipline and personal comportment upon themselves and the men who worked for them, and thus they functioned as powerful social controls. But there was more to it than that. For the belief that every man was spiritually free and self-governing enabled masters to present a relationship that denied human interdependence as the realization of Christian ideals.[42]

In other words, revivals allowed masters and factory owners to justify to themselves and others their abandonment of paternal care for their workers. Sinking or swimming was now entirely an individual matter. In the 1830s, revivals eventually spread from New York to New England and then into the upper Midwest, and everywhere "enthusiasm struck first among the masters and manufacturers, then spread through them to the ranks of workers."[43]

The preacher who sparked the great Rochester Revival of 1831 was Charles Finney. He argued that "man is not innately corrupt but only corruptible." Hence "there was no need to hold employees or anyone else in direct dependence." Rather, such dominance "prevented underlings from discovering the infinite potential for good that was in each of them."[44] What was revolutionary in these revivals was the almost complete repudiation of the Calvinistic notion of predestination. In Finney's formulation, conversion required only that sinners abandon their rebellion against God. "To repent, now to give themselves up to God, now to say and feel Lord here I am take me, it's all I can do. And when the sinner can do that," Finney concluded, "his conversion is attained."[45] In religion, as in the world of the competitive market, it was as an individual that one found reward. Indeed, the sense of agency among the prominent citizens of Rochester and Lockport came partly from the experience of their own roles in rapidly changing towns, and from the success of the canal that had also been wrought by human hands. The corollary of this salvation-as-choice idea was that social evils were also the result of individual human failing. In the temperance movement and other movements that came in the wake of the revivals, reformers attacked all social problems in this way.

In her investigation of religion in Utica, Mary Ryan also finds the roots of revivals in the anxiety brought on by "the opening of the Erie Canal, the doubling of the city's population, and the accompanying proliferation of grog shops, boarding houses, and brothels" and the inability of any traditional social organization "to oversee public morality."[46] Ryan shows that women often converted to new churches without their husbands and relatives, hence exhibiting a degree of autonomy during revivals. Her analysis not only reveals that women were usually the first to convert, but that they often bought their husbands and children along later. Women gained some moral leverage from their confinement to the domestic sphere, now often seen as elevated and pure in relation to the selfish public world of men. Women's roles in revivals, ironically, "played a central part in creating a narrowly maternal role and image for their sex." But women were also able to use this moral capital to gain their voices as reformers.

The year after the canal's completion, as "Finneyism blazed with the first flood of commerce," Charles Sellers argues that the United States was entering "the decisive phase of the market revolution, cultural as well as economic." The new middle class was constructed less on relations of production than on an ideology that "atomized society into a marketplace rewarding each according to effort," assuaging "rising anger over the class reality of bourgeois exploitation." Anyone could join this class who embraced "self-repressive norms, competitive consumption, and middle-class mythology."[47] Yet the struggle over temperance was also a matter of coercion.

> Pious employers banned drinking and hired only the sober. Churchgoing business elites rewarded churchgoing, teetotaling mechanics, clerks and

> shopkeepers with the patronage that made success possible—advice, recommendation, contracts, customers, above all, credit and capital. The churches became the gatekeepers of opportunity, and the teetotal pledge a badge of deserving middle-class respectability.[48]

Unlike Johnson, Sellers sees devout businessmen as fraught with "intolerable anxieties about their goodness as they made the difficult transition from traditional ethics to tough-minded capitalist egotism." Their sense of power "to affect distant persons and remote events" led them into antislavery and other reforms "partly because their market perspective and heightened sense of potency made them the first to feel responsible for a broader range of evil wrought or good left undone." At the same time, their pious Yankee upbringing created a need "to reaffirm traditional altruism" with "fantasies of selflessness and exercises in benevolence to sanction their pursuit of capitalist wealth." Charles Finney himself preached an antimarket altruism, condemning the maxim "to look out for number one," advising businessmen instead "to do good to others."[49] In Sellers's view, the awakening did not "suffuse the market with love" as Finney had hoped.

> Instead, and at fearful cost, it bifurcated the consciousness of cross-pressured Americans. Women and ministers ended up sustaining an illusory world of powerless love—actualized only in the female world of altruistic domesticity—that immunized and sanctified men for the segregated market world of calculating ego.[50]

Amy Kaplan suggests that the women's sphere was actually far from innocent and peripheral to the world of power. Rather, she convincingly demonstrates, women served the important ideological function of policing both domestic and national space, justifying expansion by their civilizing presence, and justifying male aggression though their need for protection.[51]

Revivals in Lockport had much in common with those of Rochester and Utica. Unfortunately, a series of devastating fires destroyed all newspaper, municipal, and most church records, preventing the kind of detailed analysis that Johnson and Ryan have undertaken. While it is beyond the scope of this study to answer basic questions about the causes of revivals and related enthusiasms, a brief examination of the form they took in Lockport and local reactions to them is part of the story of the canal's influence on the town where the waters met.

The records of the First Presbyterian Church are entirely intact, and give us at least a glimpse of the nature of the Lockport revivals. Session records from the time of the founding in 1823 up to 1830 show that the church leaders were struggling to discipline congregants, chastising and threatening excommunication to those who engaged in public displays of gambling, Sabbath breaking, swearing, and especially drunkenness. Later church leaders would change tactics and attempt to instill self-discipline. Charles Finney's revival had begun in Rochester's Third Presbyterian Church in the fall of 1830. By the spring of 1831, the fire had spread to Lockport's Presbyterians. Buoyed by a rich harvest of souls in the early spring, Pastor William Curry placed an announcement in the June 7 *Niagara Courier* proclaiming:

> A seven day's meeting will commence at the Brick Meeting House, in Lockport, on Wed., the 15th inst. [this month] Ministers who will preach, Christians who will pray, and all who are yet without hope in Jesus Christ, but will 'hear' us again on this matter are bid welcome to an Christian fellowship and hospitality.[52]

The church records show 173 conversions in 1831, including 75 on a single day.

James Hotchkin's history of the Presbyterian Church in western New York (1848) states that "it was the year 1831, that the most extraordinary displays of the power and grace of God, in reviving his work and converting souls in western New York, were exhibited." Hotchkin lists the churches of Lockport, Albion, and Niagara Falls as "most highly favored." In Lockport, "a very general effusion of the Spirit was had. Of one protracted meeting, the fruits in cases of professed conversion, were one hundred and fifty. To the church there, one hundred and ninety were added last year." Although these figures exaggerate what is recorded in the church records, clearly the Lockport revival was particularly strong. Hotchkin writes that in some congregations, "particularly in the western section," the revival has been "so general" that "the whole customs of society have been changed" including "entire abstinence from ardent spirits."[53]

Another revival commenced in 1834 with a series of "protracted meetings" that "continued nightly with increasing power." That year the church added ninety-eight new members, sixty-six on a single day. Within days after assuming the pastorate in 1842, Reverend William Wisner began "the great revival which so refreshed the church . . . perhaps the most powerful in some respects the church ever experienced," that on one Sabbath day, 150 converted.[54] Including other Lockport churches, one source indicates there were a total of 718 conversions that winter.[55] Lockport was burned over and over.

The Presbyterian records suggest that each of these revivals was general—that is, it affected most Protestant denominations. In one of the later revivals, "the churches of the city united . . . and such harmony prevailed that the windows of Heaven were opened. Hundreds of souls were converted from every class and condition of society."[56] Besides the Presbyterians, revivals were probably strongest among Lockport's Baptists—who actually used the canal at the foot of the locks for baptisms[57]—and Methodists, who reported prodigious growth in western New York in the 1830s.[58] In February of 1844 Thomas Leonard notes between fifty and seventy-five conversions at the Methodist church in a two-week period.[59] Episcopalians, on the other hand, condemned revivals with their "extravagance and wild excitement" and the "many fanatical or heretical religious movements [that] originated here in western New York amongst our staid New England ancestors," and offered instead shelter from these "emotional excesses" under the banner of "Evangelical Truth and Apostolic Order."[60] Perhaps only the Catholics were more immune to enthusiastic revivals.

In the wake of revivals, Lockport became a hotbed of almost every enthusiasm Whitney Cross identified with the Burned-Over District—temperance, Sabbaterianism, spiritualism, perfectionism, millennialism, women's rights, and abolition. Although the battle for temperance was fully engaged at Lockport, the enemy was never completely driven from the field; as the scale of industry grew, sobriety on the job became the norm, but workers' free time often continued in the old pattern.[61] Reverend Wisner of the First Presbyterian Church suggested that a temperance crusade directed partly at workers had "prepared the way" for the revival of 1842–1843. "Many of this class of our citizens," he noticed, "having broken off from a life of profligacy, are particularly susceptible to the influence of divine truth; and a large number of them have opened the door of their hearts to admit the Savior as a permanent guest."[62]

The Sabbath-keeping movement was particularly strong among Presbyterians. Deacon John Gooding of the First Presbyterian Church became a stockholder in the Pioneer Line of stages and canal boats organized by Elder Josiah Bissell of Rochester's Third Presbyterian Church. All activity on the Pioneer Line stopped on Sundays, and "at stops the Pioneer stages served hot coffee rather than the customary rum." Bissell organized a national campaign to stop Sunday mail delivery. It failed not only in Washington, but even in Rochester and Lockport. In 1828, a group of Lockport's most prominent businessmen openly opposed Gooding and Bissel, denying that "a majority of business men favor the project" and asking "that the mail be continued seven days as we have hitherto received it."[63] When elites split over basic economic issues, reformist enthusiasm could lose its legs.

By the 1840s, religious enthusiasm in the Burned-Over District began to take new forms, such as millennialism and spiritualism. Lockport was less than sixty miles from the home of the Fox sisters, who in 1848 began to report the rapping of departed souls. The canal itself contributed to a growing conviction that things venerated ancestors could never have imagined were indeed possible. A Lockport believer in spirits, Mr. Pickard, visited Rochester spiritualist A. H. Jervis in 1849. He received through Jervis a message from his deceased mother saying, "Your child is dead." On the strength of this prediction, Pickard immediately took the stage for Lockport. Returning to his home, Jervis's wife met him with a telegraph from Lockport: "Tell Mr. Pickard, if you can find him, his child died this morning." Jervis then commented to his wife: "God's telegraph has outdone Morse's altogether."[64] Through our contemporary prism, communicating with the dead may seem like one of those coups de théâtre that makes people scream at horror films, but to nineteenth-century spiritualists it was largely a source of comfort and reassurance. For those who felt their contemporary world was drifting out of control, spiritualism connected them to a timeless one. Like born-again faith, it rejected the calculating rationality of the market world by validating an intensely emotional private experience. It is perhaps no coincidence that "God's telegraph" arrived during the same era as Morse's.

The pace of economic and social transformation, combined with the surprising success of revivals, convinced most Protestants in the Burned-Over District that they were living in extraordinary times. Many believed that the millennium was at hand. This commonly took the form of postmillennialism; Christ would return at the end of a thousand years of gradual purification. It was up to Christians to work toward perfecting the world and thereby hasten Christ's return. Temperance, abolition, and many other reformist crusades were at least partly inspired by such postmillennialist beliefs. Premillennialists expected Christ to make a dramatic return, destroying evil and establishing God's kingdom. Human striving had nothing to do with this return. While postmillennialist impulses could easily be channeled into disciplined striving in the marketplace or into the national project, premillennialists rejected the bourgeois discipline that bid people defer gratification for the future, and left everything to divine intervention. The most influential premillennialist was William Miller, a devout Vermont layman who had studied Bible prophecy for fourteen years and famously predicted that Christ would come about 1843. Although he began to gain an audience for his predictions in the early 1830s, his impact on the Burned-Over District would culminate as 1843 approached. Whitney Cross argues that the idea that the Kingdom of Heaven was about to arrive, eliminating every earthly problem, was itself an expression of the "buoyant optimism" of the era. Yet Miller's followers "found the world beyond rescue, legislatures corrupt, and infidelity, Romanism, sectarianism, seduction, fraud, murder, and duels all waxing stronger."[65] Like revivals and other enthusiasms, Miller's Second Advent message was alluring to people who experienced the world around them as unsettling. Hence, Miller would find great success in Rochester and Lockport.

In January of 1843 Miller pronounced that Christ would return between March 21, 1843, and March 21, 1844.[66] In February, the nineteenth century's most brilliant comet appeared. "The comet of 1843," the *Niagara Democrat* reported on April 19, "has been frequently seen during the past week and we are told not a little apprehension has been felt by some in these parts lest it might be a forerunner of the fulfillment of Millerism."[67] Beginning in June, a group of Miller's supporters, held large services under an enormous tent. The organizers of these events, along with Reverend Elon Galusha, the pastor of Lockport's First Baptist Church, entreated Miller to come to western New York. "Should you come this way," Galusha wrote, "it may be the means of calling into the field an able Laborer tho it be at the 11th hour."[68] Galusha was among the most prominent of all American Baptists. His father had been governor of Vermont. He initiated the abolition movement among Baptists and represented them in 1840 at the world antislavery convention in London.[69]

Miller arrived in Rochester on November 12, 1843, and filled a large tent twice each day for eight days. He then proceeded to Lockport for ten days of double sessions at the First Baptist Church (March 21–30). Thomas Leonard attended both sessions on March 25. He described the second one:

> Went in the evening to hear the venerable Second Advent advocate, Mr. Miller, explain a chapter from Revelation, and prove that the final consummation of all things is to take place within five months! He closed with a very good exhortation. The Universalist minister was present and had some reply to make to some charges preferred against him by Mr. Miller.[70]

Miller later wrote that he had been "challenged to a public debate by a Universalist. I will not contend with them. It would be an admission that they *might* be right, which I cannot for a moment believe."[71] Miller's success in Lockport was electrifying. Fully one-third of the congregation of the First Baptist Church, along with Pastor Galusha, accepted his Second Advent doctrine. The historian of the church wrote that up until 1843 it had made "rapid advance" but then "evil days befell." It was a "fatal" year that brought a "shock" to the community "from which it did not recover for more than twenty years." The church finally disbanded in 1851 and reorganized as the Second Baptist Church.[72] Galusha established services in another building and often took the stump as an itinerant advocate for Millerism. At one public session in Rochester in early March of 1843, he "brought eight hundred souls to regeneration." Sparrow Sage, a Niagara County farmer, was so certain about the final consummation that he planted no crops in the spring of 1843.[73]

On March 21, 1844, Millerites gathered to await the great event. The next day, Thomas Leonard attended a Second Advent meeting. "Such a collection of fanatics," he wrote, "I have never seen before. They confessed they were disappointed yesterday being 'the day' they had pitched upon for the end of time, but they meant to 'hope on still."[74] Miller admitted his error but eventually declared that October 22, 1844, was the correct date. Leonard attended one last Millerite meeting on the evening of October 21. "Such zeal I have not seen manifested in any cause. Their *last day* is Wednesday of this week! I really felt disgusted."[75] On one hand, Leonard's reactions are indicative of the religious openness that prevailed among evangelical Protestants; most believed the end was coming and were willing to give Miller's ideas a hearing. On the other hand, Leonard, like most other Protestants, began to see them as fanatics when they persisted in their beliefs and insisted on their interpretations of scripture in the face of a disturbing world that would not go away.

The antislavery movement shook and divided Lockport even more deeply than the Second Advent preaching. The tensions it constellated revealed the town's social fractures in an explosive way. In one sense, the antislavery movement—and in particular, the abolitionist version of it—expressed a perfectionist impulse similar to that of the temperance, Sabbaterianism, prison reform, and other movements of the "benevolent empire." The perfectionism that followed the great revivals was largely a postmillennial project, but as the case of Galusha indicates, abolishing slavery was not incompatible with Millerism.

Unlike many Northern towns, Lockport had a black presence dating from the canal construction period. New York State had more "free colored" than any other Northern state until mid-century.[76] Blacks came to Lockport and other canal towns as common laborers, cooks, and barbers. Although it is difficult to assess, their presence was also related to the Underground Railroad. There were two reasons for this. The canal terminated at the Niagara River, the narrowest point on the entire Great Lakes-St. Lawrence system. A branch canal led to the Lake Ontario port of Owsego, another important point of departure for Canada. Even if one did not travel by boat, the canal marked an easy-to-follow route to Canada. Equally important, in nearly every town and village in the Burned-Over District were people sympathetic to the plight of escaping slaves. Travelers to Niagara and Upper Canada often left the canal at Lockport; the Niagara River was only twenty miles distant and Lake Ontario ten. There is every reason to suspect that the community of abolitionists in Lockport, in cooperation with the village's own black residents, helped to forward escaping slaves by numerous routes across the border.

The Underground Railroad is notoriously difficult to study, partly because it was an illegal activity in which people covered their tracks, and partly because the level of participation has been greatly exaggerated. Many whites would like to believe that their ancestors resisted slavery. The ubiquity of these legends of whites helping blacks seems to suggest that this was the central story of the slavery period, a symptom perhaps of a continuing denial of the brutal reality of slavery. Escaping slaves usually traveled on canal boats in one of two ways, hiding in the cargo holds or traveling with abolitionists posing as their masters. Caroline and William Harris, with their infant daughter, decided to leave for Canada in 1850 after living in Philadelphia for many years. Trying to pass as free colored, they bought regular tickets on a packet. They were so viciously harassed by the crew that Caroline, with her daughter in her arms, jumped through a window into the canal. She was pulled out but her daughter was never found. When the crew told the couple they would be beheaded in Syracuse, William cut his own throat and nearly died. None of the passengers intervened, especially not the three other black men who were on board. The Harris family, without their daughter, finally made it to Canada. This tragic story reminds us that not everyone in the Burned-Over District was against slavery. Indeed, some canallers and immigrant laborers saw blacks as economic competitors; they grasped the social hierarchy well enough to know that black people were the one group upon whom they could vent their frustrations.[77]

The number of blacks in Lockport is hard to ascertain since some of them were not "free coloreds" and would probably have slipped away from census enumerators. Blacks formed about 2 percent of Lockport's population in 1830 and nearly 3 percent in 1840. They also formed about 3 percent in 1855, which indicated 357 blacks in a town of 13,585. New York City had the highest percentage of blacks in the state with about 4 percent. The percentage of foreign-born residents in Lockport also steadily increased, reaching 22 percent in 1845 and 37 percent in 1855—with 2,214 of those from Ireland, Irish immigrants outnumbered blacks by about six to one.

Lyman A. Spalding was a Lockport merchant before 1825, then a mill owner and leading citizen. In April of 1836, along with a "large audience," he heard Theodore Weld speak against slavery at the Presbyterian Meeting House. Because Weld's meetings in other cities had sometimes been broken up by mobs, there had been difficulty in finding a church that would permit him to speak. At least one of the church members protested that "the meeting was being assembled for, as he said, a *political object*." Spalding was indignant at this opposition. "So far has slavery encompassed its victims through the whole U. States," he recorded in his journal, that anger is "rained against discovering the enormity of American slavery." In a recent study of the Underground Railroad, Fergus Bordewich describes Weld's visit:

> In Lockport, he was almost shouted down by hostile demonstrators, but after a marathon four-hour speech, 440 people signed the new constitution of the Niagara County Anti-Slavery Society; a year later the society would have 21,000 members, with branches in nine of the county's twelve townships.[78]

Spalding was one of the founders of the society, and during the next several years he attended antislavery meetings throughout the region.[79] As Weld's visit shows, the citizens of Lockport were far from unanimous on this subject. In April of 1836, a newspaper notice announced: "All the citizens of Lockport who are opposed to the doctrines and measures of Immediate Abolitions, are requested to meet at the Presbyterian Church . . . tomorrow afternoon to give an expression of public opinion on this subject."[80] Despite agreeing on revivals and temperance, Lockport's Presbyterians were irreconcilably divided on the issue of abolition. The dispute centered not on whether slavery was evil, but rather on whether the church should take a uniform public position on the issue. In April of 1838, forty members of the congregation sent a letter to the church elders stating that they intended to "organize a new church in Lockport, N. Y., Congregational in its government and Calvinistic in doctrine." The Congregational church was the inheritor of New England's Puritan tradition, but the 1801 Plan of Union had divided fields of effort at the New York border; all Presbyterians east of that line would become Congregationalists and all Congregationalists west of it would become Presbyterians. Lockport's renegade Presbyterians defied that agreement.[81] "So intense and bitter was the feeling on the subject of slavery," a Congregational pamphlet reports, "that a mob led by a Niagara County Judge, later to become Governor of New York State, stormed a church where an antislavery meeting was being held and a stone coming through the window struck Mrs. Whitcher, mother of Mrs. Ellen Richardson."[82] The judge leading the "mob" was Washington Hunt; it seems that middle-class people, as well as proletarians, could ignite public disorder.

During that same summer, a western New York antislavery convention commenced at Lockport's Methodist Church with "overflowing" crowds. The Methodists were just as divided as the Presbyterians; the motion to open the church to the convention had lost by a tie vote requiring its supporters "to sign a bond making them responsible for any damage." The ensuing discussions were bitter and led, once again, to the splitting of the church into two congregations.[83] The middle-class Protestants of Lockport were completely at odds over abolition.

The antislavery movement, and its opponents, continued to be active in Lockport. Spalding mentioned attending an antislavery meeting at the Methodist Church in 1840.[84] A statewide antislavery convention convened in Lockport in 1843. Black abolitionist Charles Lenox Remond, a close associate of Frederick Douglass, was one of the speakers. He commented on the "measures used to prevent people from coming out." Remond had "heard much of Lockport," and his "expectations ran high" because of "the high-minded, liberal, free and thoroughgoing abolitionism of the Liberty party advocates of this village, [where], of all places, I expected we should be welcomed." Apparently the meeting was very contentious; he called it the " 'trial hour' of truth struggling against error and sectarianism," which in the end "resulted in much good, and in the advancement of free principles."[85] Frederick Douglass came to Lockport for several days in October of 1849. Spalding heard him speak at the Congregational Church and at an "anti-slavery fair" held the next day and found him "very able," and "talented."[86] When an antislavery delegation arrived from Rochester in 1854, it could not find a church or hall to hold a public meeting; finally a local black man provided a hall by renting it himself.[87]

The antislavery Congregational Church grew from its original forty-five members in 1838 to over five hundred in 1850. Then, during a Sunday service on May 22, 1853, a violent thunderstorm gathered.

> The choir was singing when a bolt of lightening came down through the steeple which was directly over the choir loft. A large lamp in a metal frame suspended by wire was directly over the head of Deacon Luther Crocker. The lightning followed the wire down and struck and killed Deacon Crocker. There were more than a dozen other singers around him who were knocked down by the shock but he was the only fatality.[88]

Lockport's Congregational, Unitarian, Methodist, and Presbyterian churches were all on the same block of Church Street (see fig. 5.2). Although there

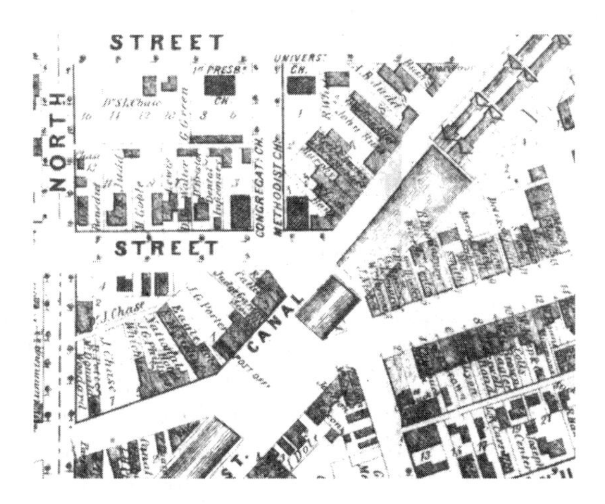

Figure 5.2. Lockport's Congregational, Unitarian, Methodist, and Presbyterian churches were all on the same block of Church Street; *Map of the Village of Lockport, Niagara Co. N.Y.*, surveyed and drawn by Bernard Callan, C.E. (New York: M. Dripps, 1851). Courtesy of Lockport City Hall.

may have been some suggestions of divine justice among the members of other churches that opposed immediate abolition, they all agreed on the immediate installation of lightning rods. The next year, a major fire destroyed both the Congregational and the Methodist churches.[89]

While the plight of slaves aroused both sympathy and action among many Northerners, the advancement of the free Negroes actually living in the North was often neglected. Moreover, supporting legal equality and freedom did not imply that one found "social intercourse between the races" acceptable. Because of this, black abolition leaders suggested in 1843 that a special report be prepared focusing on the "Condition of Colored People" targeting the black people of eight New York communities, among them Lockport.[90]

One indication of the condition of blacks in nineteenth-century Lockport was the struggle over education. Nowhere in the North could blacks vote, serve on a jury, or testify in court; transportation, churches, and even cemeteries were segregated. It was a Jim Crow world. Having no access to schooling, Negro leaders in 1835 placed a notice in the *Niagara Courier* pointing out that "the colored people in the Village of Lockport, are rapidly increasing in numbers," and that "the customs of the county do not permit us—neither indeed do we desire to join in society with those of a different complexion." They appealed "to the liberality of the citizens of Lockport" to help them build a school and place of worship.[91] Pleading did nothing, however, until Lyman Spalding donated land for a school and church in 1840. Although the village's school board administered the small segregated school, the quality of education and the physical conditions were inadequate. When Frederick Douglass returned to Lockport in 1862, he inspired a campaign to integrate the schools. As this movement gathered force, the small Negro school was burned to the ground. It was not until 1876, after decades of struggle that included boycotts, sit-ins, and civil disobedience, that Lockport's schools finally allowed black children.[92] Whether in religion, antislavery, or desegregation, it seems that there was precious little agreement in Lockport. Another equally contested issue was women's rights.

A broad transformation of the lives of women was one of the most profound changes that accompanied the commercial revolution. The old social links of the rural agricultural world "had snapped."[93] Excluded from the world of public business, having no rights of inheritance or suffrage, women were confined to the empire of the home. But during the revivals in western New York, some women began to find their voices. An official Presbyterian history, written in 1848, explained that "no audible praying by females in promiscuous assemblies" was appropriate. While on the whole the great revivals were positive, the history concluded, there were certain excesses, especially during protracted meetings in western New York churches:

"Some of the sisters, in some instances, forgetting apostolic precept, claimed it as their privilege, in public and promiscuous meetings, to lead in prayer and to address exhortation to the congregation; and there were ministers who upheld and encouraged them in this practice. . . . These irregularities, to a considerable extent, brought discredit on the revivals in Western New York."[94] Mary Ryan argues that women were crucial to the revivals in Utica.[95] Finding their voices in churches, women then began to participate in movements for temperance, prison reform, and a host of perfectionist causes. To what extent this was altruistic, and to what extent an elite attempt to create order and make the environment safe for middle-class morality, is debatable. Organizing for women's rights, although it gathered steam only gradually, was a cause closer to home.[96]

Only three years after the first Women's Rights Convention at Seneca Falls in 1848, a group of Lockport women made an extraordinary demand. The People's College Association had just been formed to offer courses of study, including "those sciences most immediately and vitally essential to agriculture and the useful arts." A group of young ladies, including Lucy Stone, demanded that women should have equal access to the college. A 1936 *Science* article concludes that "Lucy Stone and her 'female' companions were the first women students of horticulture in America—85 years ago."[97]

Lockport's most prominent women's rights advocate was Belva Lockwood. She was born Belva Bennett on a farm a few miles east of Lockport in 1830. She began to teach school in her rural district at fifteen. Soon she married, became Mrs. McNall, had a daughter, but then her husband died. She pursued further education and was among the first women graduates of Genesee Wesleyan College in 1857. Immediately she took a position as preceptor of the Lockport Union School. During her four years in this position, she lived with her daughter and younger sister in Lockport. As an active member of the Methodist Church, she worked with orphans, poor families, and "all the benevolent enterprises." It was in this position that she became a friend and ally of Susan B. Anthony and Elizabeth Cady Stanton.[98]

Women's rights emerged as a public issue in Lockport when the New York State Teachers' Association decided to hold their annual meeting there in August of 1858. Lockport native, and then New York governor, Washington Hunt welcomed the assembled teachers and asked them to "feel at home" there. Then he asked them to consider Lockport

> a new and growing town, called into being by the completion of the great link that binds the ocean and the lakes. When the Erie Canal was completed, this town was a barren wilderness. But there were higher interests than national progress. The moral and intellectual development of its people constituted the highest glory of the state.[99]

In the late 1850s, Lockport was still "new," still available as an emblem of "national progress"; by implication, the canal, education, and national progress were all expressions of a single perfecting impulse. Susan B. Anthony and Elizabeth Cady Stanton wanted to extend this impulse in a new direction. Anthony spoke on the first day, pointing out that although over 90 percent of New York's teachers were women, they were missing from the association's leadership and its important committees, adding that "she hoped much from Lockport on this matter." One by one, Anthony's initiatives were defeated—that segregated colored schools be abolished; that "sexes be educated together"; that as far as declamation and discussion are concerned, girls receive "precisely the same as boys"; that Mrs. Belva McCall, along with Miss Roundy and Miss Elizabeth Cady Stanton, be appointed to a special committee to develop a resolution on the equality of girls and boys in public speaking in the classroom. They also rejected her nomination of a woman vice president of the New York State Teachers' Association. The women delegates were equally unsuccessful in getting women onto the association's committees. Finally the officers offered positions to several women, including Anthony, on a committee to sell subscriptions for the society's periodical. "Miss Anthony begged to decline serving," the Lockport *Daily Courier* noted, "until this convention could place women upon some other committees than that of raising money for the men to spend." Belva McNall replaced Anthony on that committee.[100]

One newspaper reported that there were over one thousand in the hall for the evening sessions, and many turned away.[101] On the final day of the convention, over three hundred teachers traveled on the New York Central to Niagara Falls where Susan B. Anthony proposed a toast: "The gentlemen, dear creatures, we love them, God bless them." Several of "the gentlemen" were less generous. One was perturbed at the conference by the "eternal jangle about Women's Rights." Another described women's rights as "a stench in the nostrils of many prominent educational men."[102]

Belva McCall, after serving as preceptor of Lockport's Union School for four years, became a leading educator, taking charge of schools for girls in Hornellsville, Elmira, and Oswego, New York. She later moved to Washington, married a dentist named Ezekiel Lockwood, and began to study law at National University. Belva Lockwood practiced law in the capital for several years, but when she petitioned to practice before the Supreme Court, the court rejected her request on the grounds that there was no precedent. Lockwood was angry but focused. "Mrs. Lockwood showed," in the view of her fellow women's rights advocates, "that it is the glory of each generation to make its own precedents."[103] She then proposed a bill to Congress to change the precedent. The bill passed, and in 1879 Lockwood became the first woman authorized to practice law before the Supreme Court. In 1884 the Equal Rights Party nominated her to run for president. In a spirited campaign that attracted a great deal of media attention, Belva Lockwood became the first woman to be nominated for the nation's highest office, an accomplishment she repeated in 1888. The people of Lockport closely followed her career especially via her daughter's weekly column in the *Daily Journal*. She occasionally returned to Lockport to lecture and attend to family business.

Following one of Lockwood's lectures, the Lockport *Daily Journal* reported that "the manifest destiny idea was strongly urged. The nation was not only to spread across the continent, but the world was to be in the course of time embraced in one vast consideration."[104] The dark side of a perfectionist campaign—whether against alcohol, tobacco, slavery, or the subjugation of women—was that it implied a kind of policing that was hard to contain. If one is spreading light, those who resist have no standing. Amy Kaplan suggests that the moral authority women gained by their domesticity was sometimes used to police the boundaries of white America and to justify its expansion into the lands of other peoples. "Manifest domesticity," as Kaplan describes it, links the apparently innocent cultural work of nineteenth-century American women to the project of empire.[105]

Like antislavery, temperance, and Millerism, women's rights had strong supporters in Lockport, but each of these movements also met opposition. Public life in towns of the Burned-Over District had a contested nature. Three of Lockport's largest Protestant churches split apart between 1838 and 1843. Middle-class people, especially in manufacturing towns where a boisterous working-class culture was emerging, reacted to a world seemingly out of control in a variety of ways with consequences beyond the immediate economic and social situation.

To conclude this section on Lockport during the era of the Burned-Over District, a brief exploration of Lyman A. Spalding's writings may serve to illuminate some characteristics of the new commercial world that the canal had helped to initiate. Spalding's family, Yankees from Geneva, had owned a slave. His father was a Presbyterian, but of his father's brothers, "William, was rather a Universalist; Frederick was a Baptist, Ephraim, nearer a Christian." Later his father became a Quaker and the children followed. The Spaldings' religious searching was not unusual, even in the days before the canal. After apprenticing with several merchants in Canandaigua, Spalding moved to the Mountain Ridge and set up a store in 1822. He was able to accumulate enough capital by 1825 to purchase the land surrounding the locks on the southern side for $5,500. The state had decided to award water power rights by bid. Despite owning the crucial land, Spalding was outbid. He then bought the land on the north side and went ahead with the construction of a mill, thinking no one else could use the water, so the state could hardly refuse his bid. The state built a race for diverting water above the locks to mills, and by 1826 Spalding's flour mill was operating. In 1827, a group of local men bought a huge tract of land below the escarpment along the path of the canal and began to set up a rival village. In collusion with Canal Commissioner Henry Seymour, a group of speculators known as the Albany Company took possession, not only of the water rights, but also of most of the land in the rival village, known as East Lockport. With the aid of Seymour, the company was able to force construction of a race to bring water power through Spalding's property. In 1829, "regardless of protest," a history of the county written fifty years later records, "a party of men were sent to cut a ditch along the side of the canal. The citizens, indignant at the proceedings, assembled and drove them away."[106] The use of the term "citizens" suggests that it was not only mill hands creating this particular display of public disorder. During the affair, Spalding had the state's superintendent and his laborers arrested. One New York official stated in 1926 that the record of litigation on this lease was "longer than any other unsettled affair of the state."[107] Eventually, Spalding relinquished his claim to the water rights, but then purchased back what he needed for his mill. Washington Hunt, with some associates, took possession of all the water rights. The lessees were able to continue receiving enormous waterpower for the original fee of $200, although Spalding's mill alone was producing $600,000 worth of flour a year by 1840. Lyman Spalding had learned that it was a mistake to assume local people or "our Canal Commissioners to be honest." In Lockport, the new commercial world, on arrival, was a dog-eat-dog, selfish one.

Lyman took solace in his recent marriage to Amy Pound, and on their subsequent domestic bliss. He added "a wing on the west side of my wife's room which made a neat looking dwelling house with a fine yard in front and a good garden in the rear."[108]

While Albany speculators were conniving to deprive him of the means to develop water-powered industry, Lyman Spalding became deeply involved in the in the antimasonic movement. The controversy began in 1826 when William Morgan, a stonemason from adjacent Genesee County, threatened to expose the secrets of Masonry. A series of confrontations ensued culminating in Morgan's "arrest" by a group of Masons. He was taken to Canandaigua and soon after whisked away never to be heard from again. Apparently, Morgan moved by way of Rochester through Niagara County to Fort Niagara, at the very extremity of U.S. territory, where the Niagara River empties into Lake Ontario. Some suggested he escaped to Canada, or was drowned in the Niagara River. The story entered the emerging popular culture with expressions like "a good road to Fort Niagara, but none back"; and, as publisher of the Albany *Journal* Thurlow Weed commented when a battered body washed up on the shore of Lake Ontario, it was a "good enough Morgan" for him.[109] The idea that a group of prominent men could kidnap an innocent man and transport him across five counties was an outrage to many citizens in western New York.

It was during the agitation surrounding these events that Lyman Spalding began to publish a monthly antimasonic newspaper called *Priest Craft Exposed and Primitive Christianity Defended*.[110] Spalding's views were typical of the early advocates of antimasonry. His paper chiefly complained about elite control. He used the term "priest craft" to denounce those who would turn teaching of religious dogma into a craft, a mere way of gaining wealth. Particularly frightening was the prospect of losing religious freedom through the establishment of an official faith. The paper railed against Masons, Catholics, Episcopalians, and even Presbyterians "because we advocate the cause of the indigent and laborious poor against the systematic attacks of the harpies that would fain devour them."[111] For nearly five years, a sensational series of trials in five counties and statewide investigations kept the public aroused. At the Niagara County Courthouse in Lockport, the commandant of Fort Niagara and the County Sheriff, Eli Bruce, were tried for abduction and kidnapping. According to testimony, when Morgan's coach came near Lockport on its famous journey west on the Ridge Road, the abductors had slipped into Lockport, told Sheriff Bruce that they had Morgan, and requested his assistance in getting the prisoner to Fort Niagara. Bruce presumably then took charge of the operation during the infamous final scenes in western Niagara County. Eli Bruce was fined, imprisoned for twenty-eight months, and removed from office by order of the governor. The trials provided a host of what people like Spalding saw as outrages: witnesses whisked across the border, others bribed and threatened, and official stalling so transparent that it required another new term, "stone-walling"—derived from William Leete Stone's 1832 book *Letters on Masonry and Antimasonry*.[112] A special counsel, appointed by the legislature, reported that "these occurrences have been so numerous and various, as to forbid the belief that they were the result of individual effort alone, and they have evinced the concert of so many agents, as to indicate an extensive combination to screen from punishment those charged with participation in the offense upon William Morgan."[113] Ronald Formisano and Kathleen Kutolowski argue that, although antimasonry later attracted paranoid and fanatical people, it was republicanism that "provided the language of protest; it conditioned the fear that equality before the law—a central tenet of republicanism—was threatened by a secret society's ability to cover up dark deeds and to thwart justice."[114] Spalding did his share to stir up antimasonic feeling in Niagara County. Interestingly, the issue divided parties, churches, and even families. In the ten years following Morgan's abduction, antimasonry was a powerful political force, not only in New York, but also in Ohio, Pennsylvania, and especially New England.[115] Although antimasonry began as an outcry against elitism, the political movement evolved into what Whitney Cross calls a "businessmen's platform." Because Andrew Jackson was a mason, the antimasonic crusaders continued to oppose him and the Democratic Party. Ironically, this made it easier for antimasons to jump to the new Republican Party in the 1850s, and renew their progressive idealism.[116] Spalding provides a good example: he was appointed Niagara County's delegate to the State Republican Convention in 1855.[117]

Lyman Spalding shared a sense of idealism and social justice with many of Lockport's other Quakers. He was a good friend of Dr. Smith and his wife Edna who, during the town's earliest years, had displayed uncommon kindness to the Iroquois and wounded Irish laborers.[118] When Dr. Smith died in 1842, Spalding praised him as "a great advocate of temperance and other good reforms." Spalding provided land for the Negro school and church as well as the Catholic church, and he supported the failed Wilberforce Community for blacks in Ontario.[119] He attended meetings and lectures on abolition and temperance, and quite possibly, had a hand in the Underground Railroad. Like Thomas Leonard and so many others, he felt free to attend services and meetings at many different churches. Eventually, he became a leader of the new Congregational Church that had split from the First Presbyterian over the issue of slavery; he represented the church at the Congregational Convention in 1852.[120] In his journal he mentions all of the prominent visitors he heard speak in Lockport: Frederick Douglass, Horace Greeley, Martin Van Buren, William Miller, Henry Clay, and Theodore Weld.

Spalding was one of Lockport's most successful entrepreneurs. Reading his journal, one can see his world expanding. He frequently checked the price of flour in New York. When canal enlargement interrupted his milling operation, he complained, "Business very depressed—no $ circulating except for enlargement." He even looked into the prospects for silk-raising in Lockport. In 1840, he took a business trip to Albany. After the Lockport and Niagara Falls Railroad opened, he used it to deliver wheat to his mill—presumably from Canada—while continuing to secure western wheat via the canal. He began to make more and more extensive trips to the Midwest, always in search of wheat for his mill. In 1843 he traveled to Toledo and southern Michigan, and the following year to Cleveland, Detroit, and—by water—to Chicago. Spalding seemed to understand that, in this emerging commercial world, communication, transportation, and information were all crucial.[121]

Although Lyman Spalding was clearly not a man content to confine his altruism to his own household, he too focused more and more as time passed on the pleasures of private domestic life, apart from the roiling competition of the world beyond. It is interesting to notice how the Spaldings' celebration of Christmas evolved during the course of the 1830s, '40s, and '50s. Christmas, which had once been a public festival, then had fallen into neglect, was revived and transformed with the arrival of the new commercial world and the new bourgeois family. It became a private festival, celebrating the private utopia of the family home. Charles Dickens's visit to the Unites States in 1843, along with the popularity of *A Christmas Carol*, help spread the word about the festival's meaning. It was another antidote to market cruelty—a temporal one compared to the spatial one of domesticity—that nevertheless perfectly served the market.[122] On December 25, 1835, Spalding recognized Christmas Day, but it was a normal busy day at the mill for him and his workers. During the 1840s, his descriptions of Christmas dinner become slightly more elaborate. Then, in 1851, the full modern Christmas appears:

> A merry Christmas—Sleighing very fine & weather cold. Alice and the two boys came up yesterday & hung up stockings last night to receive Santa Claus' offerings—such lots they got this morning made them happy. Ed said Santa Claus was a good man and he would kiss him when he saw him. Turkey for dinner—no other company than Alice & boys.[123]

Every channel into which altruism fled from the cold light of the market—domesticity, reform, the Christmas festival—seems to confront us with the same question. Are these mechanisms to marginalize and effectively stifle altruism, or are they evidence of its protean and perhaps subversive survival?[124]

In the three decades that followed the meeting of the waters in Lockport, the market rushed in. Completing the canal was a crucial step in American continental hegemony, but also in the transformation of the economy. In Lockport and numerous other canal towns, people reacted to this new world in dramatic ways. For middle-class people like Spalding, with their combination of Yankee idealism and guilt, it was "only by headlong flight into domesticity, benevolence, and feeling" that they could "tolerate the market's calculating egoism." As Charles Sellers concludes, "their pessimistic piety belies our historical mythology of capitalist transformation as human fulfillment."[125]

Wedding the Waters Again and Again: Industry in Lockport

There is something at once absolutely logical and uncanny about the nature of Lockport's main industries. It is as if the imprint of the construction period continued to unfold in various ways for nearly a century. The expertise gained, especially by engineers and contractors but also by some laborers, provided Lockport with the means to transform itself from a construction site into a manufacturing town. There were people at hand who knew how to excavate stone, manipulate flowing water, and take advantage of the flow of information that the canal made possible to keep abreast of crucial economic information from both east and west.

Lockport's stone industry emerged directly out of the experience of excavating the canal. In excavating the Deep Cut and the lock basin, workers encountered many varieties of stone and became very familiar with their various properties. They also gained skill in quarrying, cutting, and transporting stone blocks. Before the canal, Lockport's stone was commercially worthless, but the canal made moving heavy material relatively cheap and easy, allowing local quarries to ship to a market that stretched from New York to Chicago. There were two kinds of quarries in early Lockport. Sandstone was first excavated commercially at the escarpment in the North End of Lockport at a place still known as Rattlesnake Hill. William Wallace Whitmore opened a quarry there in 1830 and began to market "Rattlesnake stone," a name he later suggested be changed to Medina sandstone. Whitmore shipped the stone all over western New York and eventually to Cleveland, Toledo, and Detroit. It was used in public buildings, churches, homes, sidewalks, and street pavement. Later, the firm began to buy up quarries throughout Niagara and Orleans counties and opened offices in Rochester, Buffalo, and Detroit. Workers first encountered Lockport limestone, another valuable building stone, during the excavation of the Deep Cut. A number of small entrepreneurs began quarrying almost immediately upon the canal's completion. The largest operation was Benjamin Carpenter's Lockport Limestone Quarries. This firm supplied the stone for Chicago's first courthouse, and many prominent buildings in New York, including the New York State Women's Hospital, the Presbyterian Hospital, and the Lennox Library. Carpenter's stoneworkers were very busy during the canal enlargement period laying their limestone not only for the Lockport locks but for many other structures on the canal. Even today, Lockport bears the stamp of this industry. Lockport limestone is visible in the surviving flight of locks from the first enlargement, in numerous churches, homes, and especially industrial buildings, giving the built landscape a distinctive caste.[126]

It is not an exaggeration to suggest that Erie Canal engineers possessed more knowledge about manipulating flowing water than anyone else on the continent, and during the last year of construction, a great deal of that expertise was focused on Lockport. State workers quickly constructed a tunnel and race to carry water around the locks and feed more than one hundred miles of the canal to the east.[127] When Spalding and other entrepreneurs saw the amount of water falling into the natural basin below the locks, they knew what it meant. Spalding began erecting his flour mill in the spring of 1826. The seven-story mill "was looked upon as a gigantic monument of enterprise." It was the first to use Lake Erie water, but within a few years, there were numerous sawmills, flour mills, and other manufactories. The enlarged canal demanded even greater diversions from Lake Erie, and Lockport's hydraulic potential quadrupled. Over 32,899 cubic feet of water per minute was needed for the Genesee level. Moreover, a waste weir just below the basin overflowed into a branch of Eighteenmile Creek that had been diverted under the canal. The combined power of the creek and the overflow allowed for six additional mill sites, each with its own millpond, along the creek in lower Lockport. Unlike Rochester, Lockport had no river to tap for power. Its survival as a manufacturing town was entirely the result of its artificial river.

The town's access to information made it an early adapter of new technological ideas. The Lockport and Niagara Falls Railroad (1836–1850), although essentially a tributary to the canal, was a local initiative in the first wave of railroading in the United States. When Congress built an experimental telegraph line between Washington and Baltimore in 1844, Lockport again was in the first tier of adopters. As the first line extended from Baltimore through New York, reaching Boston in 1846, another line connected New York with Buffalo (following the canal). The same year, a line from Buffalo to Lockport, and then to Queenstown, Ontario, was completed; it eventually connected to Halifax.[128] The telegraph revolutionized access to information of all kinds. Newspapers were able to get nearly instantaneous news from "the wire." Like other towns of the canal corridor, Lockport was alive with Yankee tinkerers. The Burned-Over District produced not only religious and cultural innovations, but technological ones as well. Along with New England, the region led the nation in patents in the 1830s and 1840s.[129] A few miles east of Lockport, George Pullman labored at his brother's Albion cabinetmakers shop. The sleeping car he later perfected was an improvement over the packet berths but perhaps inspired by their example. Nathan Brittan provides another illustrative example. A New Englander who graduated from Brown in 1837 and became a teacher in Rochester and Lyons, he invented a "continuous copper-strip" lightning rod that was patented, successfully manufactured, and marketed from Lockport. Naturally, Brittan "was actively engaged in religious efforts and in enterprises for social improvement."[130]

There is no dispute that Birdsill Holly (1822–1894) was Lockport's most important inventor. He was a master of flowing water in all its forms. Holly grew up in Seneca Falls, another canal town, where his reputation as an inventor had already begun to grow. He conceived and perfected the rotary pump, which he immediately applied to a steam fire engine. The fact that only one-quarter of the available waterpower in Lockport was in use may have played a role in luring him there. When he moved to Lockport in 1859, he had accumulated enough capital to join six other investors to form the Holly Manufacturing Company. Among these investors were former governor Washington Hunt and Thomas T. Flagler, who served as president of the new company. Their intention was to manufacture and market a number of Holly's inventions including a sewing machine, pumps, and hydraulic machinery.[131]

Something else immediately attracted Birdsill Holly's attention. The worst of many fires in Lockport's history had devastated ten acres of churches, hotels, shops, and houses just five years before his arrival. Holly began to work on a new system of fire protection. The key invention was a self-regulating pump that responded to the pressure in the system, automatically increasing or decreasing according to demand. Waterpower propelled the pump, which sent up to two hundred pounds of pressure into a ten-inch iron pipe, which in turn forced water into a system of smaller iron pipes beneath the city streets. The pipes carried water to businesses and homes but also provided a powerful means for fighting fires. At intervals along the pipes, he needed to provide a way to attach fire hoses. To accomplish this, Holly created a frost-proof device that could be quickly opened to allow a thick stream of water to flow out of several outlets to which hoses could be attached. This was Birdsill Holly's most familiar invention; the fire hydrant. The system was complete in 1864 and the village trustees demanded a test. Holly had promised enough pressure to force water through a 100-foot hose and then shoot it one hundred feet in the air. His first test shot water over 175 feet high. Then he attached four hoses simultaneously, each shooting over one hundred feet. Finally, he attached nine hoses that were still able to throw water over the tops of the tallest buildings on Main Street.

The economic potential of Holly's system was immediately apparent to his partners. Pumping fire engines would no longer be needed, only hoses. Lockport's insurance rates fell by 50 percent. The Holly Manufacturing Company began to build and install city water systems throughout the continent. The Great Chicago Fire of 1871, and another conflagration in Boston the following year, spurred an enormous demand. By 1875 the company had already installed systems in sixty cities, including Rochester, Buffalo, Syracuse, Columbus, Youngstown, Indianapolis, Des Moines, Minneapolis, Memphis, and Atlanta, employing 275 people in Lockport and several roving teams of installers, as well as agents in Chicago and other cities.[132]

In order to take advantage of unused water power, Holly had still another idea. The company purchased a large tract on the north side of the locks and proposed a power tunnel that would receive water above the locks and conduct it one thousand feet to the factory site, and the city's waterworks (fig. 5.3). In 1864, the Lockport Hydraulic Race Company, which had consolidated all the water lease rights from the state, dug the

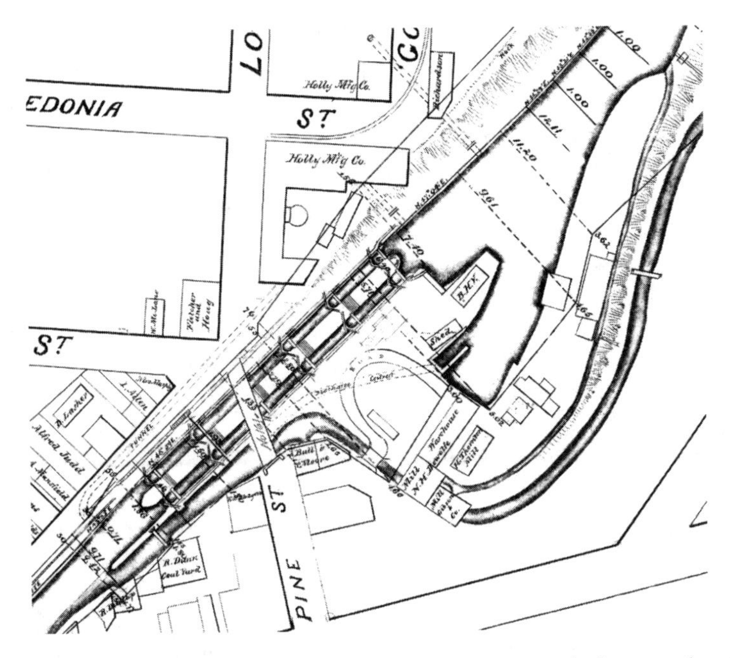

Figure 5.3. Map showing the location of Holly's power tunnel; the tunnel took in water just above the locks on the north side and delivered it to the Holly Manufacturing Company and later to several other establishments. Map of the Erie Canal at Lockport, Barge Canal sectional maps, 1896, George L. Schillner, New York State Engineer and Surveyor. Courtesy of the New York State Archives.

10.5-foot diameter tunnel through solid dolomite. By 1878, the tunnel had reached 2,643 feet, supplying power to nine different establishments, even extending power to the south side via a system of pulleys and wire cables. Since the original raceway, begun in 1825—which also passed through a short tunnel—served only industries on the south side of the canal, Holly's tunnel greatly enhanced Lockport's manufacturing capability.[133]

Birdsill Holly continued to experiment. In 1876 he began to lay five hundred feet of insulated pipe below the frost line in his own back yard and that of a neighbor and connected them to a steam boiler. When this proved successful, he ran a pipe under the street and heated a neighbor's house. He was confident that this new invention, district steam heat, would work on a larger scale. He could deliver steam to homes, schools, and businesses, just as he had delivered water. Within a year, he had created a district steam heating system for Lockport with over forty customers. By running a network of iron pipes beneath the streets, he was able to heat over one million cubic feet from a single boiler. He quickly patented his new invention, severed his relationship with his old firm, and formed the Holly Steam Combination Company, later the American District Steam Company. Once again, Holly was able to make this work because he not only had the inspiration, he also had the ability to create all the necessary valves, devices, and even meters to measure each customer's usage. Many visitors began to arrive in Lockport to investigate the system. By 1880, Holly had installed or was working on district steam systems for thirty-six North American cities including Washington, Chicago, St. Louis, Buffalo, Cleveland, Detroit, Brooklyn, New York, Toronto, and Montreal. Of greater importance perhaps, were the institutional customers—factories, asylums, sanitaria, and universities.[134]

Birdsill Holly eventually earned over 150 patents. His work came to the attention of Thomas Edison, who asked Holly to join his famous team of inventors at Menlo Park, New Jersey; but Holly, preferring to work alone, declined. [135]

Occasionally Holly's inventiveness departed from water and pipes. In 1876, he proposed a colossal 700-foot building for Goat Island, consisting of four iron towers capped with an enormous platform from which tourists would have a magnificent view of Niagara Falls. Many towers at Niagara Falls today nearly fulfill Holly's dream. Unable to finance the scheme, he traveled to New York and suggested that similar towers might relieve the city's congestion. In 1876, New York financiers could not take such an idea seriously; they dismissed him as a "farmer from the West."[136] Although students and professors at almost every university in North America are still warmed by district steam heat, and although people in cities everywhere continually pass by the humble fire hydrant, Birdsill Holly is nearly a forgotten man. On May 15, 1987, the American Society of Mechanical Engineers presented two national historic site designations in honor of Holly, recognizing Lockport as "having the first pressurized water system in 1863 and the first district steam heating system in 1877."[137]

In 1886, the Niagara, Lockport and Ontario Company secured permission from the New York State legislature to divert water via a channel from the Niagara River to Lockport, and use the drop of the escarpment there to generate an enormous amount of power. This touched off a long-standing controversy. More importantly, it spurred Thomas Evershed to make a momentous counterproposal that may have been inspired in part by the example of Holly's power tunnel (fig. 5.3). Evershed had detailed knowledge of Lockport. As a young man, he had overseen much of the lock enlargement process and had produced a rendering of the progress in 1839 (fig. 4.8). He served as a railroad and canal engineer, and after 1878, as division engineer of the Western Division of the New York Canals.[138] Evershed wrote a letter to the

Lockport *Union* on February 3, 1886, arguing that the notion of diverting water to Lockport was impractical, but then he noted: "if the people of Niagara County wish to indulge in a scheme for a magnificent waterpower, let me point one out." He proposed that a tunnel, sixteen feet in diameter, be constructed from the foot of Niagara Falls extending for one to two miles. Previously, Niagara Falls manufacturers had been supplied by a power canal that forced them to crowd near the edge of the gorge discharging their tail races into the lower river. Evershed's scheme would allow water from the upper river to be dropped over one hundred feet into the tunnel, turning machinery. "And in this way," Evershed gushed, "the line of factories and mills can be extended as far as Tonawanda [thirteen miles], if desired, and all in the County of Niagara, making it vie in population with any other county in the State of New York."[139]

Immediately a Niagara Falls manufacturer, Myron H. Kinsley, seized on the idea. Kinsley was the superintendent of the Oneida Community, the manufacturers of metal wares that had begun as the utopian experiment of John Humphrey Noyes. The refugees from the Burned-Over District were finding new enthusiasms. As Edward Dean Adams, the president of the Niagara Falls Power Company, concluded in his two-volume *Niagara Power*, Evershed's insight turned out to be one of the two keys to the momentous 1896 generation and transmission of electrical power to Buffalo, twenty-seven miles away. A. Howell Van Cleve's paean to Evershed in a 1903 address to the Buffalo Society of Natural Sciences was not atypical. Evershed "should ever be remembered" as "the man with the idea." The tunnel was "an aqueduct such as was never before built in the history of man—a conception such as could come only to a man with an imagination."[140] Twelve years after Evershed's original proposal, the electric age was about to begin. The transmissibility of electricity would render the twentieth century palpably distinct from previous eras. The second key to the harnessing of Niagara's power was Nicola Tesla's counterintuitive idea of alternating current. Evershed's experience in Lockport, Rochester, and other waterpower centers, gave him insight into the interaction of gravity, water, and rock. In his letter, Evershed mentioned the use of "turbine wheels under a head of eighty to one hundred feet, as is now done below the high falls at Rochester." Perhaps the example of Lockport's tunnel contributed to his conception. Like Niagara Falls, industrial sites in Lockport were congested. Holly's tunnel offered a solution by opening the entire north side of the locks for industry. Although the Lockport tunnel was above the drop rather than below it like Evershed's, they both allowed a string of industries to spread along a linear power tunnel. The Lockport tunnel was smaller, but still comparable at more than half the diameter and a third the length of the Niagara tunnel. Moreover, since they passed though similar rock formations, they both required long and difficult work.

Having lost a bitter battle to Niagara Falls over the right to divert Niagara River water, Lockport leaders were able to extract something in compensation; as part of the agreement to relinquish their water rights, they were guaranteed cheap electricity. Shortly after the first transmission to Buffalo, Lockport began receiving electrical power from Niagara Falls (fig. 5.4). A 1910 promotional booklet boasted that Lockport could deliver one hundred horse power for a month for only $100, while in Pittsburgh it cost $419, in Chicago $629, and even in Tonawanda $144.[141] The destinies of Niagara Falls and Lockport were still competitively intertwined. The natural wonder had been tamed by art, a force General Lafayette had first recognized on the Niagara Frontier in the artificial wonders of Lockport. Although Lockport's new power was coming though wires, it derived from the same zero-sum calculation of the St. Lawrence flowing over the Niagara Escarpment. One's loss, it seems, would always be the other's gain. At the end of the century, one could still see the trace of Lockport's inception.

As an addendum to the story of industry in Lockport, I must briefly examine the city's most successful enterprise, Harrison Radiator. Its story,

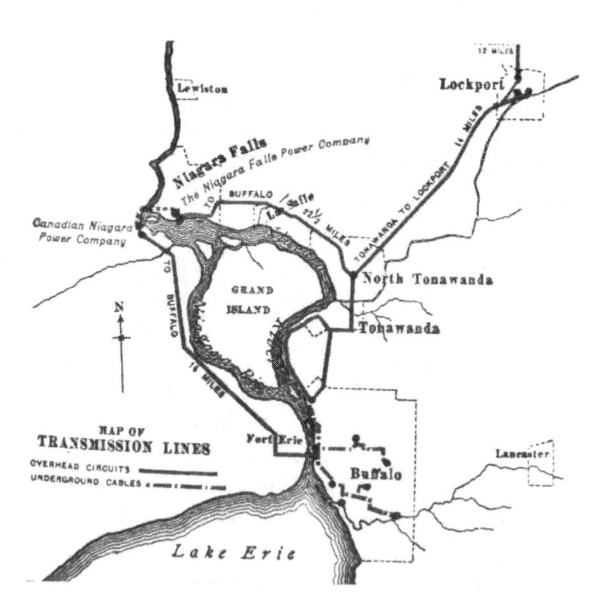

Figure 5.4. Early transmission lines from Niagara Falls; Lockport was in the first tier of cities to receive alternating current in the United States. Adapted from "Information for Visitors," a booklet published by the Niagara Falls Power Company (New York), and the Canadian Niagara Power Company (Ontario), 1910, 11. Map by Cherin Abdel Samie and Yasser Ayad.

too, emits faint echoes of 1825. Herbert C. Harrison was no ordinary immigrant when he arrived in New York in 1907. He had been born in Calcutta, graduated from Oxford, and worked as an electrochemist and as general manager of the Leeds Copper Works. Arriving in the United States, he quickly organized his own smelting company, then sold it to Union Carbide. Harrison became interested in problems holding back the infant automobile industry. No one had been able to find a solution to the inconvenient fact that internal combustion engines tended to overheat. After much experimentation, Harrison produced a "ribbon cellular" radiator featuring the "Harrison hexagon core." The honeycomb core afforded a great leap in efficiency, and it was less expensive to produce. Possibly because of cheap electricity, Harrison came to Lockport, and with the aid of some local investors, started producing the new cellular radiators. The first plant was adjacent to the locks on Richmond Avenue, formerly Canal Street. Like those locks, the new radiator was a hydraulic device. The former used fluid as a medium of transportation, the latter as a medium for dispersing heat. The thin copper hexagonal cells allowed the air—taken in through the grill of the onrushing vehicle—to mix intimately with the fluid heated by the engine's combustion, cooling it and, in turn, cooling the engine. The internal combustion engine requires an intricate balance of fire and water: portable combustion would be impossible without the hydraulic system that tames and contains it. After three years of growth, Harrison moved his company to a large stone building recently vacated by the American District Steam Company, which had relocated to Buffalo. In 1918 the plant produced 250,000 radiators and became part of the General Motors Corporation. By 1925, production had soared to over a million units, and to 2.3 million in 1929. Later they also produced automotive heaters, air conditioning units, and a host of related products. Short of space again in 1937, the company purchased a sixty-five-acre site at the western edge of Lockport where they expanded even more rapidly. In 1981, Harrison Radiator Division of GMC built its two hundred millionth radiator.[142] Although General Motors has recently spun off the company Herbert Harrison founded nearly a century ago, and it has fallen on hard times, Harrison Radiator was the mainstay of Lockport's economy for a century, sustaining Lockport's industrial character—and its class structure—into the twenty-first century.[143]

Between 1821 and 1825, the skill of engineers and the brawn of laborers made the waters of Lake Erie flow through the Mountain Ridge toward the east. The liquid streaming through the center of the town, and then through races, tunnels, waterwheels, pumps, pipes, hydrants, and finally the tiny copper cells of radiators, has continued to fertilize this artificial port, and at the same time, to perpetuate aspects of the social relations present at its inception.

Natural City

In 1865 the New York State legislature approved a charter that officially transformed the village of Lockport into the city of Lockport. Although cities are preeminent achievements of human artifice—sites where it is possible to defy nature's diurnal and seasonal cycles—many of Lockport's citizens seemed eager to veil the artifice that had built their city at the very moment of its recognition. Most interpretations of Lockport's landscape before 1850, like those of the Erie Canal in general, described nature and art as adversaries, with Lockport itself serving as a symbol of the triumph of art over nature. Sam Woodworth had proclaimed at the canal's opening, "Proud Art o'er Nature has prevailed," but gradually evidence of a struggle—to create either the canal or the town—began to disappear from memory. Recall that Holland Land Company surveyors had imposed order on their recently conquered land by laying out a grid of square townships, oriented to the cardinal directions, and numbering the creeks by distance from the Niagara River. Lockport's leaders, perhaps taking a cue from the Manhattan grid plan of 1811, also numbered the village streets. By 1845, new street names replaced numbers with trees: Pine, Locust, Elm.[144]

A local interpretation of Lockport's landscape, a poem that appeared in the *Niagara Democrat* in 1843, also elided the struggle that visitors such as Mrs. Trollope had emphasized:

> Oh! Rock enthroned! Fair daughter of the West,
> The Wilderness is gone and thou art here;
> So full of commerce, busy, bustling, gay;
> The splashing of a hundred water wheels;
> The hum of industry, of a thousand hammers;
> The sharp clink, and all the varied forms
> Of manly, wealth producing industry—that one
> May almost turn the ear incredulous
> At the great changes. Thy fathers even, they
> Who cut the proud old forest down, are here,
> Stand proudly up, almost unbent with years;
> See! Through thy streets, how Erie's waters come,
> Wafting the golden wealth of the great west.
> And dashing down thy mountain, bear away
> A tide of plenty on to Hudson's tide.
> Fair Daughter of the West,
> Thy circumjacent forest still retires,
> And o'er the golden fields which mark the toil
> Of husbandmen, rings merrily the harvest home;

> Increasing plenty pours along thy streets;
> The wilderness is gone, and thou art here.

In a manner familiar from other comments on the "immaculate conception" of Lockport, this poem contrasts the bustling urbanity of 1843 with the wilderness that existed twenty-two years earlier when the "fathers" began to "cut the proud old forest down." Although it is crowded with images of human toil, the poem's vocabulary erodes any sense of toil: waterwheels are "splashing," industry has a "hum," and commerce is "gay." The Mountain Ridge, the "Rock," is no longer an obstacle; it has become "enthroned." The waters of Lake Erie, seemingly without human effort, waft "the golden wealth of the great west" and burst through the mountain. Gone are the human costs of "dashing down" the mountain. Lockport, the emblem of art and progress, is simply, inevitably, here—precisely because "the wilderness is gone." In this poem, we can see the beginning of the pastoralization of Lockport: the town was becoming a natural thing in two senses: first, in the sense that it is inevitable rather than the contingent result of forces and decisions and that might have produced a world without it; and second, in the sense of the transfer to an urban scene of a vision of a virtuous, harmonious, rural world where the farmer "rings merrily the harvest home." This process of pastoralization accelerated at Lockport, especially after the Civil War, burying deeper and deeper all traces of human decision, struggle, and cost. Before examining these efforts to render Lockport a natural city, I will briefly explore how the same process affected perceptions of the Erie Canal and of the United States as a nation.

The notion that nature had destined a New York canal, leaving art the mere task of completing a natural design, had preexisted the Erie's construction.[145] The canal itself had helped to spur the growth of an urban middle class—"released from the toils of working the land and the frustrations of confronting nature as an unrelenting antagonist"[146]—who could travel the canal in search of natural wonders, culminating at Niagara Falls. Commenting on an extensive exhibit on the art of the Erie Canal in 1984, Patricia Anderson divided the works into three periods according to the first three paintings in Thomas Cole's series, *The Course of Empire*. The earliest paintings, representing the "Savage State," depicted an awesome sublime nature, but by midcentury artists were depicting an "Arcadian or Pastoral State." The canal itself, the triumph of "Art o'er Nature," had become "a fine thread of civilization woven into the variegated fabric of nature."[147] Images such as William Bartlett's 1840 view of Lockport looking east (fig. 4.12), as we have already seen, tended to deemphasize the artifice and effort that had overcome the Mountain Ridge. Bartlett's prints were included in N. P. Willis's *American Scenery* (1840),[148] the success of which had convinced many artists to travel west on the canal. James Crawford argues that many artists saw the canal as a conduit to nature that could transport people "away from the industrial east to an unsettled" west.[149] James Machor suggests that because the canal had spurred the growth of New York City, and invigorated the economy of the entire country, by midcentury it was "eliminating the distinctions between country and city." Because of the canal, he continues, "America was experiencing a union of the agricultural interior and the cities of the East—an urban-pastoral linkage spanning half the continent."[150]

In Carol Sheriff's words, the Erie Canal had become "second nature." In response to requests to close the canal on the Sabbath, the state had decided in 1858 that the canal was more like a lake than a road, and "just as 'no one' would call for a halt to oceangoing navigation on Sunday, so it was unthinkable to close the Erie. The emphasis in the term 'artificial river' had shifted from 'artificial' to 'river,' from man-made to natural."[151]

In the works of Hudson River writers like Irving, Cooper, and Bryant, and painters like Thomas Cole and Frederic Church, nature became entwined with nation. In his "Essay on American Scenery," Cole sought to distinguish the United States from Europe. Though Europe had a landscape marked by "heroic deeds and immortal song," he proposed that "nature has shed over *this* land beauty and magnificence . . . features, and glorious ones, unknown to Europe" (emphasis in original). The "most distinctive, and perhaps most impressive, characteristic of American scenery is its wildness."

> For those scenes of solitude from which the hand of nature has never been lifted, affect the mind with a more deep toned emotion than aught which the hand of man has touched. Amid them the consequent associations are of God the creator—they are his undefiled works, and the mind is cast into the contemplation of eternal things.

In this nationalistic vision, Americans are blessed with a direct connection to the divine via the natural landscapes surrounding them. Here was a people who could read its destiny in its landscape. Switching from the third to the first person plural, Cole then offered this paean to Niagara:

> In gazing on it we feel as though a great void had been filled in our minds—our conceptions expand—we become part of what we behold. At our feet the floods of a thousand rivers are poured out—the contents of vast inland seas. In its volume we conceive immensity; in its course, everlasting duration; in its impetuosity, uncontrollable power.[152]

The "we" of this passage clearly implies an "imagined community" that is immense, enduring, and powerful.[153] Despite his enthusiasm, Cole was alert to the contradictions of a nation that was eliminating the very natural landscape that provided its difference. He clearly preferred *The Pastoral or Arcadian State*, represented in the second painting of *The Course of Empire* (1834–1836); the final paintings in the series—showing, in turn, *The Consummation of Empire, Destruction, and Desolation*—dramatized his warning against "progress" beyond this state of balance.[154]

Why did such vigorous cultural nationalism first appear in the Hudson River Valley? The glimpse of an awesome potential via connections to the West and its "vast inland seas," may have motivated people like Cole. The Knickerbocker writers and painters emerged in conjunction with the national ascendancy of New York City, which in turn was deeply connected to the Erie Canal. They wrote and painted for a New York audience as the canal was artificially extending the Hudson—and New York's reach—into the heart of the continent. Describing the Hudson, the Connecticut journalist and essayist George William Curtis wrote that "no European river is so lordly in its bearing, none flows in such a state to the sea." He preferred it "over all the rivers of the world" because the Hudson "implies a continent behind."[155]

The conflation of canal and nature was not innocent. In the pastoral vision, the future of the nation and its destiny of expansion were justified by the millennial grandeur and beauty of the balanced landscape that was to be created. Art historian Angela Miller suggests that the pastoral served as a bastion against the social tensions of the new commercial and industrial order. It eliminated tensions by subordinating details—and differences—to a vision of the whole. It was a vision, Miller points out, that was "only accepted and understood as such by a certain aesthetically and ideologically conditioned sector of the public capable of making the leap beyond literalism."[156] Appreciating and promoting the pastoral vision marked one's status as a member of the Northern middle class: nationalism, social distinction, and aesthetic reverie converged in an anxious imperative.

A group of prominent Lockport citizens, in the midst of the Civil War, initiated a project that would demonstrate their own "leap beyond literalism" by creating a romantic pastoral landscape, just as the state was about to recognize Lockport as a city. They called it Glenwood Cemetery. After forming an association, they purchased a site north of the business district, halfway down the escarpment, along a road that had replaced the path of the old Lockport and Niagara Falls Railroad (fig. 5.5). An 1878 county history suggested that "the grounds seemed to have been formed in all their picturesqueness by some convulsion or upheaval from beneath, or a violent sundering of masses of earth from the brow of the mountain, to shape the hills and valleys with which they are broken, and which afford an opportunity for the display of improvement." The grounds were in fact part of a glacial moraine. The association hired Frederick E. Knight, "the

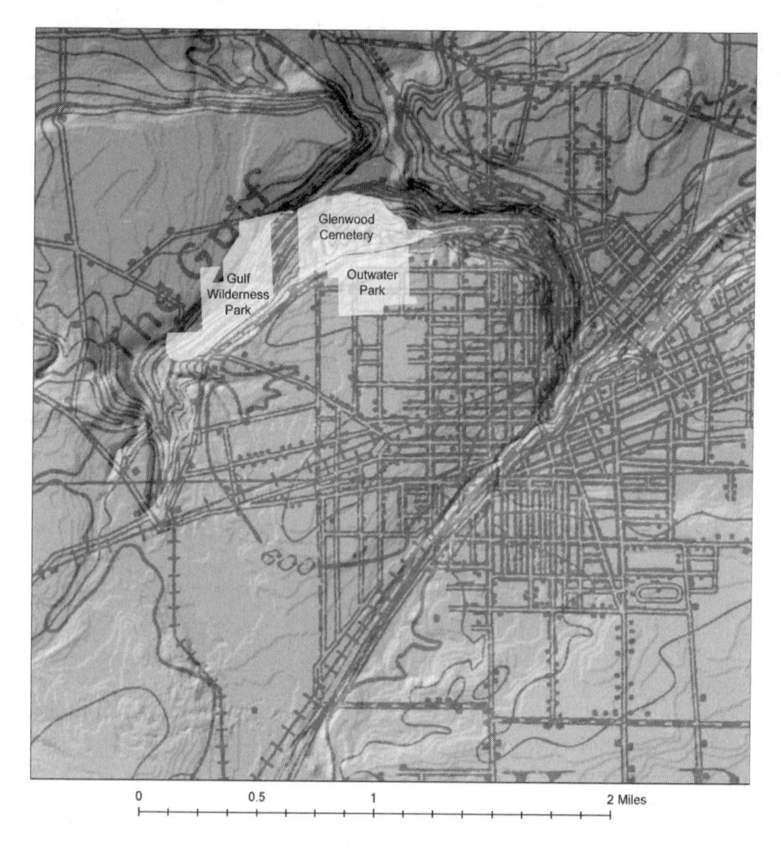

Figure 5.5. Map indicating location of Lockport parks superimposed on a topographic map and a digital elevation model (United States Geological Survey). Map by Cherin Abdel Samie and Yasser Ayad.

accomplished civil engineer who laid out the Central Park grounds in New York" to design the cemetery's plan. Like Frederick Law Olmsted and Calvert Vaux's Central Park, Knight's Glenwood Cemetery featured curvilinear patterns of roads and pathways (fig. 5.6). Central Park and Glenwood Cemetery were naturalistic landscapes meant to form an antithesis to and a respite from the grid of city streets. Yet in their design and construction, the hand of artifice was everywhere. Knight planned subsurface drainage and water systems for Glenwood and even provided fire hydrants to protect the necropolis from the perils of conflagration.[157]

The success of Boston's Mount Auburn Cemetery (1831) had helped to inspire the urban parks movement. It was a new kind of burial ground,

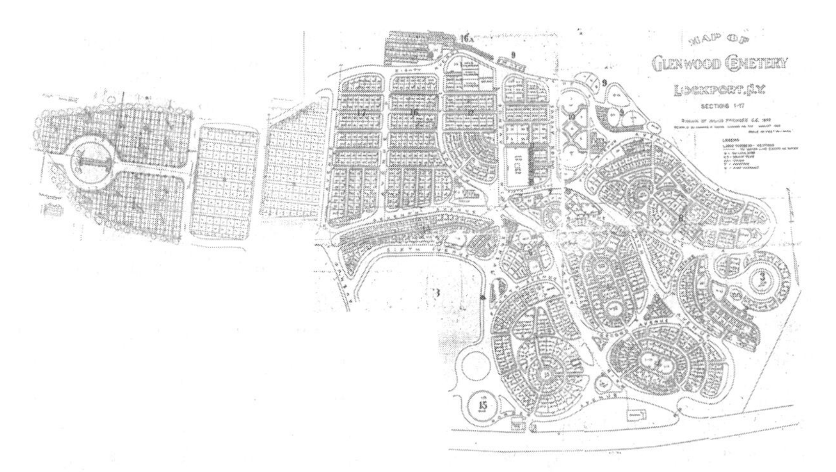

Figure 5.6. *Map of Glenwood Cemetery, Lockport, N.Y.*, drawn by Julius Frehsee C.E., 1892. Courtesy of the Niagara County Historical Society.

located at the edge of the rapidly growing city and designed to offer the weary bourgeois visitor a green and picturesque venue for picnics and leisurely carriage rides, a form of environmental art that resonated with the literary and graphic creations of Cooper and Cole. Andrew Jackson Downing, along with Vaux and Olmsted, gradually began to apply this idea to urban parks. Like Cole, Olmsted believed that natural beauty could actually improve people. If sublime surroundings could lead Americans "into the contemplation of eternal things," why not bring nature into the heart of the city? He hoped that Central Park would help to civilize the throngs of immigrants then crowding into New York, instilling within them the aesthetic predilections of the new bourgeoisie. Later, Olmsted began to design curvilinear suburbs where the bourgeoisie could at least escape the urban masses in their own parklike districts.[158]

A newspaper article that appeared in 1864 as Glenwood was nearing completion described the aspirations of the cemetery's promoters. The site presented "those natural undulations and variety of surface necessary to make it a beautiful and desirable resort for our people—where the busy struggling world can commune with nature, and be recalled by the associations of the place, if even for an hour, to considerations of more importance." Glenwood was "a most romantic place" planned to give "breadth and shadow to the landscape," rendering "this spot the chief attraction of the city and neighborhood." It was a credit "to this people, that whilst busily engaged in manufacturing and other enterprises necessary for the living, . . . that they have not forgotten what is due to the dead," but "are preparing a last resting place" that "will rob death of half its terrors."[159] It is clear that "this people" saw Glenwood as a bastion against the struggles and tensions of their urban industrial world. They believed "natural" beauty had the power to erase these tensions: it could even remove the terror of death.

Glenwood was officially open to corpses of all classes, sects, and religions; there was even a small area set aside for Jewish burials. Unofficially, it replaced the Protestant Cold Spring Cemetery that had become full. The cost of burial plots, moreover, acted as a class filter. For the members of the association, the creation of Glenwood was a social gesture that demonstrated their elite taste. The numerous massive monuments were also displays of social status. An 1897 article described "the huge granite and bronze shaft of the late Governor Hunt," the Van Vleet monument with its "32 ft granite shaft obelisk," and the Gridley monument, "a huge Grecian dome supported by four granite pillars." The visitor could not help but notice "the many modern and costly monuments" that "instantly impress one with the fact that here lies the remains of people who in life were high in the social and financial world."[160]

In the decades following the Civil War, urban leaders throughout the United States replicated the Glenwood Cemetery Association's gesture, creating parks, parkways, and cemeteries, often with equally sylvan names. Although these open spaces have often become cherished by all classes, this does not change the fact that they were elite responses to profound social and economic tensions.

Contractor Daniel Price once proposed an Olmstedian park system for Lockport with a "Central Park" near Main Street. In 1920, Dr. Samuel Outwater donated over one hundred acres for a park on the brow of the escarpment in Lockport's North End (fig. 5.5). By this time, however, the meaning and purpose of parks had evolved in a more democratic direction; sports and other active uses began to overshadow the contemplation of natural beauty.[161] In 1972, Lockport opened Gulf Wilderness Park, one of the most unusual urban parks in the state (fig. 5.5). Three years after the first Earth Day, the appreciation of nature had become more democratized. The park occupied a portion of the Lockport Gulf that had survived in a wild enough state to attract hikers interested in wildflowers, shrubs, trees.[162]

Although not directly related to parks or other elements of Lockport's landscape, pastoral rhetoric surged in 1910 when the city secured preferential electrical rates. A publication aimed at industrial investors began with this paean:

> A few words concerning a city set on a hill: Where the devastating disturbances of the elements are strangers, but the grateful gifts of

> bounteous orchards are ever-present guests; where the wheels of industry whir under the tireless and exhaustless energy of Niagara's electric current, undimmed by the smoky clouds which begrime other manufacturing centers.[163]

The year before, Lockport had begun to draw its water supply via a pipeline directly from the Niagara River, and now the falls itself was sending clean power to Lockport through wires. Another echo of this pastoral boosterism appeared at the 1925 celebration of the canal's centennial. The cover illustration of the commemorative booklet combined the canal with trees and smokestacks (fig. 5.7).[164] The pastoral canal leads directly to Lockport's pastoral smokestacks with all contradictions thoroughly aestheticized.

Perhaps the most intriguing instance of Lockport's pastoralization is the 1883 enthusiasm to explore and develop the legendary Lockport Cave. The cavern had been well known since one of the original Quaker settlers, John Comstock, built his log home "near the mouth of the cave" in 1816, but it had attracted very little notice until the 1850s.[165] The cave entrance

Figure 5.7. Collage of Lockport scenes that seamlessly blends the canal, smokestacks, and trees; cover illustration form George R. Worley, "Old Home Week: Celebrating the Hundredth Anniversary of the Erie Canal," Lockport Old Home Week Committee, Lockport, New York, July 19–25, 1925.

was located in the west bank of the branch of Eighteenmile Creek, just above the water line, east of Lockport's center where Cave Street intersects the old Indian trail that became Main Street (fig. 5.8). Dr. A. Leonard and a companion provided the first published account of exploration in 1858. As they cleared the cave's entrance of debris, a number of young boys gathered about them and related various legends including one that "the cave came out near Niagara Falls because the 'constable had chased a man clear through.' " The explorers crawled through tiny, watery passages, down several subterranean cataracts, frequently encountering pottery and other debris carried by freshets, until they arrived at a large limb blocking their way. "The sound of another waterfall," they reported, "is heard echoing along the dark throat of the passage beyond." A few days later, another party managed to remove the limb and penetrate to a bifurcation with "one branch running in the direction of the Gulf underneath the railroad bridge [the natural basin below the locks] and the other crossing diagonally underneath Main Street, in the direction of the locks."[166] A key element in the legend of the Lockport Cave is that it underlies the very center of the town, a natural counterpart to the locks. The following year, another

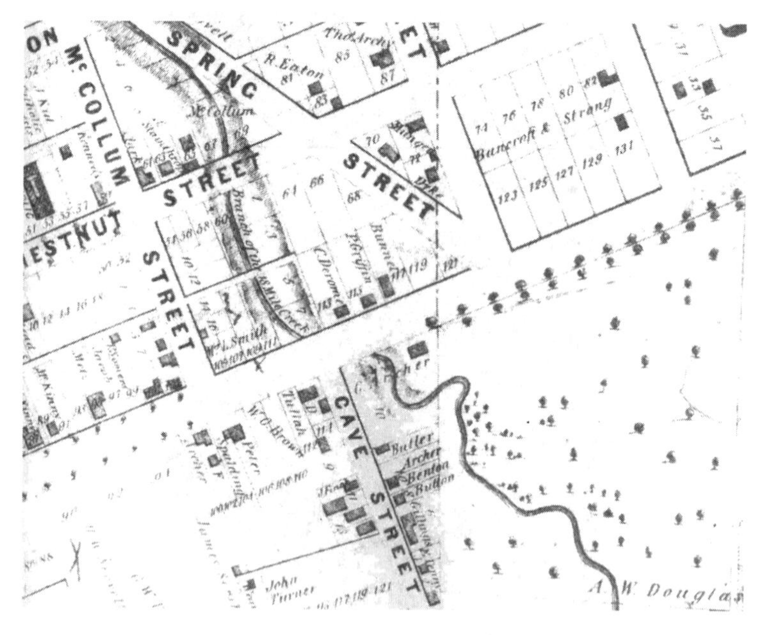

Figure 5.8. The entrance to the Lockport Cave was adjacent to Cave Street, on the west side of the creek just south of where it disappears beneath Main Street; "Map of the Village of Lockport, Niagara Co. N. Y.," surveyed and drawn by Bernard Callan, C.E. (New York: M. Dripps, 1851). Courtesy of Lockport City Hall.

exploration account described a torturous journey, "kneeing and elbowing and shinning and crawling over, under and between sharp rocks." One of the two male adventurers was a reporter for the Lockport *Chronicle*. They encountered several small chambers where it was possible to at least sit up, and reported hearing "the rumbling noise of carriages passing over us." The explorers panicked when their candle went out, but eventually emerged "at starlight after having spent six hours under ground."[167]

In February of 1883 cave fever erupted among some of Lockport's most prominent citizens. The spark was a February 1 lecture on the geology of Lockport. The following day, a man who lived near the cave entrance claimed he had entered it on several occasions to a point where "it separates into three branches." A February 3 article titled "That Questionable Cave" reported that "quite a little band of explorers issued forth from the Opera House yesterday to prosecute further the investigation as to the opening of the cave, which has gained through the newspapers a considerable notoriety for Lockport of late." Four days later, a group of prominent men formed the Lockport Cave Company at the Opera House to explore what the *Chronicle* referred to as "Lockport's alleged cave." The company quickly issued $100,000 in stock, mostly to local investors. Dr. Simeon T. Clarke, the company president, believed the cave would "rival the Pyramids," attracting tourists from the whole world to see "the eighth wonder."[168] There was a wonder of nature beneath the center of Lockport—the city with a giant mechanical device at its heart, born of a struggle with "nature's strongest barrier," the embodiment of American artifice, possessed by what Mrs. Trollope called the "demon of machinery." The natural landscape was no longer an adversary: it had become the town's salvation. In the ensuing days, explorers penetrated hundreds of feet into the cave's narrow passages and installed electric lights. When an "unparalleled flood from melting snow and heavy rains" arrived in early March, filling the cave with debris and silt, the Lockport Cave Company dissolved.[169] The *Daily Journal* reported three years later that the cave's "gaping jaws were hermetically sealed, thus shutting out forever the possibility of returning wondrous riches which this cave was supposed to possess." A culvert now conducted the creek beneath Main Street, and the ravine—filled with earth—disappeared along with the cave's entrance.

No longer accessible, the Lockport Cave persisted and waxed as a legend. A 1921 article mentioned "numerous stories as to the cave being used during the anti-Slavery days as the hiding place for escaped slaves on their way to Canada, and other stories to the effect that a band of desperadoes made the cave their rendevouz [*sic*] for a number of years. Still other stories deal with Indians." Niagara County historian Clarence O. Lewis renewed interest by writing a series of articles beginning in 1950. Like the 1883 enthusiasts, mid-twentieth-century promoters assumed the colossal cave of legend was a fact. Indeed, it seemed as if a great many others on the Niagara Frontier wanted to believe it too. A 1958 article in the Buffalo *Evening News* reported that Lockport officials were "considering a proposal to allow scientists from the Buffalo Museum of Science to explore the immense cave that honeycombs the city." The project "could mean a great new tourist attraction for the city and the region as a whole, and offer an appropriate counterpoint to the nearby magnificence of Niagara Falls." A year earlier, Niagara County Civil Defense director Roger Winner had sent letters to Albany and Washington pleading that the cave would make an ideal bomb shelter and claiming it "could accommodate most of Niagara County's 200,000 population." The discourse on the Lockport Cave was now responding to the new tensions of the atomic age and the Cold War; the city's natural landscape was to offer a literal bastion. Nonetheless, it was still a discourse about the city's distinction, and Niagara Falls still provided the counterpoint.[170]

"There is no file in our office consulted more by young and old alike," the Niagara County Historian's Office revealed in 1966, "as our voluminous file on the Lockport Cave."[171] Fifteen-year-old Thomas P. Callahan opened the next epidemic of cave fever in 1969. With a group of young friends, he brought a petition before the city's Common Council to encourage exploration and development of the lost cave. What made the project "one of the most unusual in the world," Callahan suggested, "is that the caves lie directly underneath the main business district of the city." Mayor Rollin T. Grant called it "sort of a romantic idea . . . something we have all thought about." "If we can explore the moon," Alderman Benjamin Donner suggested, "the least we can do is allow exploration of the Lockport Cave." People began to contact Callahan with stories of cave entrances and holes in cellars. One man remembered two passageways under the Erie Canal Big Bridge where he used to play as a boy. Callahan speculated that "the cave may run under the Erie Barge Canal, and possibly to the Lockport Gulf." Memory and imagination rendered the line between natural and artificial passages blurry.[172]

In mid-January 1970, Callahan found a small lateral cave that intersected the tunnel now housing the creek north of the cave entrance explored in 1883. It was a tortuous little passage, averaging about eighteen inches in height, that extended about five-hundred feet westward before pinching off.[173] For several months, a rival group of spelunkers had been searching for the Lockport Cave. On April 7, a Lockport *Union-Sun & Journal* headline announced "Long Lost Caverns Located by Searchers." James Boles Jr., a recently discharged Army veteran, discovered a way into "the maze of tunnels that makes up the Lockport Cave." He led a news team into the passages to a room, ten feet high and eight feet wide, decorated with draperies and stalactites. Boles kept the entrance a secret. The newspaper later called this "the most significant discovery."[174] The Lockport Caves were formed during the period when the waters of Lake Tonawanda had poured over the escarpment carving the two gorges or gulfs. The channel that followed the present path of the canal had provided the erosional power to create small subsurface channels, some of which extended to the ravine near Cave Street. The duration of this flow was not sufficient to erode a large cavern in the soluble, but resistant dolomite.[175]

Despite the rediscovery of the Lockport Caves—Boles suggested the plural form—there was no colossal cavern, no "eighth wonder of the world," no commercially exploitable salvation beneath Lockport's streets. Grasping this deflating fact, Thomas Callahan proposed in 1971 "to construct the world's largest man-made cave to be located in the area of the famed locks."[176] If you cannot find the cave of legend, perhaps you can create it.

James Boles and his team of spelunkers made one other "discovery" in 1970. They led a news team who announced that a "third cave, this one man-made, has been discovered beneath the City of Lockport, a huge, water-filled cavern more than a half-mile long."[177] Boles had rediscovered Birdsill Holly's hydraulic power tunnel, abandoned for seventy years, and nearly forgotten. Here was the large artificial cave Callahan had dreamed of, adjacent to the locks, but on the opposite bank of the canal. Five years after the "discovery," Thomas Callahan created the Lockport Hydraulic Race Company and began conducting tours of the tunnel, billed as "Lockport Cave Raceway Tour." Because of its age, and the amount of water it had accommodated, the tunnel had begun to resemble a natural cave. The company's brochure pointed out that "the cave features Stalactites, Flowstone, Rock and Geologic formations." The tours lasted only until 1980, but were reopened by Callahan in 1996 under the name "Lockport Cave & Underground Boat Ride."[178] Are these obvious references to the legendary cave, and the remarkable blurring of the natural and the artificial, merely postmodern marketing strategies? What has been lost and what found? What remains hidden? The final section explores a forgotten part of Lockport that might have been its very center.

The Gulf

The 1843 poem "Oh! Rock!" quoted earlier, repeats the line: "The wilderness is gone and thou art here." Although the poem renders the "dashing down" of the mountain as a destined—almost natural—process, it retains a vision of Lockport as a counterpart to wilderness. But was the wilderness completely gone? And if it was gone, where did it go? In the discourse of travelers and local elites concerned with defining Lockport, the town's wild counterpart was clearly Niagara Falls. One mile west of the locks, however, was another place many described as wild: the Lockport Gulf. As late as 1904, a Buffalo newspaper claimed that the area north and west of the city center looked "as though it had never been molested by man."[179]

North of the Mountain Ridge was a prehistoric delta that had burst through the even more ancient beach that became the Ridge Road. During the War of 1812, a corduroy road temporarily spanned the swampy district. A Tuscarora chief, in an 1848 interview, remembered taking "salmon from the Eighteenmile Creek, where the Lewiston Road crosses near Lockport, with my hands, three feet long." The Senecas from Tonawanda Creek had also made a yearly trip to the Eighteenmile Creek for fishing.[180] Silas Hopkins, a fur trader who lived in Lewiston in the late 1780s, bought beaver, otter, muskrat, and mink from Tuscaroras who hunted in "the marshes along the Ridge Road," where they killed at least one panther.[181] When canal construction began to attract settlers and laborers, most came via the Ridge Road and were forced to pass through a "howling wilderness" from Wright's Corners.[182] Crossing the ancient delta continued to annoy stage travelers for many decades. A road running north through the delta connected Lockport with the Ridge Road in 1823. While the delta area was gradually tamed, the rugged, swampy, and densely forested western valley of the Eighteenmile Creek—in local parlance, simply the Gulf—remained remarkably wild. It was as if the wild connotations originally associated with the entire area below the escarpment now became concentrated there. Those connotations, on the one hand, could be desolation and danger: when early villagers worked to eradicate rattlesnakes, they proceeded from the natural basin north and west, leaving the Gulf as the snakes' last stronghold. On the other hand, the connotations could be romantic, as in Edna Smith's retrospective description of the natural basin in 1821, when presumably it closely resembled the head of the western valley. At its center was a pond "fed by numerous springs and rivulets flowing from the hillside"; the slopes of "the surrounding precipices were covered with trees and shrubs, mossy rocks, ferns and wild flowers," and the "autumnal brilliance of crimson and gold . . . made all the hillside like a gorgeous carpet."[183]

Although the Gulf itself continued to resemble the 1821 natural basin for most of its history, a certain amount of human activity has always clustered around its head. Orasmus Turner, the editor of Lockport's first newspaper, wrote that "at the head of a deep gorge, a mile west of Lockport (similar to the one that forms the natural canal basin, from which the combined Locks ascend), in the early settlement of the country, a circular raised work, or ring-fort, could be distinctly traced. Leading from the enclosed area, there had been a covered way to a spring of pure cold water that issues from a fissure in the rock, some 50 to 60 feet down the declivity."[184] Holland Land

Company surveyors sketched the contours of the Gulf's box canyon (fig. 5.9), and Geddes's survey of 1816 that routed the canal through the western gorge depicted a dwelling labeled "O. Comstock" at the northern finger of the Gulf, one of only two dwellings depicted in the vicinity (fig. 5.10).[185]

Had the Geddes route remained unchanged, the locks would have climbed the Mountain Ridge at the head of the Gulf, where presumably a town, perhaps called Lockport, would have arisen. Because rock excavation in the Deep Cut would have been considerably reduced, such a counterfactual Lockport might have hosted a canal celebration in 1824 instead of 1825.

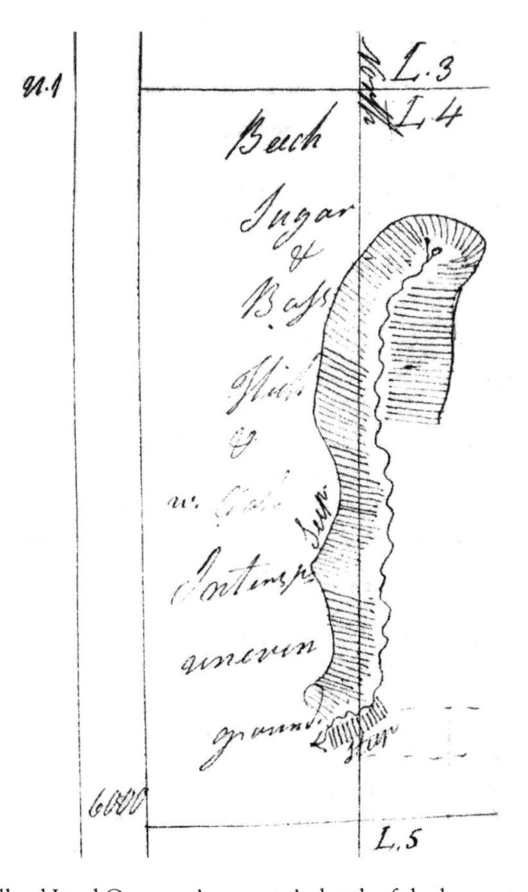

Figure 5.9. Holland Land Company's surveyor's sketch of the box canyon in the Gulf; Field Books from Holland Land Company maps, field notes, and deeds, 1788–1899; these surveys were used to prepare the Holland Land Company Townships Maps (1800–1819), the Mountain Ridge site is on the maps of T.14.R.7. and T.14.R.6. Courtesy of the New York State Archives.

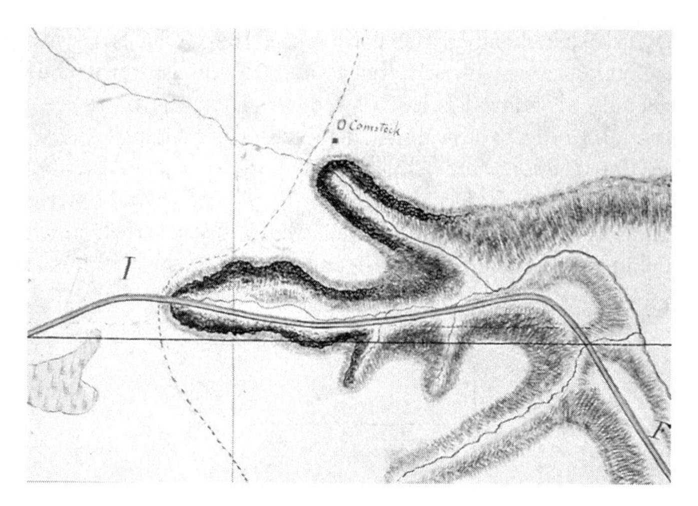

Figure 5.10. Detail from Geddes's 1817 map showing the Gulf. From James Geddes, Original maps of surveys for the Erie Canal, 1817. Courtesy of the New York State Archives.

The dwelling Geddes marked "O. Comstock" was situated at the edge of the Gulf's western finger where a small stream poured into a box canyon forming Indian Falls. This was the most impressive of three cataracts that plunged over the Mountain Ridge in 1821 near the future site of Lockport; the others were in the two other branches of the same creek—one that flowed past the cave entrance, and the other that followed the path of the canal and plunged near the present site of the locks. Indian Falls is presumably the one marked on many precanal maps such as Eddy's 1818 map of New York State, where it is the only named feature in the vicinity.[186] Its name may refer to the builders of the circular fort, but to nineteenth-century Euro-Americans the word "Indian" coincided with the site's wild reputation.

Indian Falls was well known enough to appear on maps partly because it was one of the first sites precanal settlers found useful. Perhaps as early as 1810, Jabez Pomeroy employed the creek's waters to operate a carding mill. The 1878 *Illustrated History of Niagara County* mentioned that in 1818 Pomeroy was still in "the business of cloth dressing at the head of the Gulf."[187] The Lockport and Niagara Falls Railroad (1836–1850) passed just south of the Gulf, as did the New York Central after 1851. Like cloth dressing, a number of industries that residents probably would have considered noxious, gravitated to the Gulf in the nineteenth century. In 1867, Seth M. Bornea announced the completion of "one of Page's Perpetual Lime Kilns at the Gulf, one mile west of the City of Lockport," and offered "to furnish any quantity of lime of a superior quality, either at the kiln or delivered in any part of the city."[188] A rare image of the district in the *Illustrated History* shows the Bornea Lime Works atop the ravine with a New York Central locomotive pulling cars through the landscape (fig. 5.11). The 1875 *Atlas of Niagara and Orleans Counties, New York,* also shows the Akron Water Lime works at the site. There was enough activity in the region for Lockport to establish the "Gulf School" near the kilns.[189]

In the mid–nineteenth century, Luman H. Nicholls and John St. John began to market water from one of the mineral springs in the Gulf and to promote their site as a health resort. A May 1862 advertisement proclaimed that "Spring has come, and excursionists are talking of country resorts, watering places etc." The Lockport Mineral Spring "is most romantically located, and the water for mineral and medical properties, will compare favorably with that of the most noted watering places in the world."[190] This venture, which began only a few years before the construction of Glenwood Cemetery, shows that Lockport's wild gorge was evolving, in the minds of the promoters of the Lockport Mineral Spring at least, from the abode of rattlesnakes to a potential haunt of elite "excursionists."

While in one counterfactual scenario, Lockport itself would have developed at the western Gulf, in two other scenarios, the city's center of gravity might have shifted to the Gulf and perhaps its size expanded to metropolitan proportions. Both of these scenarios involve canals, one for ships, the other for power.

There have been at least sixteen attempts to create a ship canal between the Niagara River and Lake Ontario through Niagara County. Since Upper Canada's Welland Canal first connected the two lakes in 1829, proposals for an all–American canal arose from arguments that the Welland was inadequate or—especially during times of international tension such as 1836 and 1861—that U.S. shipping was in peril from British naval power. There were many possible routes for such a channel: from the Niagara River to Lewiston or to any number of points along the Ontario shore. Routing a ship canal through Lockport's western Gulf shared an advantage with the canal

Figure 5.11. Sketch of Bornea Lime Works at the head of the Gulf; from *Illustrated History of Niagara County, New York* (New York: Sanford and Company, 1878), 148.

itself: a preexisting gorge already provided part of the channel. Moreover, the valley of Eighteenmile Creek could be followed northward to the lake. Ship canal proposals in 1853–1854 and 1862 listed several possible routes, but the Gulf route was most secure because it was farthest from the border. The latter proposal came during the Civil War when President Lincoln was concerned about British intervention on the part of the South. He supported the construction of a canal around the obstruction of Niagara to get U.S. gunboats into the upper Great Lakes. Congress refused Lincoln's request, but the possibility of a ship canal through the Gulf lingered, appearing again in 1888, 1892, 1896, 1916, 1918, 1926, and 1960. Had any of these schemes been implemented, the Gulf, and Lockport, might have developed into very different places.[191]

The idea of constructing a power canal through the Gulf arose only once, but it came much closer to fruition than any of the ship canal schemes, and—in the eyes of its supporters at least—the potential impact would have been far greater. Alexander Holley first proposed the idea in 1886. In fact, it was to Holley's proposal that Thomas Evershed was responding when he outlined, in a visionary letter to a Lockport newspaper, the power tunnel for Niagara Falls. Nonetheless, Holley's proposal—backed by a group of Lockport capitalists called the Niagara, Lockport and Ontario Power Company—gained a state charter to divert an unlimited amount of water from the Niagara River for power generation at the Gulf.[192] The Niagara Falls Power Company was able more quickly to attract financial backing from New York investors. So, while the Lockport group was able to successfully renew their charter in 1890 and 1894, the Niagara Falls project came to fruition in 1896. Under the provisions of the renewed charter, all diversion rights would be forfeited if work was not begun by May 21, 1904. As the fateful date approached, there were rumors that important investors were taking an interest. The immediate concern was to secure another extension.[193]

A heated public debate developed, with supporters and opponents of the proposal vying to influence state legislators. The March hearings in Albany attracted a number of prominent speakers, including Charles M. Dow, the president of the Niagara Falls Preservation Commission. Opponents warned that additional water diversions would destroy the beauty of Niagara Falls, and in particular, the American Falls whose channel was not as deep as that of the Horseshoe Falls. Supporters countered that the Niagara Falls Power Company was attempting to create a monopoly and destroy competition.[194] Once again Lockport was pitted against Niagara Falls in the zero-sum game of water flowing over the escarpment. Almost everyone was surprised when the Niagara, Lockport and Ontario Power Company announced on April 8 that a contract had been signed to finance and start building the power canal at once. The source of the sudden influx of capital remained a secret, and the company stressed that it was not abandoning "all hope of passage of the renewal bill." The *Buffalo Courier* reported that the Westinghouse Company was interested in the project that would make "Lockport a great manufacturing center."[195] Work on the canal commenced a few days later. One day after the announcement of the contract, the company received more good news: the renewal bill had passed the State Senate by a three-to-one majority and had "gone to the governor for his signature."[196]

Now attention finally turned to the site where the power was to be generated. The *Buffalo Enquirer* noted that "anyone who rides into Lockport on a New York Central train from the west may observe as the train approaches the western limits of the city a deep gulf or a natural river bed which terminates abruptly within a few feet of the railroad tracks." A short distance north of the tracks, "are the Indian Falls, and at this point the Lockport Ontario power canal finds its outlet, and there is where great water power will be obtained." Although the place was "picturesque," the *Enquirer* pointed out that it was only "in the flood seasons of the year that any considerable water is seen there." This "precipitous ledge" had another "natural" purpose.

> That it is the natural outlet for water brought from the Niagara River is at once evident and the possibilities that exist for making power there and turning the gorge into a series of mill sites are so extensive that no one at this time do aught but forecast a great future for all that undeveloped territory and in imagination see great factories looming up where now nothing of an industrial order exists.[197]

One has to wonder how long the antique topographical designation—turned proper noun—"Gulf" would have lasted in such a domain of urban artifice. Marcus Ruthenberg, a Lockport industrialist, had recently set up the first electromagnetic furnace for steel making. He pointed out that the location of steel production was more tied to energy sources than to the location of iron ore. He believed that electricity would soon replace coal and was therefore prepared to contract for a large amount of electricity at Lockport in order to "immediately establish an immense steel manufactory here."[198] Finally, the *Lockport Journal* relayed a report from Cleveland that "the Vanderbilt-Andrews interests were involved heavily in the project to build the power canal to this city and construct here the largest and costliest power generating plant in the whole world." [199]

All of this buoyancy may have been masking a certain panic among Lockport's leaders about their city's prospects and its civic identity. Twelve days prior to the state Senate's passage of the renewal bill, the *Buffalo Courier* stated that "Mayor William H. Baker had no hesitancy in saying that it will be a dark day for Lockport if the extension is refused." Calling such an outcome "a hard blow for Lockport," the mayor elaborated:

> We have a good city here. . . . What we need is some special advantage, like cheap power. The proposed canal would give it to us. It would mean that the city was to be born anew and to enter upon a future compared to which the past is insignificant, great as that has been. I hesitate to say what our future will be without the canal.[200]

Governor Odell soon dealt Lockport exactly the "hard blow" Mayor Baker had feared: he refused to sign the renewal bill.[201] Lockport did manage to trade its water rights for some inexpensive electricity, but the city's metropolitan dreams could only be bitterly repressed.

The article reporting the mayor's apprehensions carried a photograph of the Gulf covered in snow. The caption read " 'The Gulf'—Panoramic View of a Picturesque Section of Lockport's Proposed Power Canal. If the Canal is Constructed This Great Ravine, Two Miles Long and a Mile From Lockport's City Line, Will Be Used to Connect the Excavated Portion With Eighteen-Mile Creek—Otherwise It May Some Day Become a Pretty Part of Lockport's Park System." Figure 5.5 shows that, a few decades latter, a green necklace of parks and cemeteries nearly completed the arc around the Mountain Ridge as if poised for a culmination at Indian Falls, but other things were in store. The 1904 "dark day" of disillusionment must have made many Lockportians eager to forget about the Gulf, or worse.

Alexis Muller Jr., writing an official history for Lockport's celebration of the "150th Anniversary of the Grand Erie Canal," surprised many of his neighbors when he broke from the normal mode of local history and instead offered a jeremiad. He reminisced about playing as a youth at the twenty-five-acre Boy Scout Camp that later became Hickory Park in the middle of the Lockport Gulf. A small dam in the creek created a pool for swimming. Even for adults, it had been a pleasant place for a nature walk. "That all stopped dead," he lamented, when Harrison Radiator opened its new West Plant in 1937 and

> began dumping raw sewage into the creek at the head of the gulf. Soon the whole bottom of the valley was covered with a greasy slimy mess which killed vegetation, ruined widely known sources of fossils studied by neighboring college students and totally polluted the creek. The city of Lockport, itself added to this destruction when it used open areas of the gulf as a city dump. Why? Simply because this beautiful and almost unspoiled natural valley offered a cheap and easy spot to dump a truck full of junk—junk which was a breeding ground for mammoth rats—junk which inevitably caught fire, burned for days on end, smelled to high heaven and polluted the air in all directions.[202]

The notion that the Gulf was "beautiful and almost unspoiled" was partly a continuation of the old notion that the Gulf was the wild counterpart to Lockport's center. Throwing trash into Lockport's ravines, caves, and watercourses was not a new problem. The 1865 city charter contained specific ordinances to "preserve the course of Eighteen-mile creek in and through the city" and "to prevent the casting therein of any dead animal, offal, filth or any foul or offensive substance or thing, or any earth, stones or rubbish of any kind." An ordinance on "Noxious trades" warned: "nor shall any offensive or deleterious waste substance, gas, tar, sludge, refuse or injurious matters be permitted to accumulate upon the premises or be thrown or allowed to run in any public waters, stream, watercourse, street or public place."[203] The existence of these ordinances indicates that such practices were already common enough to merit attention. As the city grew, the rationalization of disposal meant that trash had to be concentrated in some place. The Gulf dump, it seems, was the "natural" extension of the previous informal practice of simply using the nearest ravine.

The industries clustered at the head of the Gulf in the nineteenth century certainly used the valley for their industrial waste, but it was during the twentieth century that the place was systematically exploited as a garbage dump. Even before the construction of Harrison's West Plant, the city began to use a site adjacent to the Gulf's easterly ravine for trash disposal. A 1983 environmental study found that this "original landfill was constructed on the edge of the escarpment with filling proceeding into the ravine" with the "base of the landfill" extending into the Gulf.[204] In the early 1950s, the city established a new dump north of the original one, but again on the edge of the escarpment (fig. 5.12).

There were a few dissenting voices. Lockport's Planning and Zoning Commission argued in the late 1940s that the entire Gulf was "ideal for a rustic park."[205] The most energetic defender of the Gulf's natural purity during the 1950s and 1960s was County Historian Clarence O. Lewis. He proclaimed in 1961 that "Lockportians for 100 years have dreamed of the Lockport Gulf as a park area." Much of the Gulf, he admitted, had been "ruined by two city dump sites and the water flowing through the winding creek looks more like milk than water yet there are long stretches of the gulf still in its natural state and the stream could be alleviated." Later that year Lewis complained that this natural "paradise" had been made into a "public dump," and while "every city has to have such an unsavory place," plans "can still be made to preserve the lower half of 'The Gulf' for a park." Lewis, in full Jeremiah mode by 1963, continued to rail against "cutting down all the trees and covering the area with garbage." He noted that a small ski slope had been created in the Gulf, a harbinger he hoped of things to come "if pollution of the brook which meanders through the gulf could be stopped and perhaps a little more water from some source diverted into this section of the creek." Later in 1963, Lewis pleaded that "no one should be allowed to dump industrial waste" and the "garbage disposal area should not

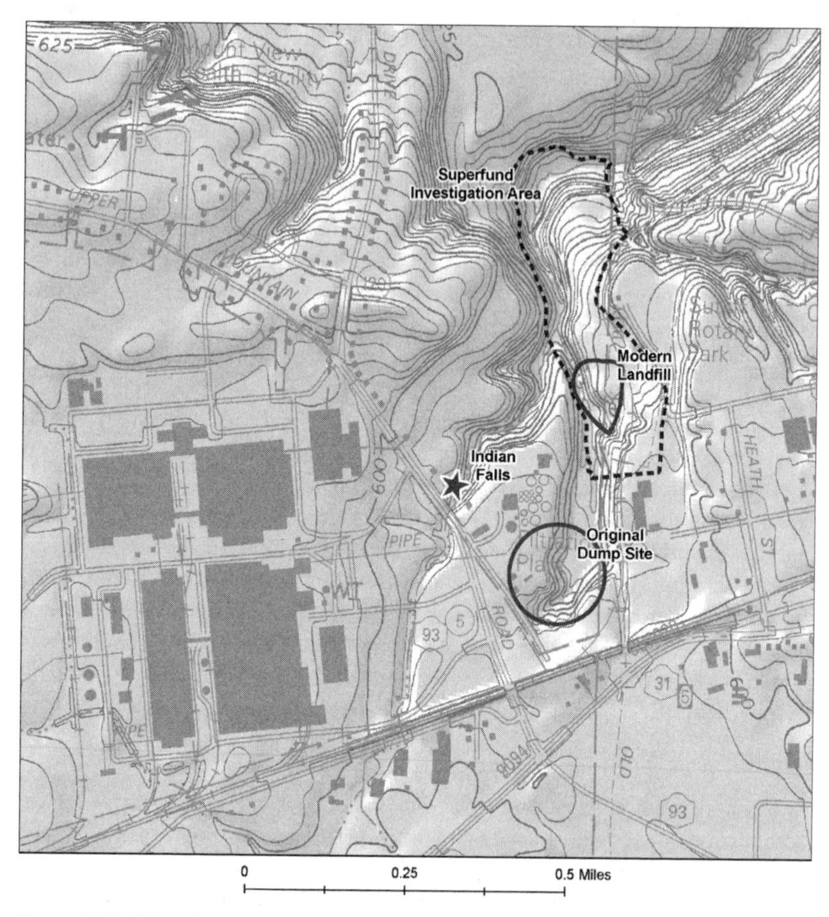

Figure 5.12. Map showing landfills and Superfund site in the Gulf, superimposed on a topographic map and a digital elevation model (United States Geological Survey). Map by Cherin Abdel Samie and Yasser Ayad.

be allowed to pass" into the lower Gulf. With his words seemingly falling on deaf ears, Lewis's rhetoric inflated in 1966: the city had begun "to desecrate this magnificent creation of nature by dumping its garbage there." Apart "from the lower end its attractiveness has vanished," and soon "the entire Gulf will be past redemption."[206] Lewis was aware of Indian Falls, although there is some doubt about whether he ever visited it. He described it as "a picturesque waterfall of some 10 feet or more"—a vast underestimation of the thirty-five- to forty-foot cataract. At any rate, whatever charms it once possessed had "vanished." The Lewis proposal of a park in the lower Gulf, nonetheless, came to fruition with the establishment of the Gulf Wilderness Park in 1972.

"Realizing the problems associated with the control and correction of water pollution," a Harrison Radiator publication reported, the company "built a multi-million dollar water treatment plant, which was put into operation in 1969 in the west Lockport Complex. The treated water, which is not reused, is discharged into the Gulf, a Tributary of Eighteenmile Creek—meeting all government standards."[207] When a local conservation organization, a few years later, "publicly commended the corporation" for "cleaning up its own mess," Alexis Muller asked: "if someone entered your house and spilled a gallon of yellow paint in the middle of your living room rug, would you reward him for carrying the pail away?"[208] The Harrison filtration facility occupied the ridge between the two fingers of the Gulf gorge.

By the early 1980s, it had become clear that Lockport Jeremiahs like Lewis and Muller, if anything, had underestimated the severity of pollution in the Gulf. The New York State Department of Environmental Conservation (NYSDEC) ordered an investigation, and a consulting firm prepared the Superfund Phase I Summary Report in 1983. The immediate concern was the second city landfill, in operation from the 1950s until 1976 which "was a continuation of operations that had originally been located south of the Harrison filtration plant" (fig. 5.12). The report recommended that "the original landfill should be investigated for remediation" as well. An earlier NYSDEC report had noted that the second landfill had received "unknown quantities" of industrial pollutants including organic peroxides. In addition, "Harrison Radiator was suspected of shipping metal sludges to the facility."[209] It seems that no records were kept of either wastes dumped or areas filled. The landfill area was supposed to be three to four acres, but investigation of historical aerial photography revealed that it covered an area of at least twenty to thirty acres. The boundaries of the Superfund site enclosed not only the landfill but also the slope and valley floor of the Gulf (fig. 5.12), an area visibly strewn with piles, drums, and tires. NYSDEC classified the Lockport City Landfill as "Class 2, which is a site posing a significant threat to public health or the environment and requiring action."[210]

The landfill operations had consisted of a series of trenches, as much as eighty feet deep. According to city employees, "whenever a drum was received containing liquid, the shipper was instructed to drain the liquid into the trench" so the drums could be crushed. Employees also stressed "that probably every industrial facility in Lockport and that portion of Niagara County disposed of wastes there," and noted that "a large amount of dumping occurred while no city personnel were present." Harrison Radiator delivered two truckloads of wastes daily, "including drums of paint sludge and acid." Two electric companies regularly disposed of transformers, which "were broken open and the copper salvaged." The liquids containing PCBs were then dumped into the trench. VanDeMark Chemicals disposed of two types of wastes:

> One (1) was a powdery solid that emitted a chlorine odor. When drums of this material were ruptured, clouds of powder formed irritating worker's [*sic*] skin eyes and lungs. Due to the extreme discomforts associated with this material, the barrels were kept separate from the other refuse being disposed of and covered. The second material was extremely reactive when exposed to air or water. The city workers thought this material contained phosphorus. When drums of this material were punctured, fires resulted. According to the workers, fires were frequent occurrences at the landfill.[211]

VanDeMark Chemicals, most of which was purchased by the French Groupe SNPE in 1999, was the subject of a feature article in *Chemical and Engineering News* in 2001 entitled "The Craft of Living Dangerously." The firm's particular niche was in "taking the risk to work with extremely dangerous chemicals that other companies decline to handle" including explosives and the poison war gas phosgene, which is used in pharmaceutical manufacturing. VanDeMark is the "only merchant producer of phosgene in the United States."[212]

In 1998, the rupture of a chlorine pipe at the VanDeMark plant forced police to cordon off a half-mile square "advisory zone" within which residents were told to close windows and remain indoors. Later, children were evacuated from DeWitt Clinton Elementary School, about one mile from VanDeMark's. Several students complained of "nausea and eye irritation," and two policemen were treated at the Lockport Memorial Hospital for throat irritation. Assistant Police Chief James Donner commented that "the situation could have been a lot worse, with the potential of 300 pounds of chlorine being released."[213] A year earlier, an industrial fire at a Lowertown plastic firm spawned concern over the release of PCBs. High levels of PCBs had been found not only in Eighteenmile Creek, but also in the canal.[214] In 1991 the New York State Health Department launched a study of autoimmune diseases of present and former residents of Newfane, about eight miles downstream on the Eighteenmile Creek. There were twenty-seven cases of lupus, multiple sclerosis, thyroid disease, and rheumatoid arthritis in the vicinity. Despite the high level of disease among people who lived near the creek, it was difficult to demonstrate that waterborne pollutants were the culprit. That the Eighteenmile Creek was severely contaminated, nonetheless, was beyond dispute. The Department of Environmental Conservation identified fifteen dump sites in Lockport and Newfane—all within the creek's watershed, and the State Health Department advised against eating any fish from the creek because of elevated PCB levels.[215]

A *USA Today* story in 2000 revealed that Lockport's specialty steel mill, Simonds Saw and Steel, had participated in nuclear weapons production between 1948 and 1956. "Simonds rolled 25 million to 35 million pounds of uranium and 30,000 to 40,000 pounds of thorium into billets for fuel rods used in nuclear reactors." Apparently, workers were not properly informed that they were working with radioactive materials. A number have since died of cancer and other diseases. Joy Christie, whose husband Gary died of kidney cancer in 1994, admitted: "for a long time after Gary died, I would drive by the plant and I wanted so badly for someone to take bright orange paint and spray 'murderer' on the outside of that dam [*sic*] building."[216]

Harrison Radiator and Simonds had both arrived in Lockport at the time of the cheap electricity deal with Niagara Falls. VanDeMark Chemicals may have come for power advantages, but perhaps stayed because their dangerous and dirty work was somehow tolerated, or not noticed, in the polluted valley of the Eighteenmile Creek. Niagara Falls was the source of the power, its natural fury transmuted into the juice that also fed a large cluster of chemical manufacturers there. The Love Canal, and perhaps other landscapes in the city of Niagara Falls unfit for human habitation, show how inept and shortsighted the human use of electrical power could be. From the beginning, there had been many in Lockport who wanted to divert some of Niagara's power, but when it began arriving through transmission lines, those who wielded it created similar landscapes of environmental devastation. One question remains: Why should the Gulf be the locus of that devastation?

Joyce Carol Oates has often compared her feelings about Lockport with those of James Joyce toward Dublin. "In a sense, I'm always thinking about it," she remarked on a visit to the city of her birth in 1987; "James Joyce has his Dublin, and I have my Lockport."[217] Both authors lived most of their lives away from the city that provided the setting for so much of their fiction. "Lockport," she admitted, "was luminous to me because I was young."[218] Born into a modest working-class Irish family, the childhood Oates remembers was not tragic. "I recreate it," she said in 1980, "as it was, as it could have been, as I wanted—and didn't want—it to be."[219] Oates sees always that things could have been otherwise, and she revels in exploring the possible worlds that human decisions and actions might have brought into being. In her 1987 novel, *You Must Remember This*, she recreated the city of her childhood as Port Oriskany "almost as if Lockport had the heavy industry Buffalo had." Oates imagined the city that might have emerged had the power channel to Lockport become a reality in 1904, with "great factories looming up where now nothing of an industrial order exists,"[220] but the promoters of the power canal might be disappointed to discover that this counterfactual city, in Oates's rendering, is streaked with as many shadows as the Lockport of history. "While writing the novel," she recalls, "I had a map of Port Oriskany taped to my wall so that . . . I could simply stare at it," and like the protagonist Enid, "the contours of whose soul so resemble my own, traverse its streets, ponder its buildings and houses and vacant lots, most of all the canal that runs through it, as it runs through Lockport, New York . . . that canal that, in Enid's heightened and often fevered imagination, as in my own, seemed an object of utter ineffable beauty. (It must be remembered that beauty does not mean mere prettiness but something more brutal, possessed of the power to rend one's heart.)"[221]

At the time of the Oklahoma City bombing, Oates learned that Timothy McVeigh had been born near her own home south of Lockport, and that, like her, he had attended school in Lockport. Oates described the city in a 1993 *New Yorker* article. "It's a city of vertiginously steep hills built on the banks of the Erie Barge Canal—Lockport's predominant feature—which cuts through it in a deep swath and divides it approximately in two." The canal's "beauty," for Oates, could not be separated from its power to rend. "To walk along the canal's high banks," she confided, "on cracked and littered pavement, gazing down at the foaming, black water below, is mesmerizing. Framed by dizzyingly steep, stark stone walls, the canal has the look of a nightmare domesticated by frequent viewings."[222] In her 1973 poem, "City of Locks," Oates again presents an image of the canal as piercing the city and its people, and blurring the boundary between the human and the nonhuman:

> evilly, an odor rises from the bubbling fall
> of one artificial level to another
> old municipal buildings at the canal's edge
> sink down to its silent strata
> as if to a past before history
>
> water flows through us today, warmly
> through the streets of Lockport, New York
> and through us
> in a small foaming series of falls.[223]

Joyce Carol Oates crafted her most sustained meditation on the city of her birth in her 1971 novel *Wonderland*. She precedes the novel with an author's note assuring the reader that "all the settings—especially Lockport, New York; Ann Arbor, Michigan; and Toronto, Ontario—are fictional. Any resemblance to reality is accidental and should be resisted." This epigraph follows:

> We . . . have dreamt the world. We have dreamt it as firm, mysterious, visible, ubiquitous in space and durable in time; but in its architecture

> we have allowed tenuous and eternal crevices of unreason which tell us it is false.
>
> —Borges, *Labyrinths*

According to the *Oxford English Dictionary*, the root meaning of "fiction" is "to fashion or form." It shares this root with "fictile": "molded into form by art." Oates does not give the Lockport of *Wonderland* a pseudonym. Yet her declaration that it is "fictional" might be taken in both senses—that is, as an invented setting for her narrative and as a place formed by human art. The latter sense would resonate with her suggestion that the subject of Joyce's fiction was not just his characters but Dublin itself, even the Dublin precisely on June 16, 1916. Oates happened to be born on June 16, 1938. All things formed by art—all fictions—rely on exclusions. The Lockport Oates experienced as a child was there because other possible Lockports were not. The city's total absence, of course, was another possibility. Oates's forming of Lockport in *Wonderland*, and the shaping of the Lockport of history, are both fictional processes, matters of human agency, though not necessarily of self-conscious decisions. Both cities are imagined, and shaped, by definitional exclusions and oppositions. What is excluded does not vanish, even if it vanishes from consciousness. It remains to haunt the creation. So though the city may be imagined as a "firm" and "durable" fact of nature, "crevices of unreason" creep in to reveal the seams of artifice.

The protagonist of *Wonderland*, Jesse, experiences such a recognition in one of Oates's most penetrating renderings of Lockport. Jesse is a boy with a tragic past, the sole survivor of his father's murderous rampage against his own family and himself. Jesse's life improves when he moves to Lockport and is adopted by a well-off family. His healing—or more accurately, his stubborn decision to persist—is linked to a growing interest in the canal and its history. The passage begins with specificity of time and place—as if, like Joyce, her subject were the Lockport of history and of her own experience.

> One morning in August, 1940, Jesse was standing at the railing of the largest bridge in Lockport, high above the Erie Canal. He was gazing down at the locks. He held a library book tightly, as if fearful it might somehow fall; he was fascinated by the depth of the bridge, the steep damp sides of the canal, the different levels of water. So he was here at last, standing here alone. . . . Everything in sight was illuminated with a hazy, pearl-like glow because of the humidity and the brilliant, glazed sunlight. . . . *He had survived. He was here* [emphasis in original].

We learn later that the library book is titled *Clinton's Ditch: A History of the Erie Barge Canal.* In the presence of the canal, he holds "tightly" to the book about the canal and to his own existence: "*He was here*." He looks down: "Beneath him, a long dizzying drop to a pit of water that was dark and fairly still." When his thoughts drift to his biological family and his rural life before Lockport, "he erased the thought." Jesse was creating a new self through exclusions as well. He understands his own shocking contingency: an accident had saved him from the terrible fate the rest of his family had suffered. He also notices the contingency of Lockport itself, and the fact that almost no one else does:

> Behind him traffic moved as usual, not very much of it on a weekday morning in Lockport. Women strolled downtown to shop, in no hurry. In the distance there was a church steeple, hazy in the sunlight. What was so fascinating about this, Jesse thought, was its ordinary nature—the canal, the locks, the noisy water; the town itself ordinary and quiet, as if it had existed for centuries, with a profound certainty of its right to exist, no awareness of the fact that it had no reason for existing, no guarantee of its right to exist. It was here; it moved in a slow timed orbit. Already he could define himself against it: *Jesse Pedersen on the big bridge, waiting for a barge to come through the locks* [emphasis in original].

Turning his eyes back downward,

> he stared, fascinated. Hypnotized. A dank, fetid odor rose from the deepest pit of water, directly below him. Jesse leaned over the railing. It was important for him to see everything, as much as he could see. Along the canal's banks buildings had been built, decades ago, that seemed to descend into the foundation of gray, dreamlike rock itself, their peaks and arches and chimneys rising to the sky, their lower parts descending into the smudged, rain-washed gray rock, as if going back to a time when there was no distinction between human life and the life of rocks. Had they slept for thousands of years before being wrenched out of the earth, dug up to make way for the canal . . . ? Jesse frowned and thought of the calm, wide, muddy canal that existed away from the sequence of locks, winding into the distance. No one would suspect, approaching the locks, that the water could turn so violent and dangerous.[224]

At Lockport, the calm pastoral canal is revealed to be a lie. The violent wrenching "out of the earth" seems to hint at the forgotten, excluded realities of the canal's construction. And the city's "certainty of its right to exist," "as if it had existed for centuries," indistinguishable from the rocks themselves, is exactly the sort of naturalness that Oates takes pleasure in exploding. For her, as for Jesse, there are myriad, sometimes shocking possibilities just beneath the surface of what seems natural and ordinary.

Had a few individuals made different decisions, Lockport would never have existed, and most obviously, had David Thomas not altered the canal's route at the last moment, Lockport would have developed at the head of the Gulf, a place that remains strikingly excluded from local consciousness. As many Lockportians dreamed of a subterranean wonder of nature beneath the heart of their city, some Lockportians slowly covered and contaminated that other valley with its waterfall. Was its presence an irritation that threatened to reveal the contingency of the town's very existence—like a foreign substance that the oyster covers with layer upon layer of secretion until no longer noticeable? Is the Gulf with its waterfall a doppelganger to the natural basin with its locks, a disturbing mirror that threatens to expose the "crevices of unreality" beneath the "City of Locks"?

My modest suggestion is that there is nothing inevitable about Lockport. It is a human creation, molded by choices made within contexts no individual or group could easily control—choices based upon reason and interest, but also upon half-recognized desires and fears. Humans deserve the credit for overcoming the Mountain Ridge and everything that followed from it, from commercial prosperity to the fire hydrant: the canal that flows through Lockport, for Oates, is "an object of utter ineffable beauty." Yet this beauty is not "mere prettiness but something more brutal, possessed of the power to rend one's heart." The canal cuts deep: humans are responsible for the complete creation, the full beauty.

The picture of Lockport I have presented is one of many that might be sketched. I have examined Lockport's moment of birth, so celebrated in narratives of progress and nationalism. I have also attempted to attend to the stories excluded from those narratives, stories that reveal the struggles, the costs, and the often unseen implications of the conquest of the Mountain Ridge: the dispossession of native peoples, the horrors of common laborers' experience, the shocking impact of the market revolution the canal helped to ignite, and the environmental costs of the progress the canal once symbolized. Moreover, I have sought to emphasize that, while celebratory narratives were nearly always dominant, they have been frequently contested with words and deeds. Indeed, this book is a small contribution to this ongoing dialogue about nationalism, progress, capitalism, and empire that made the Erie Canal such a salient point of departure during the antebellum era.

To refer to the United States as an empire may seem misplaced to some, yet nineteenth-century New Yorkers had no hesitation boasting of the Empire State. Between 1800 and 1850, the United States expanded from a coastal community to a continental power, invading and absorbing a large part of Mexico and the lands of the continent's First Nations. The completion of the Erie Canal played a key role in facilitating that hegemony. In that sense, the Deep Cut and the Lockport locks that overcame the Mountain

Ridge and completed the link between New York City and the Great Lakes were indeed a stairway to empire. Narratives of America's "natural" and "inevitable" hegemonic expansion, justified by assertions of innocence and moral superiority, have been called upon again and again to serve new and wider arenas of action. There are always other choices.

The locks, the Deep Cut, and Lockport itself, were once celebrated symbols of progress. When the forgotten stories of canal construction and the city are remembered, Lockport also seems a representative example of the costs of progress. I have no desire to denigrate the human achievements of the canal, the city, or technological progress. Prometheus is a hero because he acts to benefit the human world; progress must be judged in similar terms.

In response to Joyce Carol Oates's *New Yorker* article about Timothy McVeigh, a number of letters appeared in the Lockport *Union-Sun & Journal* complaining about what writers perceived as her critical tone. Her response represents my own aspirations in writing this book and my own sentiments about the city of both our births: "Writers try to focus on what, from their necessarily limited perspective, is real. They mean to be honest, not propagandistic. And often their purpose is to show that is it possible to love what is imperfect."[225]

Appendix

Mountain Ridge Contractors

The year-by-year summary of contracts is based on original contracts, vouchers, and other records in the New York State Archives ("Contracts and accounts for construction and repair," [ca. 1817–1828] New York [State] Office of the Canal Commissioners; and "Accounts of monies paid to contractors and others for construction, repair, and enlargement of the Erie and Champlain canals, 1817–1871," New York [State] Office of the Auditor of the Canal Department).

1821 (all contracts issued on May 21, 1821)

Section 1.
Half mile (40 chains), including locks; Claudius V. Boughton, Oliver Culver, John Maynard, and Joseph Comstock

Section 2.
One mile and thirty-five chains (1.44 miles); Joseph Comstock and Otis Hathaway

Section 3.
Two miles and thirty-five chains (2.44 miles); Ebenezer F. Norton, Lyhman Collins, and Philo Norton

Section 4.
Two miles and fifty-five chains (2.69 miles); Jared Boughton, Jacob Lobdell, Jonathan Smith, and Joseph Maynard

1822 (new contracts issued on June 15, 1822)

Section 1.
Boughton, Culver, and Maynard

Section 2. (old section 2 now divided into new sections 2, 3, and 4)
East (or northerly) subdivision (37.5 chains or .47 miles); Jonathan Child and Elijah Hamlin

Section 3.
Central subdivision (40 chains or .5 miles); Darius Comstock

Section 4.
West (or southerly) subdivision (37.5 chains or .47 miles); John Gilbert

Section 5. (1822 section 3)
Norton and Collins

Section 6. (1822 section 4)
Boughton, Lobdell, Smith, and Maynard

1823

Section 1.
Culver, Boughton, and Maynard continue on entire section until July 26, 1823, when Oliver Phelps assumes the southerly portion (about .285 miles), leaving about .215 miles to the original contractors. Oliver Culver and John Maynard then take over that northerly part on October 1, 1823. An additional contract for this portion is issued to Culver alone but it is not clear that this is a further subdivision. Similarly, a contract is issued to Orange Dibble and Jonathan Olmsted for work from October to December 1823 but it is not clear if this is a further subdivision.

Section 2.
East subdivision (1822 section 2)

Child and Hamlin continue with their 1822 contract until October 4 when they divide the section into equal parts (.235 miles each) with Child assuming the more northerly part.

Central subdivision (1822 section 3)

Darius Comstock, new contract October 4, 1823.

West subdivision (1822 section 4)

John Gilbert, new contract December 1, 1823.

Section 3 (1822 section 5).
Norton and Collins continue with entire section until the autumn when the commissioners announce that original sections 3 and 4 were divided into eleven contracts by the end of 1823 (presumably five of these were in section 3). They also divided the entire Deep Cut from the Main Street Bridge (the northerly end of Phelps's section) to the Tonawanta Creek into 184 stations of 3 chains (198 feet). Only three of the new sections were actually contracted in 1823:

> Ebenezer F. Norton, Stephen Bates, and William House, December 10, 1823, for the northerly .49 miles of section 3 (stations 47–59).
>
> Rueben B. Heacock, payments beginning September 17, 1823, stations 91–100 (.375 miles).
>
> Seymour Scovell, December 15, 1823, stations 101–111 (.41 Miles).

Section 4 (1822 section 6).
Boughton, Lobdell, Smith, and Maynard continue with their contract, but decide to subdivide most of it (stations 121–184, about 2.4 miles) themselves into eight sections (.3 miles each), with Smith and Maynard taking charge of sections 1, 3, 5, and 7 and Boughton and Lobdell taking 2, 4, 6, and 8). Then, on December 17, 1823, in accordance with the commissioners' new divisions, three new contracts are issued:

> Daniel Brown, for stations 137–142 (.225 miles).
>
> Wallace Bell and William Richardson, for stations 148–155 (.26 miles).
>
> Jared Boughton and Jacob Lobdell, for stations 167–184 (.675 miles).

1824

Section 1.
Culver, Boughton, and Maynard continue on in their .215-mile northerly part of the section, with Culver again carrying out work on the same section under a separate contract.

Oliver Phelps continues on the southerly .285 miles, but again Dibble and Olmsted carry on work in the same section; then Olmsted takes over this work alone in mid-1824.

Section 2.
East subdivision (1822 section 2)

Child continues with northerly part and Hamlin with the southerly part (.235–mile each) under separate contracts.

Central subdivision (1822 section 3)

Darius Comstock continues.

West subdivision (1822 section 4)

John Gilbert continues.

Section 3 (1822 section 5).
Norton, Bates, and House continue with the northerly .49 miles of section 3 (stations 47–59).

E. F. Norton assumes stations 60–70 (.41 miles) on December 26, 1824.

Wallace Bell, William Richardson, and Thomas Griffith assume stations 84–90 (.26 miles) on December 27, 1824).

Bell, Richardson, and Griffith assume stations 91–94 (.15 miles) on October 1, 1824.

Rueben B. Heacock continues on stations 91–100 (.375 miles) until October 1; then he continues only on stations 95–100 (.225 miles).

Seymour Scovell continues on stations 101–111 (.41 miles).

Three additional contractors held parts of section 3 in 1824, although the exact stations are not mentioned anywhere. In the contracting chart, they are tentatively placed where they might reasonably have worked, based on the order of reporting expenses by Bouck, which generally goes from north to south; the contractors are:

> Norton and Collins, who worked all of 1824.
>
> Horace S. Turner, who worked all of 1824 and 1825.
>
> Orange Dibble, who worked from September 24, 1824, to the end of 1825.

Section 4 (1822 section 6).
Daniel Brown continues on stations 137–142 (.225 miles).

Bell and Richardson continue on stations 149–155 (.26 miles).

Elihu Evers, receiving first payments on December 15, 1824, on stations 156–166 (.41 miles).

Boughton and Lobdell continue for stations 167–184 (.675 miles).

In addition, the following three held for parts of section 4 throughout 1824; although there is no exact evidence of the stations they worked, they are tentatively placed in the contracting chart according to the order Bouck mentions them:

Jenks and Putney.

Snyder and Lane.

Pratt and Simpson.

1825

Section 1.
Completed by the end of 1824.

Section 2.
East subdivision (1822 section 2)

Completed by the end of 1824.

Central subdivision (1822 section 3)

Completed by the end of 1824.

West subdivision (1822 section 4)

John Gilbert continues throughout 1825.

Section 3 (1822 section 5).
Norton, Bates, and House, continue with the northerly .49 miles of section 3 (stations 47–59).

E. F. Norton continues on stations 60–70 (.41 miles) throughout 1825.

Bell, Richardson, and Griffith continue on stations 84–90 (.26 miles) and 91–94 (.15 miles) throughout 1825.

Rueben B. Heacock continues on stations 95–100 (.225 miles) throughout 1825.

Seymour Scovell continues on stations 101–111 (.41 miles) throughout 1825.

As in 1824, three additional contractors held parts of section 3 in 1825, although the exact stations are not mentioned anywhere. In the contracting chart, they are tentatively placed where they might reasonably have worked, based on the order of reporting expenses by Bouck; the contractors are:

Brown and Campbell, who worked all of 1825 (perhaps taking over from Norton and Collins).

Horace S. Turner, who continues from 1824 and throughout 1825.

Orange Dibble, who continues from 1824 and throughout 1825.

Section 4 (1822 section 6).
Daniel Brown continues on stations 137–142 (.225 miles) throughout 1825.

Bell and Richardson continue on stations 148–155 (.26 miles) throughout 1825.

Evers continues on stations156–166 (.41 miles) throughout 1825.

Boughton and Lobdell continue for stations 167–184 (.675 miles) throughout 1825.

As in 1824, the following three held parts of section 4 throughout 1825; although there is no exact evidence of the stations they worked, they are tentatively placed in the contracting chart according to the order Bouck mentions them:

Jenks and Putney.

Snyder and Lane.

Pratt and Simpson.

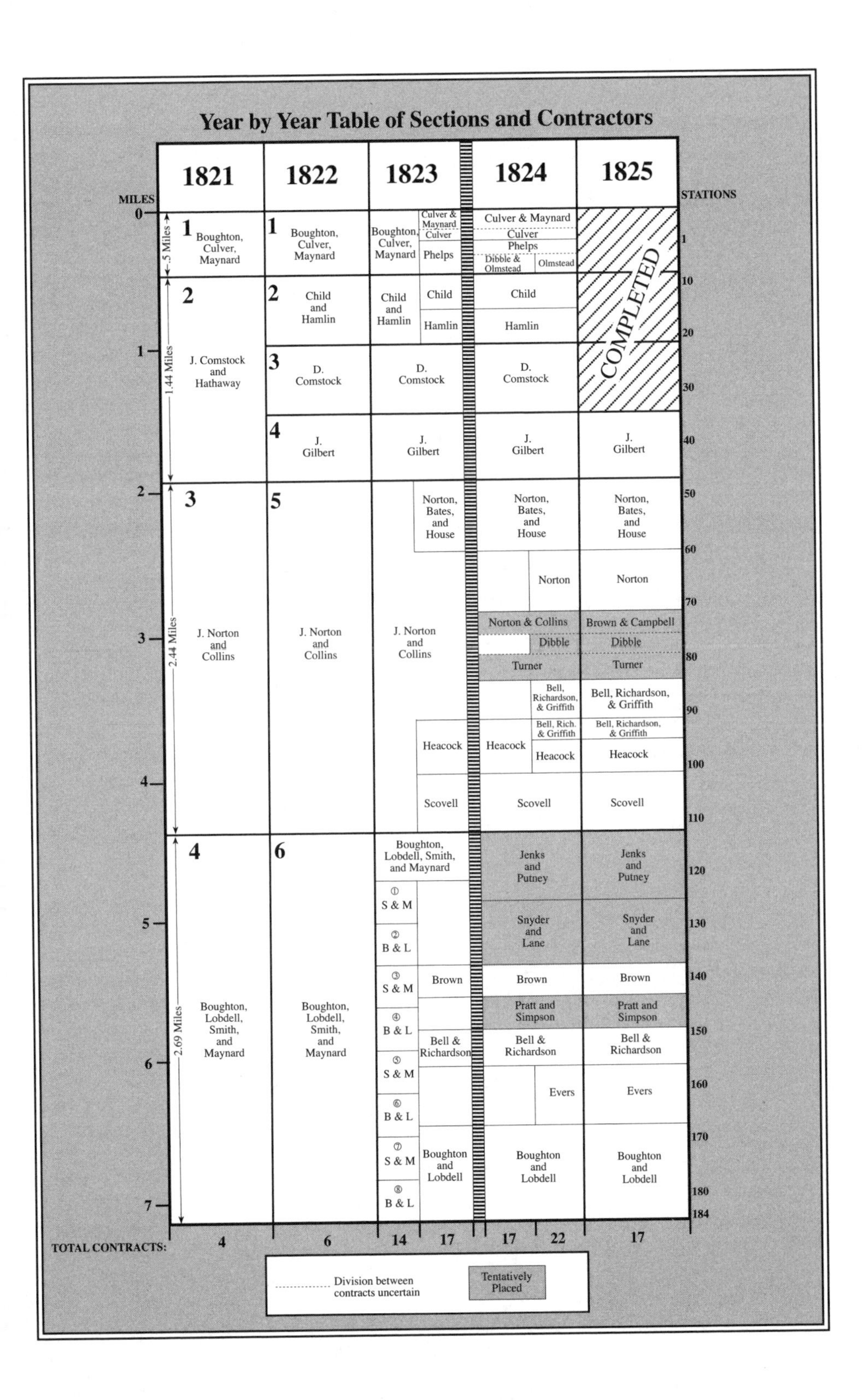

Year by Year Table of Sections and Contractors
1821
1822
1823
1824
1825
MILES
STATIONS
0
1
2
3
4
5
6
7
.5 Miles
1.44 Miles
2.44 Miles
2.69 Miles
1
Boughton, Culver, Maynard
2
J. Comstock and Hathaway
3
J. Norton and Collins
4
Boughton, Lobdell, Smith, and Maynard
1
Boughton, Culver, Maynard
2
Child and Hamlin
3
D. Comstock
4
J. Gilbert
5
J. Norton and Collins
6
Boughton, Lobdell, Smith, and Maynard
Boughton, Culver, Maynard
Culver & Maynard
Culver
Phelps
Child and Hamlin
Child
Hamlin
D. Comstock
J. Gilbert
Norton, Bates, and House
J. Norton and Collins
Heacock
Scovell
Boughton, Lobdell, Smith, and Maynard
① S & M
② B & L
③ S & M
④ B & L
⑤ S & M
⑥ B & L
⑦ S & M
⑧ B & L
Brown
Bell & Richardson
Boughton and Lobdell
Culver & Maynard
Culver
Phelps
Dibble & Olmstead
Olmstead
Child
Hamlin
D. Comstock
J. Gilbert
Norton, Bates, and House
Norton
Norton & Collins
Dibble
Turner
Bell, Richardson, & Griffith
Bell, Rich. & Griffith
Heacock
Heacock
Scovell
Jenks and Putney
Snyder and Lane
Brown
Pratt and Simpson
Bell & Richardson
Evers
Boughton and Lobdell
COMPLETED
J. Gilbert
Norton, Bates, and House
Norton
Brown & Campbell
Dibble
Turner
Bell, Richardson, & Griffith
Bell, Richardson, & Griffith
Heacock
Scovell
Jenks and Putney
Snyder and Lane
Brown
Pratt and Simpson
Bell & Richardson
Evers
Boughton and Lobdell
1
10
20
30
40
50
60
70
80
90
100
110
120
130
140
150
160
170
180
184
TOTAL CONTRACTS:
4
6
14
17
17
22
17
Division between contracts uncertain
Tentatively Placed

Notes

Prologue

1. Rochester *Telegraph*, June 28, 1824.

2. Jesse Hawley, *Ontario Messenger*, January 27, 1841.

3. C. F. Briggs, "Lockport," in *The United States Illustrated; In Views of City and Country. With Descriptive and Historical Articles*, ed. Charles A. Dana (New York: H. J. Meyer, [1853]), 162–64.

Chapter 1. Leaving Lockport: And Returning to the Canal's Beginning

1. Herman Melville, *Moby Dick, or The White Whale* (New York: Library Publications, 1930), 11.

2. Joyce Carol Oates, *Wonderland* (New York: Vanguard Press, 1971), 11.

3. *Illustrated History of Niagara County, New York* (New York: Sanford and Company, 1878), 173; Clarence O. Lewis, *The Erie Canal, 1817–1967*, Occasional Contributions of the Niagara County Historical Society (Lockport, NY: Niagara County Historical Society, 1967), 36.

4. New York State Department of Environmental Conservation, "Lockport City Landfill," New York State Superfund, Phase 1 Summary Report, prepared by Recra Research (Albany, NY: New York State Department of Environmental Conservation, 1983), 7–10; New York State Department of Environmental Conservation, "Remedial Investigation at the Lockport City Landfill," NYSDEC Site No. 9-32-010, prepared by URS Consultants (Albany, NY: New York State Department of Environmental Conservation, 1992).

5. Charles Sellers, *The Market Revolution: Jacksonian America, 1815–1846* (New York: Oxford University Press, 1991).

6. Roland Barthes, *Mythologies* (New York: Hill and Wang, 1972), 128–29.

7. James Joyce, *Ulysses* (New York: Vintage Books, 1961), 394.

Chapter 2. Locating Lockport: The Canal and the Continent's Last Barrier

1. Moustafa Bayoumi and Andrew Rubin, eds., *The Edward Said Reader* (New York: Vintage Books, 2000), 110, from "The Scope of Orientalism," in *Orientalism* (New York: Vintage Books, 1979).

2. Ramón de la Sagra, "Five Months in the United States of America: From the 20 of April to the 23 of September of 1835" (Paris: Printing Shop of Pablo Renouard, 1836), 260–61, translated and partially reprinted in Watt Stewart, "A Pilgrimage through New York State in 1835," *New York History* 19 (1938): 407–18, 416–17.

3. Ibid., 417 (261 in original).

4. Edward Said, in Bayoumi and Rubin, *Said Reader*, 113.

5. Recently, critical geographers have questioned the concept of scale, particularly when levels such as "national" and "international" become naturalized. In the 1820s, concepts of scale were fluid, although the notion of the national was in the process of entering public discourse as a commonsensical concept; see Andrew Herod, "Scale: The Local and the Global," in *Key Concepts in Geography*, ed. S. L. Holloway, S. P. Rice, and G. Valentine (London: Sage Publications, 2003), 229–47; Sally Marston, John Paul Jones, and Kevin Woodward, "Human Geography without Scale," *Transactions of the Institute of British Geographers* 30, no. 4: 416–32.

6. Quoted in Robert Greenhalgh Albion, *The Rise of New York Port [1815–1860]* (Hamden, CT: Archon Books, 1961), 84.

7. D. W. Meinig, *The Shaping of America: A Geographical Perspective on 500 Years of History*, 4 vols., vol. 2 *Continental America, 1800–1967* (New Haven and London: Yale University Press, 1993), 311–23; James E. Vance, Jr., *Capturing the Horizon: The Historical Geography of Transportation* (New York: Harper and Row, 1986), 101–45.

8. Elkanah Watson, *Men and Times of the Revolution*, ed. Winslow C. Watson (New York: Dana and Company, 1856), 280–81.

9. Robert Fulton, *A Treatise on the Improvement of Canal Navigation* (London; I. and J. Taylor at the Architectural Library, 1796), 11–15.

10. The National Road is the single important counterexample of federal internal improvements; see Ralph H. Brown, *Historical Geography of the United States* (New York: Harcourt, Brace and World, 1948), 185–86.

11. Thomas Jefferson, *Notes on the State of Virginia* (Chapel Hill: University of North Carolina Press, 1955), 15–16.

12. Ellen Churchill Semple, *American History and Its Geographic Conditions* (Cambridge: Riverside Press, 1903), 263–78, provides the classic narrative of this argument. See also: Carville Earle, "Beyond the Appalachians, 1815–1860," in *North America: The Historical Geography of a Changing Continent*, second edition, ed. Thomas F. MacIlwraith and Edward K. Muller (New York: Rowman and Littlefield, 2001), 165–88.

13. D. W. Meinig, *Continental America*, 318–322; Vance, *Capturing the Horizon*, 110–45.

14. Parker E. Calkin, "Late Pleistocene History of Northwestern New York," in *Geology of Western New York Guide Books*, New York State Geological Association, 38th Annual Meeting, ed. E. J. Buehler (Buffalo: New York State Geological Association, 1966), 58–68.

15. Elkanah Watson, *History of the Rise, Progress, and Existing Conditions of the Western Canals in the State of New York* (New York: D. Steele, 1820), 342; see also Christopher Colles, *Proposals for the Speedy Settlement of the Lands in the Western Frontier of . . . New York, and for the Improvement of the Inland Navigation Between Albany and Oswego* (New York: Samuel Louden, 1785).

16. Paul Johnson, " 'Art' and the Language of Progress in Early-Industrial Paterson: Sam Patch at Clinton Bridge," *American Quarterly* 40 (1988): 433–49, 440.

17. John Seelye, *Beautiful Machine: Rivers and the Republican Plan; 1755–1825* (New York: Oxford University Press, 1991), 9.

18. Seelye, *Beautiful Machine*, 40; Hildegard Binder Johnson, *Order Upon the Land: The U.S. Rectangular Land Survey and the Upper Mississippi Country* (New York: Oxford University Press, 1976).

19. For recent perspectives on the early American imperial gaze, see Daniel K. Richter, *Facing East from Indian Country: A Native History of Early America* (Cambridge, MA: Harvard University Press, 2001); and Malino Johar Schueller and Edward Watts, eds., *Messy Beginnings: Postcoloniality and Early American Studies* (New Brunswick: Rutgers University Press, 2003); see also Mary Louise Pratt, *Imperial Eyes: Travel Writing and Transculturation* (London: Routledge, 1992).

20. Watson, *History of the Western Canals*, 353–54.

21. Edwin G. Burrows and Mike Wallace, *Gotham: A History of New York City to 1898* (New York: Oxford University Press, 1999), 450.

22. This depends on how one calculates; see, for example, Nobel Whitford, *History of the Canal System of the State of New York*, 2 vols. (Albany: State of New York, 1906), 808–927; Diane Lindstrom, *Economic Development in the Philadelphia Region, 1810–1850* (New York: Columbia University Press, 1978).

23. Rohit T. Aggarwala, "The Rise of New York, 1790–1815: A Reassessment of Sources of Growth," presented to the Conference of New York State History, Buffalo, June 5, 1998, manuscript.

24. Burrows and Wallace, *Gotham*, 335–37; Aggarwala, "The Rise of New York," 2–4.

25. Burrows and Wallace, *Gotham*, 442–43.

26. Burrows and Wallace, *Gotham*, 443, Thomas Kiernon, *The Road to Colossus* (New York: William Morrow and Co., 1985).

27. Albion, *Rise of New York Port*, 83.

28. Burrows and Wallace, *Gotham*, 419–51, 450; see also Albion, *Rise of New York Port*, 76–94; Whitford, *History of the Canal System*, 909–27; Ronald E. Shaw, *Erie Water West: A History of the Erie Canal, 1792–1854* (Lexington, KY: University of Kentucky Press, 1966), 277–92; Charles Sellers, *The Market Revolution: Jacksonian America, 1815–1856* (New York: Oxford University Press, 1991), 40–48, Kiernon, *Road to Colossus*.

29. Soldiers' observations of central and western New York during the war also attracted many to seek land later. William Wyckoff, *The Developer's Frontier: The Making of the Western New York Landscape* (New Haven: Yale University Press, 1988), 106; Royal Lovell Garff, "Social and Economic Conditions in the Genesee Country, 1787–1813" (PhD dissertation., Northwestern University, 1939), 64.

30. A typical example of this rhetorical use of Indians is Cadwallader D. Colden, *Memoir at the Celebration of the Completion of the New York Canals* (New York: The Corporation of New York, 1825; facsimile reprint: Ann Arbor: University Microfilms, 1967), 3–4; Laurence M. Hauptman, *Conspiracy of Interests: Iroquois Dispossession and the Rise of New York State* (Syracuse: Syracuse University Press, 1998).

31. D. W. Meinig, "A Geography of Expansion," in *The Geography of New York State*, ed. John H. Thompson (Syracuse: Syracuse University Press, 1966), 140–71.

32. Phillip Lord, "Heading West: 200 Years Ago, Inland Navigation in the 1790s" (Albany: University of the State of New York, State Education Department), brochure compiled from data collected by the Durham Project, New York State Museum, n.d., c. 1994); see also Russell Bourne, *Floating West: The Erie and Other American Canals* (New York: W.W. Norton, 1992), 71–89; Seelye, *Beautiful Machine*, 286–87.

33. Lyman A. Spalding, *Recollections of the War of 1812 and Early Life in Western New York*, Occasional Contributions of the Niagara County Historical Society, no. 2 (Lockport, NY: Niagara County Historical Society, 1949), 5.

34. John Melish, *Travels in the United States of America in the Years 1806 and 1807, and 1809, 1810 and 1811* (Reprint edition, New York: Johnson Reprint Corporation, 1970; original: 1812), 513; see also Wyckoff, *Developer's Frontier*, 102; Orasmus Turner, *Pioneer History of the Holland Purchase* (Buffalo: Jewett, Thomas and Co., 1849), 623.

35. Wyckoff, Developer's Frontier, 102.

36. Spalding, *Recollections*, 6–9.

37. Whitford, *History of the Canal System*, 48–68; Shaw, *Erie Water*, 22–30; Colden, *Memoir*, 28–40.

38. The phrase that Gouverneur Morris purportedly was the first to use.

39. Reported by Merwin S. Hawley in *Publications of the Buffalo Historical Society*, vol. 2, 243–44, quoted in Whitford, *History of the Canal System*, 44–45.

40. Jesse Hawley, *Ontario Messenger*, January 27, 1841; quoted in Shaw, *Erie Water*, 24.

41. Alfred Wegener, *Origin of Continents and Oceans*, John Biram, translator (New York: Dover, 1966; original: Berlin: 1915).

42. Seelye, *Beautiful Machine*, 275–76; Bourne, *Floating West*, 91–92; George E. Condon, *Stars in the Water: The Story of the Erie Canal* (Garden City, NY: Doubleday, 1974), 19–21; Turner, *History of the Holland Purchase*, 621–22, 628–30; Gouverneur Morris may have mentioned (and therefore independently invented) the conception in 1803 or earlier: see Letter from Simeon De Witt, esq. To Mr. William Darby (Albany, February 25, 1822), *Laws of the State of New York, in relation to the Erie and Champlain Canals, together with the Annual Reports of the Canal Commissioners, and other Documents . . .*, 2 vols. (Albany: Published by the Authority of the State, E. and E. Hosford, Printers, 1825), I: 38–42.

43. Joshua Forman also claimed to have independently discovered the interior route, but this has been disputed; see Whitford, *History of the New York Canals*, 56–58; Shaw, *Erie Water*, 30.

44. Assembly Journal, 1808, 58.

45. The Seneca word for this watercourse was eventually softened to "Tonawanda"; alternative spellings of the Seneca word include "Tonnewanta," "Tonewanta," "Tonnewonta," "Tonnawanta"; I use "Tonawanta," the spelling that appears on the New York State map in Colden, *Memoir*, following p. 6.

46. *Laws*, I: 31–44.

47. David Hosack, *Memoir of DeWitt Clinton: With an Appendix, Containing Numerous Documents, Illustrations of the Principal Events of His Life* (New York: 1829), 346.

48. Clinton kept an interesting journal of his travels in 1810, "His Private Canal Journal 1810," in William W. Campbell, *Life and Writings of DeWitt Clinton* (New York: Baker and Scribner, 1849).

49. John H. Eddy, "Map of the Western Part of the State of New York, Showing the Route of the Proposed Canal from Lake Erie to Hudson's River," in Whitford, *History of the New York Canals*, following p. 66.

50. Seelye, *Beautiful Machine*, 287–89; Bourne, *Floating West*, 91–93; Turner, *History of the Holland Purchase*, 621–26; Meinig, "Geography of Expansion, 1705–1855," 159.

51. See, e.g., Samuel Woodworth, "Ode for the Canal Celebration," in Colden, *Memoir*, 250–53.

52. Wyckoff, *Developer's Frontier*, 22, n. 57.

53. Paul D. Evans, *The Holland Land Company* (Buffalo: The Buffalo Historical Society, 1924), 186–95; Wyckoff, *Developer's Frontier*, 30; Hauptman, *Conspiracy of Interests*.

54. Joseph Ellicott, with William and John Willink and others, "Map of the Morris Purchase or West Geneseo in the State of New York," Holland Land Company, 1804.

55. See also L. L. Pechuman, "Niagara County and Its Towns" (Lockport, NY: Niagara County Historical Society, 1958), 13.

56. Brown, *Historical Geography of the United States*, 203.

57. Johnson, *Order Upon the Land*, 228.

58. Wyckoff, *Developer's Frontier*, 24–26.

59. Wyckoff, *Developer's Frontier*, 75–82.

60. James Geddes, Report on Canal Surveys, 1811, *Laws*, I: 36, 49.

61 Turner, *History of the Holland Purchase*, 630.

62. Map and Profile of the Proposed Canal from Lake Erie to the Hudson River in the State of New York Contracted by the Direction of the Canal Commissioners from the Maps of the Engineers in 1817 (Albany: New York State Canal Commissioners, 1817).

63. *Laws*, I: 32.

64. Holland Land Company Field Books, vol. 666, notes, New York State Archives.

65. Holland Land Company Townships Maps (1800–1819), the Mountain Ridge site is on the maps of T.14.R.7 and T.14.R.6, New York State Archives.

66. Frank K. Seischab, "Forests of the Holland Land Company in Western New York, circa 1798," in P. L. Marks, Sana Gardescu, and Frank K. Seischab, *Late Eighteenth Century Vegetation of Central and Western New York State on the Basis of Original Land Survey Records, New York State Museum Bulletin* no. 484. (Albany: University of the State of New York, State Education Department, New York State Museum, 1992).

67. Turner, *Holland Purchase*, 312, 315.

68. James Geddes, Survey Maps of 1817, Map 1 and Map 2, Original Maps of Surveys for the Erie Canal, 1817, Office of the Canal Commissioners, New York State Archives; the Niagara Road was also called the Lewiston-Queenston Road.

69. This image is a composite of elevation data (10-meter grid) from two 7.5 minute United States Geological Survey quadrangles, Lockport and Pendleton, New York, 1:24,000 series, United States Geological Survey.

70. Report of the Canal Commissioners, March 2, 1811 (based on Geddes's 1810 survey), in *Laws of the State of New York, in relation to the Erie and Champlain Canals, together with the Annual reports of the Canal Commissioners, and other Documents* . . . (Albany: Published by Authority of the State, E. and E. Hosford, Printers, 1825), I: 56.

71. James Geddes's Statement 1816, appended to Mr. Van Rensselaer's Report (March 21, 1816), *Laws*, I: 144. To these figures must be added the 4-foot depth of the canal; see Whitford's profiles of the Deep Cut in *History of the New York Canals*, vol. 2, 1046; one chain equals sixteen feet, 330 chains equal one mile.

72. Report of the Commissioners under the Act of April 17, 1816, issued February 17, 1817, *Laws*, I: 145.

73. "Minutes of the Canal Commissioners and Superintendent of Public Works, 1817–1821," 7 March 1820, New York State Archives.

74. Report of the Canal Commissioners, March 12, 1821, *Laws*, I: 7–27, 9.

75. Report of the Canal Commissioners, 1821, 8–9.

76. Clarence O. Lewis, "The Erie Canal: 1817–1967" (Lockport, NY: Niagara County Historical Society, 1967), 9; Ruth Higgins, *Expansion in New York* (Columbus: Ohio State University Press, 1931), 137.

77. Whitford, *History of the Canal System*, II: 1168.

78. As evidenced in his "Miscelanies," a small handwritten treatise on canal cutting and construction, David Thomas Papers, 1820–1829, New York State Archives.

79. Myron Holley to David Thomas (Albany and Union Springs, Cayuga County, respectively) April 1, 1820, David Thomas Papers, New York State Archives.

80. Benjamin Wright to David Thomas (Pittsford to Murray, Genesee County) June 20, 1820, David Thomas Papers, New York State Archives.

81. Benjamin Wright to David Thomas (Rome to Scipio, Cayuga City), January 4, 1821, David Thomas Papers, New York State Archives.

82. Benjamin Wright to David Thomas (Rome to Manchester [later the village of Niagara Falls]), October 16, 1820, David Thomas Papers, New York State Archives.

83. Benjamin Wright to David Thomas (Canandaigua to Manchester), October 20, 1820, David Thomas Papers, New York State Archives.

84. *Laws*, I, Report of the Commissioners, March 12, 1821, 68; Turner, *History of the Holland Purchase*, 631.

85. Benjamin Wright to David Thomas (Rome to Scipio, Cayuga City), January 4, 1821.

86. *Laws*, I: 160, Report of the Canal Commissioners of February 20, 1824; Report of the Commissioners, February 24, 1823, 98.

87. Donald H. Zenger, *Stratigraphy of the Lockport Formation (Middle Silurian) in New York State*; Bulletin no. 404, New York State Museum and Science Service (Albany: University of the State of New York, the State Education Department, October 1965.

88. E. M. Kindle and Frank B. Taylor, *Geologic Atlas of the United States: Niagara Folio* (Washington, DC: United States Geological Survey, 1913), Physiographic Map of the Vicinity of Lake Ontario, fig. 2, 2.

89. Kindle and Taylor, *Niagara Folio*, 19, fig. 10; other studies that contributed to current knowledge include: John P. D'Agostino, "Lake Tonawanda: History and Development," MA thesis, University of Buffalo, February 1958; Parker E. Calkin and Carlton E. Brett, "Ancestral Niagara River Drainage: Stratigraphic and Paleontologic Setting," *Geological Society of America Bulletin*, 89 (August 1978): 1140–54; Edward J. Buehler and Parker E. Calkin, eds., "Guidebook for Field Trips in Western New York, Northern Pennsylvania and adjacent, Southern Ontario," Annual Meeting of the New York State Geological Association, Oct. 8–10, 1982, Amherst, New York (New York Geological Society, 1982); Parker Calkin and Thomas A. Wilkinson, "Glacial and Engineering Geology Aspects of the Niagara Falls and Gorge," 247–77 in Buehler and Calkin, "Guidebook"; Parker E. Calkin, "Late Pleistocene History of Northwestern New York," 58–68, in "Geology of the Western New York Guidebook," E. J. Buehler, ed., New York Geological Association, 38th Annual Meeting, 1966.

90. Calkin, "Late Pleistocene History," 62.

91. Kindle and Taylor, *Niagara Folio*, 11.

92. Calkin and Brett, "Ancestral Niagara River Drainage," 1153.

93. The feature is known as the Barre Moraine, *Niagara Folio*, 17.

94. D'Agostino, "Lake Tonawanda," 55.

95. Calkin, "Late Pleistocene History," 61–62.

96. Calkin, "Late Pleistocene History," 62–63.

97. Calkin and Brett, "Ancestral Niagara River Drainage," 1153.

98. DeWitt Clinton, "His Private Canal Journal 1810," in *Life and Writings of DeWitt Clinton*, ed. William W. Campbell (New York: Baker and Scribner, 1849), 142–45.

99. James Hutton, *Theory of the Earth*, 2 vols., 1795; John Playfair, *Illustrations of the Huttonian Theory of the Earth* (Edinburgh: W. Creech, 1802).

100. Whitford, *History of the New York Canals*, 79.

101. *Laws*, II: 7–24, 9; Turner, *History of the Holland Purchase*, suggests the decision was effectively made late in 1819, 630.

102. Report of the Canal Commissioners in February 1822, *Laws*, II: 60–78.

Chapter 3. Cutting Lockport: Laborers, Citizens, and Five Years at the Mountain Ridge

1. Joyce Carol Oates, *(Woman) Writer: Occasions and Opportunities* (New York: Dutton, 1988), 379–80.

2. Report of the Canal Commissioners, February 1822, *Laws of the State of New York, in relation to the Erie and Champlain Canals, together with the Annual Reports of the Canal Commissioners, and other Documents* . . . 2 vols. (Albany: Published by Authority of the State, E. and E. Hosford, Printers, 1825), II: 60–78, on 67, 68, and 64 respectively.

3. Both of these levels were given a slight gradient to facilitate the eastward flow; this was proposed by Benjamin Wright in a letter to David Thomas (Rome to Manchester) October 6, 1820.

4. Report of the Canal Commissioners, February 1822, in *Laws*, II: 65.

5. L. T. C. Rolt, *From Sea to Sea: Illustrated History of the Canal du Midi* (Grenoble: Euromapping, 1994; original: Allen Lane, 1973), 12, 90.

6. Reported in Benjamin Wright to David Thomas (Rome to Scipio, Cayuga County), January 4, 1821.

7. Ronald B. Shaw, *Erie Water West* (Lexington: University of Kentucky Press, 1966), 148; Nobel Whitford, *History of the Canal System of the State of New York* (Albany: State of New York, 1906), 116.

8. Early twentieth-century local historian Joshua Wilbur lists every one of them, and then year by year adds all newcomers by name, "Town and City of Lockport," in *Souvenir History of Niagara County New York* (Lockport: Niagara County Genealogical Society, 1986, 2nd edition; original: Lockport: Pioneer Association of Niagara County, 1902), 144–50.

9. Orasmus Turner, *Pioneer History of the Holland Purchase* (Buffalo: Jewett, Thomas and Company, 1849), 551, 654–55; other sources include: *Illustrated History of Niagara County, New York* (New York: Sanford and Company, 1878), 165–70; Clarence O. Lewis, *The Erie Canal, 1817–1967*, Occasional Contributions of the Niagara County Historical Society (Lockport: Niagara County Historical Society, 1967), 21–23; Wilbur, "Town and City," in *Souvenir History*, 146–55.

10. Wilbur, "Town and City," in *Souvenir History*, 148; *Illustrated History*, 165; George C. Lewis, "Early History of Lockport and Vicinity," n.p., n.d., mimeograph, Niagara County Historian's Office, 4.

11. *Illustrated History*, 165.

12. Quoted in full in Turner, *Pioneer History*, 654–55.

13. Edna Deane Thomas (Smith), "Recollections of an Early Settler," transcribed by Laura P. Colton, and published in five parts in the *Lockport Daily Union* beginning with part I (August 11, 1873), part II (August 14), part III (August 21), part IV (August 28), and part V (September 6), part IV. The author was Edna Smith during the construction period; later she remarried.

14. Edna Thomas (Smith), "Recollections," part III.

15. Turner, *Pioneer History*, 655.

16. Edna Thomas (Smith), "Recollections," part I.

17. *Niagara Democrat*, January 2, 1851; Lockport *Union-Sun & Journal*, July 17, 1965.

18. Elizabeth Wayman, *Middleport Herald*, March 8, 1851.

19. *Illustrated History*, 171.

20. Edna Thomas (Smith), "Recollections," part I.

21. Report of the Canal Commissioners, February 1822, *Laws*, II: 66.

22. Wilbur, "Town and City," in *Souvenir History*, 151–52.

23. Turner, *Pioneer History*, after p. 562.

24. Shaw, *Erie Water*, 92.

25. Annual Report of the Canal Commissioners, February 1822, in *Laws*, II: 64–65.

26. Although the original Mountain Ridge contracts are straightforward, a growing crisis led to multiple subdivisions. Sources include: "Contracts and Accounts for construction and repair," Office of the Canal Commissioners, contains original contracts and vouchers; Commissioner William Bouck's monthly and yearly vouchers from the Mountain Ridge for the period 1823 to 1825; "Accounts of monies paid to contractors and others for construction, repair and enlargement of the Erie and Champlain Canals, 1817–1871" are ledger books that were kept in Albany in which payments to contractors were recorded; they are listed by contract number and contractor and sometimes mention the section where the work was being done; combined, these records do not allow us to unambiguously identify every contract and section and assign it a definite location. Hence the maps in this volume for 1823 and 1824, and the corresponding information in the appendix, provide an approximate picture based on certain assumptions that I am careful to name. Less precise information comes from Turner, *Pioneer History*, 655–56 (he erroneously mentions five instead of four contracts in 1821) and The *Niagara County Business Directory*, c. 1870, which mentions six original contracts.

27. Lewis, *Erie Canal*, 10.

28. Cadwallader D. Colden, *Memoir at the Celebration of the Completion of the New York Canals* (Albany: State of New York, 1825), puts the figure at one thousand to fifteen hundred, 368; Carol Sheriff, *Artificial River: The Erie Canal and the Paradoxes of Progress, 1817–1862* (New York: Hill and Wang, 1996), suggests one thousand to eleven hundred, 43; Shaw, *Erie Water*, cites Rochester *Telegraph*, May 6, 1823, and estimates one thousand at that date, 130.

29. Peter Way, *Common Labour: Workers and the Digging of North American Canals, 1780–1860* (Baltimore: Johns Hopkins University Press, 1993), 102.

30. Tyler Anbinder, *Five Points: The Nineteenth-Century New York City Neighborhood That Invented Tap Dance, Stole Elections, and Became the World's Most Notorious Slum* (New York: Plume, 2001); see also Edwin G. Burrows and Mike Wallace, *Gotham: A History of New York City to 1898* (New York: Oxford University Press, 1999).

31. Way, *Common Labour*, 106–7.

32. Letter from Myron Holley, Canandaigua, June 20, 1821, to his father, Luther Holley, Esquire, Salisbury, Connecticut, Papers of Myron Holley, New York State Archives.

33. The description by Morris H. Tucker, Esquire, of Lockport in the summer of 1821 is printed in full in Turner, *Pioneer History*, appendix, 654–55.

34. The Niagara County Historian's Office has appended a number of miscellaneous items to Mrs. Smith's "Recollections." One is an obituary published in the Union Springs, New York, *Advertiser*, undated.

35. Edna Thomas (Smith), "Recollections," part I.

36. Marcus Moses, description of Main Street in 1823, reprinted in *Illustrated History*, 167–68.

37. See Edna Thomas (Smith), "Recollections," parts I and V; and *Illustrated History*, 168–69.

38. Edna Thomas (Smith), "Recollections," part V.

39. Report of the Canal Commissioners, February 1822, *Laws*, II: 60–78, 66. The report states that "the grubbing and clearing is nearly all done and they have excavated about fifteen thousand cubic yards of rock and probably seventy thousand of earth." Nathan Roberts advertised in early June of 1822 for one thousand hands at Lockport, *Souvenir History*, 151.

40. Report of the Canal Commissioners, February 1822, *Laws*, II: 66.

41. D. Clinton (Albany) to D. Thomas (Union Springs) March 16, 1822, David Thomas papers, New York State Archives.

42. Report of James Geddes, February 1822, *Laws*, II: 529.

43. William Wyckoff, *The Developer's Frontier: The Making of the Western New York Landscape* (New Haven: Yale University Press, 1988), 144–47.

44. See Shaw, *Erie Water*, 148–65, 164–80.

45. *Laws*, II: 519–43, contains all the resolutions related to the terminus question; see also Shaw, *Erie Water*, 147–63.

46. Shaw, *Erie Water*, 152.

47. Report of the Canal Commissioners, February 1822, *Laws*, II: 66; Annual Report of the Canal Commissioners, February 24, 1823, *Laws*, II: 99.

48. Report of the Canal Commissioners, February 24, 1823, *Laws*, II: 98.

49. Ibid.

50. Ibid.

51. Shaw, *Erie Water*, 130.

52. Minutes of Canal Commissioners, meeting of June 10, 1822, Buffalo; see also 1822 contract and 1822 expense vouchers of Darius Comstock.

53. Report of the Canal Commissioners, February 24, 1823, *Laws*, II: 98.

54. Contracts of 1822 for Jonathan Child and Elijah Hamlin (new Mountain Ridge section 2), Darius Comstock (new section 3), and John Gilbert (new section 4).

55. William Bouck's monthly vouchers, 1823–1825.

56. Report of the Canal Commissioners, February 24, 1823, *Laws*, II: 100.

57. Ibid., 99.

58. Benjamin Wright (Rome) to David Thomas (Scipio), January 4, 1821.

59. Report of the Canal Commissioners, February 24, 1823, *Laws*, II: 99.

60. Ibid., 97.

61. Chipman P. Turner, brother of newspaper editor Orasmus Turner, was the rider, *Souvenir History*, 151; *Illustrated History*, 167.

62. *Souvenir History*, 150–51.

63. Edna Thomas (Smith), "Recollections," part II.

64. *Illustrated History*, 174, the printer was Bartimus Ferguson; see also *Souvenir History*, 150.

65. Edna Thomas (Smith), "Recollections," part II; see also *Illustrated History*, 167; and *Souvenir History*, 151.

66. *Souvenir History*, 151.

67. See, for example, Turner, *Pioneer History*, 618.

68. Turner, *Pioneer History*, 655–56; see also *Illustrated History*, 174; *Souvenir History*, 150.

69. Lewis, *Erie Canal*, 13; the original paper no longer exists.

70. Lyman A. Spalding's journals, 9 volumes, as well as his correspondence and other papers, Syracuse University, Special Collections; the Manuscript Collection at Cornell University has additional correspondence and papers; among his published works are *Recollections of the War of 1812 and Early Life in Western New York*, Occasional Contributions of the Niagara County Historical Society, no. 2 (Lockport: Niagara County Historical Society, 1949); Spalding also briefly published a partisan newspaper, *Priest Craft Exposed*, 1828–1830; see also *Illustrated History*, 176.

71. Spalding, *Recollections of the War of 1812*, 3–12.

72. Shaw, *Erie Water*, 135–39.

73. Spalding, *Recollections of the War of 1812*, 16.

74. Spalding, *Recollections of the War of 1812*, 16–17.

75. Quoted in Lewis, *The Erie Canal*, 12–13.

76. Quoted in *Souvenir History*, 150; the date indicated, December 7, 1821, must be wrong since the Lockport *Observatory* did not yet exist; it must be 1822, although there are no newspapers records to verify this inference.

77. Rochester *Telegraph*, December 31, 1822.

78. *Illustrated History*, 167.

79. George E. Condon, *Stars in the Water: The Story of the Erie Canal* (Garden City, NY: Doubleday, 1974), 91.

80. *Niagara Sentinel*, January 17, 1823; reprinted in Rochester *Telegraph*, January 28, 1823: the eight were Louis Berque, William McKinstry, James Kelly, James Josslin, Michael Moran, Patrick Looney, Francis Buck, and Michael Flemming.

81. Rochester was founded by William's father, Nathaniel Rochester.

82. *Niagara Sentinel*, October 10, 1823, announced the beginning of the trial "Tuesday last"; *Niagara Sentinel*, March 12, 1824 announced Kelly's acquittal.

83. *Niagara Sentinel*, April 2, April 30, and May 14, 1824.

84. Rochester *Telegraph*, December 31, 1823.

85. Blake McKelvey, "The Erie Canal: Mother of Cities," *New York Historical Society Quarterly* 35 (1951): 55–80, 64.

86. Report of the Canal Commissioners, February 24, 1823, *Laws*, II: 116–17.

87. *Laws*, II: 374–75; Colden, *Memoir*, 67–68; Shaw, *Erie Water*, 157–61.

88. Shaw, *Erie Water*, 162.

89. DeWitt Clinton, letter of February 23, 1822, is quoted in J. J. Thomas, "Memoir of David Thomas," a paper read before the Cayuga County Historical Society, December 17, 1878, 8.

90. DeWitt Clinton to David Thomas (Albany to Union Springs, New York), March 16, 1822.

91. Thomas signed the contract appointing Roberts as chief engineer, dated December 19, 1823. He also signed Comstock and Hathaway's account of expenses on the same day. William Bouck's yearly report shows Thomas at work for all of 1823, but not 1824. Whitford, *History of the New York Canals*, states that Nathan Roberts served as chief engineer from 1822 to 1825, 1163. Moreover, Thomas continued to receive some correspondence at Lockport in the fall of 1823, see Clinton (Albany) to David Thomas (Lockport), September 7, 1823; and Charles E. Dudley (Albany) to David Thomas (Lockport), both in David Thomas papers, September 29, 1823. Turner, in his *Pioneer History*, states: "His sudden removal from a sphere of great usefulness, in which no blemish or wrong doing was shown, with another memorable instance, must always be passed over by the historians, with the conclusion that the times, and not the men, were of fault," 632. On the antagonism between Thomas and his two rivals, Geddes and Roberts, see Shaw, *Erie Water*, 149, 156–57.

92. Shaw, *Erie Water*, 165.

93. Report of the Canal Commissioners, February 20, 1824, *Laws*, II: 160–61.

94. Ibid., 161.

95. Russell Bourne, in *Floating West: The Erie and Other American Canals* (New York.: W.W. Norton, 1992), uses this suggestive term, but his version of events at the Mountain Ridge is decidedly at odds with the surviving evidence. Condon, *Stars in the Water*, also speaks of "the miracle of Lockport," 85–92.

96. In section 2, Child and Hamlin's payments increased from $25,500 in 1822 to $46,000 in 1823; in section 3, those of Darius Comstock increased from $30,000 to $54,353, and in section 4, Gilbert's increased from $16,500 to $40,550; see Bouck's yearly report of vouchers for 1823, and Albany Record Book of Contracts. In at least one contract, the 1823 increase was even greater; see voucher for Boughton, Lobdell, Smith, and Maynard, December 16, 1824; their expenses quadrupled in 1823.

97. Lockport *Observatory*, October, 1822, quoted in Lewis, *Erie Canal*, 13.

98. This assumes between $4.75 and $5.00 per keg, Bouck's yearly expense vouchers 1823.

99. See, for example, Walter Licht, *Industrializing America: The Nineteenth Century* (Baltimore: Johns Hopkins University Press, 1995), 42–44; and Richard D. Brown, *Modernization: The Transformation of American Life, 1600–1865* (New York: Hill and Wang, 1976), 122–43.

100. See Bourne, *Floating West*, 111–12; also Eric Brunger and Lionel Wyld, "The Grand Canal: New York's First Thruway," Publications of the Buffalo and Erie County Historical Society, vol. XII, 1964, Buffalo, 10.

101. Lewis, *The Erie Canal*, 10; Condon, *Stars in the Water*, 87.

102. Brunger and Wyld, "The Grand Canal," 10; Lewis, *The Erie Canal*, 10.

103. Catlin's *Process of Excavation, Lockport* in Colden, *Memoir*, opposite p. 298; see Lewis, *The Erie Canal*, 10–11; and Condon, *Stars in the Water*, 90–91; while the exact timing of these innovations is uncertain, Lewis dates the crane to 1824.

104. Colden, *Memoir*, 68.

105. Perhaps anticipating the coming subdivision, he ordered the printing of one hundred contracts on August 2, 1822, Canal Contracts and Vouchers; Myron Holley's last Mountain Ridge signature is on a voucher dated May 11, 1822.

106. Turner, *Pioneer History*, 632.

107. Report of the Canal Commissioners, February 20, 1824, *Laws*, II: 161.

108. Contracts and expense vouchers for 1823, 1824, 1825.

109. Expense voucher for work done from May 1822 to December 1824, Bouck's vouchers, canal contracts.

110. Bouck's contracts, dated July 26, 1823; Phelps also received 20 cents a day for board and deducted from his expenses $100.87 for "the use of Shanties at Lockport on my canal contract." Did the state own these shanties? Perhaps, if they were built under the stipulations of the 1822 contracts.

111. See Oliver Phelps's voucher dated October 30, 1824.

112. Wyckoff, *Developer's Frontier*, 16.

113. The executive decision to give part of Boughton, Culver, and Maynard's section to Phelps seems a breach of the former's contract. It may have raised some hackles since they finally agreed, in a voucher of February 27, 1824, to turn over this work to him as a subcontractor. Perhaps these contractors, who were building the double flights of combined locks, were challenged enough in that task.

114. Report of the Canal Commissioners, February 20, 1824, *Laws*, II: 161–62.

115. These were contracts to: Norton, Bates, and House; Reuben Heacock; Seymour Scovell (all on original section 3); and to Daniel Brown; Bell and Richardson; and Lobdell and Boughton (all on original section 4); all contracts were signed by Bouck.

116. Contract to Boughton, Culver, and Maynard, December 19, 1823.

117. Contract dated December 1, 1823; his completion date was May 1, 1825; they offered $1.84 per cubic yard for completing 11,000 yards, and $1.82 for 10,000 yards.

118. Thomas received his last payment on his engineering contract from Bouck on May 3, 1824; "Accounts of monies paid to contractors and other for construction, repair and enlargement of the Erie and Champlain Canals, 1817–1871."

119. *Souvenir History*, 152; Culver and Maynard's lock contract of October 6, 1823, specifies a price of $874.50 per foot of lift. That amounts to only $52,470.

120. *Illustrated History*, 168.

121. Marcus Moses, "Main Street in 1823," *Illustrated History*, 167–68.

122. Holmes Hutchinson (Utica) to Lyman Spalding (Lockport), December 16, 1823; Cornell University Library.

123. Quoted in J. P. Merritt. *Biography of Hon. W. H. Merritt, M.P.* (St. Catherine, ON: E. S. Leavenworth, 1875), 57; see also Roberta M. Styran and Robert R. Taylor, with John Jackson, *The Welland Canals: The Growth of Mr. Merritt's Ditch* (Erin, ON: Boston Mills Press, 1988), xvii and 27.

124. Shaw, *Erie Water*,164–80; Whitford, *History of the New York Canals*, 118–20; Holley was later cleared of wrongdoing.

125. Report of the Canal Commissioners, March 4, 1825, *Laws*, II: 244.

126. Report of the Canal Commissioners, February 20, 1824, *Laws*, II: 173.

127. Report of the Canal Commissioners, February 20, 1824, *Laws*, II: 172.

128. Report of the Canal Commissioners, February 20, 1824, *Laws*, II: 165; the Champlain Canal was also completed in 1823, 151.

129. Rochester *Telegraph*, May 17, 1824, repeats the Buffalo *Journal* article dated May 10, 1824.

130. Rochester *Telegraph*, Tuesday, July 27, 1824; reprinted from Lockport *Observatory*, July 13, 1824.

131. Ibid.

132. Ibid.

133. *The Niagara Sentinel*, Lewiston, N.Y., July 16, 1824.

134. Ibid.

135. Way, *Common Labour*, 287, table 17.

136. There would be about twenty-five more military interventions during construction of North American canals, Way, *Common Labour*, table 17, 278.

137. Shaw, *Erie Water*, 132.

138. Quoted in Walter J. Walsh, "Religion, Ethnicity, and History," in *The New York Irish*, ed. Ronald E. Bayor and Timothy J. Meagher (Baltimore: Johns Hopkins University Press, 1996), 48–69; discussion of Greenwich Village Riot, 61–64; see also Burroughs and Wallace, *Gotham*, 491–92.

139. Burroughs and Wallace, *Gotham*,192.

140. The church was founded on the "third Sabbath of January 1823"; although the Quaker Meeting House had been built in 1819, the Quakers refused, "as a matter of conscience" to accept the Holland Land Company's offer of one hundred acres to the first church in the village for support of its pastor; the Presbyterians then accepted the offer; see Gertrude Strauss, *Our First One Hundred and Fifty Years* (Lockport: First Presbyterian Church, 1973), 6, 9.

141. Session Records, January 1823 to August 1838; while the Presbyterian records are remarkably intact, many other church records, like Lockport's early newspaper records, were destroyed in fires.

142. Way, *Common Labour*, 113.

143. He renewed this practice in 1822; see Minutes of the New York Canal Commissioners July 16, 1821 (vol. I: 23) and February 7 1822 (vol. I: 23); Shaw, *Erie Water*, 129; Way, *Common Labour*, 99.

144. Way, *Common Labour*, 193.

145. Ibid.,150, 160.

146. Ibid.,160–61.

147. Ibid., 98.

148. Whitford, *History of the New York Canals*, 91; Way, *Common Labour*, 99.

149. Way, *Common Labour*, 112.

150. Ibid., 62.

151. Stephen John Hartnett, *Democratic Dissent and the Cultural Fictions of Antebellum America* (Urbana: University of Illinois Press, 2002), 40–92; Sheriff, *Artificial River*, 27–51.

152. Shaw, *Erie Water*, 91.

153. Way, *Common Labour*, 115–30.

154. Not surprisingly, there is no surviving proof of prostitution during Erie Canal construction; see Way, *Common Labour*, 85, 171; we do know that an 1832 visitor who encountered two prostitutes at Lockport was told that the "neighborhood abounded with similar characters"; Rev. Isaac Fidler, *Observations on Professions, Literature, Manners, and Emigration, in the United States and Canada, Made during a Residence There in 1832* (Reprint: New York: Arno Press, 1974; original: London: Whittaker, Treacher, and Co., 1833), 203–4.

155. Shaw, *Erie Water*, 133.

156. *Niagara Sentinel*, March 12, 1824.

157. Frederick Marryat, *Diary in America* (Bloomington: Indiana University Press, 1960; orig. ed. 1839), 92–93; quoted in Way, *Common Labour*, 144.

158. Charles Dickens, *American Notes, and Reprinted Pieces* (London: Chapman Hall, 1868), 126; quoted in Way, *Common Labour*, 144; see also Sheriff, *Artificial River*, 42.

159. Edna Thomas (Smith), "Recollections," part III.

160. Liston Edgington Leyendecker, *Palace Car Prince: A Biography of George Mortimer Pullman* (Boulder: University of Colorado Press, 1992).

161. Edna Thomas (Smith), "Recollections," part IV.

162. At the St. Mary's Canal at Sault Ste. Marie, one hundred workers occupied a structure one hundred by twenty-two feet; Way, *Common Labour*, 146.

163. Expense voucher from Bouck to Otis Hathaway, for "services as overseer," May 6, 1825.

164. Expense voucher, from Bouck to Boughton and Lobdell, December 1824.

165. Voucher to from Bouck to Seymour Scovell, December 3, 1825; it is probable he was getting his whiskey much cheaper than $0.50 per gallon.

166. Way, *Common Labour*, 183; Condon, *Stars in the Water*, 68.

167. Expense voucher from Bouck to Boughton and Lobdell, December, 1824: $16 a barrel. An undated receipt, found with Bouck's 1822 vouchers, shows Bouck paying $109.50 for 789 gallons ($0.28 per gallon).

168. Richard D. Borgeson, "Irish Canal Laborers in America 1817–1846," MA thesis, History, Pennsylvania State University, 1964, 63, says the price varied between $0.05 and $0.25 a quart; given the inflated prices caused by Lockport's isolation the higher figure seems more appropriate.

169. Quoted in William Hodge, *Canal Enlargement in New York State and Papers on the Barge Canal Campaign* (Buffalo: Buffalo Historical Society, 1909), 309.

170. See Borgeson, "Irish Canal Laborers," 63; Marvin A. Rapp, *Canal Water and Whiskey: Tall Tales from the Erie Canal Country* (New York: Twayne Publishers, 1965), 40; *New York Folklore Quarterly* 11 (winter 1955): 297; see also Condon, *Stars in the Water*, 92. This story is preposterous if we consider that one hundred men would have been a very large crew on this job and for them to drain the entire barrel (thirty-one gallons), they would each have had to drink forty ounces, an amount that probably would have prevented them from standing let alone continuing to the next barrel.

171. Borgeson, "Irish Canal Laborers," 62.

172. W. J. Rorabaugh, "Estimated U.S. Alcoholic Beverage Consumption, 1790–1860," *Journal of Studies on Alcohol* 37 (March 1976): 357–64; cited in Way, *Common Labour*, 183.

173. Report of the Canal Commissioners, February 24, 1823, *Laws*, II: 105–7; Report of the Canal Commissioners, February 20, 1824, *Laws*, II: 177–79. The report also listed beer, ale, cider, and several kinds of brandy. Other common beverages included rum and blackstrap (a mixture of rum and molasses), see Borgeson, "Irish Canal Laborers," 62.

174. William Rorabaugh, *The Alcoholic Republic: An American Tradition* (New York: Oxford University Press, 1981), see also Sellers, *Market Revolution*, 259–65.

175. Way, *Common Labour*, 181–87.

176. Peter Way says that "the common explanation of this period" was to "attribute labor disputes to the bottle," *Common Labour*, 186.

177. Way, *Common Labour*, 133, 134.

178. Rochester *Telegraph*, April 13, 1824.

179. "Fatal accident," a Lockport *Observatory* story reprinted in the *Niagara Sentinel*, April 23, 1824.

180. The fires were in 1843, 1849, 1854, 1881, and 1928; Lockport *Union-Sun & Journal*, July 17, 1965, 12–B.

181. Spalding, *Recollections*, 16.

182. Turner, *Pioneer History*, 656.

183. Edna Thomas (Smith), "Recollections," part IV, "The Dangers of Blasting."

184. Colden, *Memoir*, after p. 298.

185. Victor Lefebure, *The Riddle of the Rhine: Chemical Strategy in Peace and War* (New York: The Chemical Foundation, 1923), 58.

186. Dr. Thomas L. Nichols, *Forty Years of American Life*, 2 vols. (London: John Maxwell and Company, 1864), vol. 1, 120.

187. Way, *Common Labour*, 173–76.

188. Ibid., 90.

189. Edna Thomas (Smith), "Recollections," part IV.

190. Ibid.

191. Giorgio Agamben, *Homo Sacer: Sovereign Power and Bare Life* (Stanford: Stanford University Press, 1998).

192. Edna Thomas (Smith), "Recollections," part V.

193. Steven Mintz, *Moralists and Modernizers: America's Pre-Civil War Reformers* (Baltimore: Johns Hopkins University Press, 1995).

194. Way, *Common Labour*, 173–74; on laborers' entertainments, see Borgeson, "Irish Canal Laborers," 62.

195. Way, *Common Labour*, 172.

196. Ibid., 195.

197. Ibid., 194.

198. Ibid., 199; another view is presented by Sean Wilentz, *Chants Democratic: New York City and the Rise of the American Working Class, 1788–1850* (New York: Oxford University Press, 2004); Wilentz highlights workers' agency, although his focus is on the artisans and skilled laborers rather than unskilled common laborers.

199. Edna Thomas (Smith), "Recollections," part III.

200. Ibid.

201. "1 Killed, 12 hurt In Canal Rioting," Lockport *Union-Sun & Journal*, July 17, 1965, 10–B.

202. "Murder," Lockport *Observatory*, February 6, 1823, reprinted by the Rochester *Telegraph*, Tuesday, February 11, 1823.

203. Clarence O. Lewis, "The Lockport Story," in *Souvenir Program Commemorating the Lockport Centennial, 1865–1965* (Lockport: 1965), 23.

204. David R. Roediger, *Wages of Whiteness: Race and the Making of the American Working Class*, 2nd ed. (London and New York: Verso, 1999).

205. Noel Ignatiev, *How the Irish Became White* (New York: Routledge, 1995), 5–10.

206. "Kennedy Spoke Here During '60 Campaign," Lockport *Union-Sun & Journal*, July 17, 1965, 14–D.

207. Edna Thomas (Smith), "Recollections," part V.

208. Turner, *Pioneer History*, 655.

209. *Souvenir History*, 148–49.

210. Voucher from Davis Hurd to I. J. Thomas, September 11, 1824, for 271 days as assistant engineer at $2 a day, $542; approved by Bouck.

211. Buffalo *Journal*, January 15, 1822; see also Shaw, *Erie Water*, 131.

212. Advertisement quoted in *Souvenir History*, 150.

213. Lockport *Observatory*, Saturday, October 29, 1825, describes such a celebration at Lockport.

214. Edna Thomas (Smith), "Recollections," part II; a similar celebration ignited when the Lockport Hotel completed it's planed floor and held a dancing party, see Turner, *Pioneer History*, 656.

215. *Souvenir History*, 154.

216. Moses, "Main Street in 1823," in *Illustrated History*, 168.

217. Edna Thomas (Smith), "Recollections," part V.

218. Patrick McGreevy, "Place and the American Christmas," *Geographical Review* 80, no. 1 (January 1990): 32–42; Christopher Lasch, *Haven in a Heartless World* (New York: Basic Books, 1977).

219. Edna Thomas (Smith), "Recollections," part IV; the man's name was John Allen.

220. Sellers, *Market Revolution*, 237–39.

221. Ibid., 237.

222. Quoted in Sellers, *The Market Revolution*, 238.

223. Sellers, The Market Revolution, 237.

224. Edna Thomas (Smith), "Recollections," part III.

225. Mariam Fraser and Monica Greco, Introduction in *The Body: A Reader*, ed. Mariam Fraser and Monica Greco (London: Routledge, 2005), 1–42, 9.

226. See, for example, Mintz, *Moralists and Modernizers*, for an overview; chapter 5 of this book deals with these issues in more detail.

227. Sheriff, *Artificial River*, 50.

228. Paul E. Johnson, *A Shopkeeper's Millennium: Society and Revivals in Rochester, New York, 1815–1837* (New York: Hill and Wang, 1978); Whitney R. Cross, *The*

Burned-Over District: The Social and Intellectual History of Enthusiastic Religion in Western New York, 1800–1850 (Ithaca: Cornell University Press, 1950).

229. Report of the Canal Commissioners, March 4, 1825, *Laws*, II: 245.

230. There were also 12,000 yards of hardpan, a mixture of rock and earth, to be excavated; Report of the Canal Commissioners, February 20, 1824, *Laws*, II: 245.

231. There were two contracts on both divisions of original section 1; the additional contracts were to Oliver Culver, and to Orange Dibble and Jonathan Olmstead, both originated in 1823. The appendix provides additional details.

232. Rochester *Telegraph*, September 14, 1824; the Oak Orchard Feeder provided the water.

233. "Erie Canal Packet Boats," Rochester *Telegraph*, May 10, 1824.

234. Report of the Canal Commissioners, March 4, 1825, *Laws*, II: 245–46; they also noted that the line to Black Rock was nearly completed.

235. Contract of Horace Turner and Ebenezer Jackson, December 28, 1824, $1,000, "to be completed by May 1, 1825"; Assistant Engineer Alfred Barnett, who certified the voucher, reduced the payment to $900 because "the road was not completed as directed."

236. Horatio Gates Spafford, *A Gazetteer of the State of New York* (Albany: B. D. Packard, 1824; reprinted: Interlaken, NY: Heart of the Lakes Publishing, 1982), 290.

237. *Niagara Sentinel*, July 2, 1824.

238. Reprinted in the Rochester *Telegraph*, July 6, 1824.

239. Spafford, *Gazetteer*, 290.

240. In an expense voucher for December 27,1825, for example, contractors Bell and Richardson request payment on a particular job for 2,448 days of labor, and they provide exact beginning and ending dates indicating the job was finished in seventeen working days. To complete 2,448 days of labor in seventeen days would require 144 men. By similar calculations we can estimate several additional crews as follows: Jared Boughton and Jacob Lobdell, in a voucher of December 1824, listed 1,975 days in thirty working days indicating a crew of sixty-six; Reuben B. Heacock, in a voucher of December 3, 1825, listed 2,926 excavating days in twenty-six work days Heacock indicating a crew of 113; and Scovell (voucher is dated December 3, 1825) listed his fifty-four men by name. The average size of these crews was ninety-four. Since contractors were generally housing these workers and paying them on a monthly basis, crew sizes were presumably somewhat stable. Generalizing this average figure to all twenty-two contracts that were active at least until July of 1824 yields a total workforce of 2,068.

241. Although Bouck continued to issue large disbursements to Darius Comstock on his contract for section 3.

242. Bell and Richardson, voucher of December 3, 1824.

243. Listed on both Bouck's monthly and yearly vouchers for 1825.

244. John Gilbert, the contractor for section 4 (1822 numbering), continued to receive large disbursements until August. Bouck directed the largest amount of effort to section 5, particularly the contracts of Bell and Richardson, E. F. Norton, and Reuben Heacock. These contractors received large disbursements through August. Some contractors on section 6 also received payments through late summer though they were not as large; Bouck's monthly and yearly vouchers, 1825.

245. Whitford, *History of the New York Canals*, 123; the line reached Buffalo in August of 1825.

246. Andrew Burstein, *America's Jubilee* (New York: Alfred A. Knopf, 2001), 6.

247. Benedict Anderson, *Imagined Communities: Reflections on the Origin and Spread of Nationalism* (London: Verso, 1983).

248. John Seelye, *Beautiful Machine: Rivers and the Republican Plan; 1755–1825* (New York: Oxford University Press, 1991), 293; see also Burstein, *America's Jubilee*, 10, 32;.

249. Auguste Levasseur, *Lafayette in America in 1824 and 1825; or, Journal of a Voyage to the United States*, 2 volumes (Philadelphia; 1829). Facsimile version by Research Reprints, New York, 1970), Volume II: 213; see also, Lewis, "The Story of Lockport," in *Lockport Centennial*, 9–11; and Condon, *Stars in the Water*, 89.

250. Way, Common Labour,174.

251. Levasseur, *Lafayette in America*, II: 213–14.

252. Levasseur, *Lafayette in America*, II: 214.

253. Rochester *Telegraph*, June 14, 1825.

254. Lockport *Observatory*, Thursday, July 8; reprinted in the Rochester *Telegraph*, July 13, 1825; and in the *Niagara Sentinel*, July 16, 1825.

255. Lewis, *The Erie Canal*, 7.

256. There were two separate announcements in the Rochester *Telegraph* on June 7, 1825; and on June 14, 1825.

257. Vouchers to George Whipple for painting railings (December 7, 1825); Joseph Latheld for drilling holes for railings (December 7, 1825); Spalding and Rogers for building a wharf (December 7, 1825); Darius D. Watterhouse for lighting lamps (December 24, 1825); Bell and Richardson for "19 days work done about the locks" (December 29, 1825); all approved by Bouck.

258. The receipt survives among Bouck's vouchers at the New York State Archives. On May 21, 1825, Bouck paid Thomas Morgan $66.75 for the tablets. Morgan instructed him to pay Spalding and Rogers. A second date (May 28, 1825) is also listed on the voucher, and Spalding and Rogers receive $74.06 on July 23, 1825. The difference may represent the cost of transport. At least some of the stone for the locks came from the quarry of Alman Millard just northwest of the village; voucher to Alman Millard for $230.98 for "cutting stone from his own quarry for Lockport Locks," 1824; Millard, who built Lockport's first framed house, was also a major land speculator; see Holland Land Company parcel maps, Township 14, Range 7; and *Souvenir History*, 148.

259. They referred to the flight as "Neptune's Staircase," Styran and Taylor, *The Welland Canals*, 30 and fig. 3.1, "The Welland Canal, 1837," 42.

260. L. T. C. Rolt, *From Sea to Sea* (Seyssinet, FR: Euromapping, 1994; original: by Allen Lane, 1973), 12, 90; British examples included the five combined locks at Bingley in Yorkshire and "Neptune's Staircase" of eight chambers on the Caledonian Canal in Scotland, Henry Dale and Rodney Dale, *The Industrial Revolution* (New York: Oxford University Press, 1992), 32.

261. Whitford, History of the New York Canals, 797–98.

262. John Prindle, "Tales of the Towpath," *Canal Times*, 1990, 3, 12; the long-lost tablet was discovered at the Schoharie Crossing Visitor's Center by Tom

Grosso, who also had the detailed knowledge and presence of mind to compare it to Thomas Evershed's 1839 sketch of the locks, which shows the placement of the distinctive oval tablet.

263. Lewis, *The Erie Canal*, 23.

264. Rochester *Telegraph*, June 21, 1825.

265. Sheriff, *Artificial River*, 35, 45; Shaw, *Erie Water*, 406–7; Way, *Common Labour*, 76–78.

266. Way, *Common Labour*, 78.

267. Sheriff, *Artificial River*, 45.

268. "Lockport Irish Dedicate Plaque," Lockport *Union–Sun & Journal*, Friday March 17, 1989.

269. Rochester *Telegraph*, June 28, 1825.

270. Sheriff, *Artificial River*, 50.

271. Reverend F. H. Cuming, "An Address Delivered at the Laying of the Cap-Stone, of the Ten Combined Locks at Lockport, on the Anniversary of St. John the Baptist," June 24, 1825 (Lockport: By request of Ames's Chapter of Royal Arch Masons, 1825).

272. Rochester *Telegraph*, June 28, 1825.

273. Karl Bernhard, the Duke of Saxe-Weimar, *Travels Through North America, During the Year 1825 and 1826*, vol. 1 (Philadelphia: Carey, Lea and Carey, 1828), 67–68.

274. Colden, *Memoir*, 133; Turner, *Holland Land Purchase*, 632–33.

275. Whitford, *History of the New York Canals*, vol. 1, 123.

276. Letter from J. G. Bond (Lockport) to Joseph Ellicott (Batavia), September 29, 1825, on behalf of the committee; in addition to the invitation, the letter also contained a copy of the announcement "Meeting of the Waters," D. Comstock, chairman, dated September 23, 1825, Lockport.

277. Colden, *Memoir*, 143–44.

278. Turner, *Holland Purchase*, 633.

279. Walter Edmonds, *Erie Water* (Boston: Little, Brown and Company, 1933), 497.

Chapter 4. Writing Lockport: A Dialogue on American Progress

1. John Locke, *An Essay Concerning Human Understanding*, ed. P. H. Nidditch (Oxford: Oxford University Press, 1975; rev. 1982) book II, chapter 11, 17; S. H. Clarke, " 'The Whole Internal World His Own': Locke and Metaphor Reconsidered," *Journal of the History of Ideas* 59, no. 2 (1998): 241–65.

2. Lionel D. Wyld, *Low Bridge: Folklore and the Erie Canal* (Syracuse: Syracuse University Press, 1962), 176.

3. C. F. Briggs, "Lockport," in *The United States Illustrated; In Views of City and Country, With Descriptive and Historical Articles*, ed. Charles A. Dana (New York: H. J. Meyer [1853]), 162–64, included an illustration, *Erie Canal at Lockport, NY*, following p. 162.

4. Cadwallader D. Colden, *Memoir at the Celebration of the Completion of the New York Canals* (Albany: State of New York, 1825); *Laws of the State of New York, in relation to the Erie and Champlain Canals, together with the Annual reports of the Canal Commissioners, and other Documents . . .* , 2 vols. (Albany: Published by Authority of the State, E. and E. Hosford, Printers, 1825); Colden was also the biographer of Robert Fulton, see John Seelye, *Beautiful Machine: Rivers and the Republican Plan; 1755–1825* (New York: Oxford University Press, 1991), 322.

5. Colden, *Memoir*, 9; later Colden explicitly calls the canal "the most impressive example . . . since the adoption of the Federal Constitution, of the benefits of a free government, upon the character of a community," 70.

6. The idea that rapids and falls, and particularly Niagara, served as "an indirect means of conveying the energies of the opening and expanding republic" (Seelye, *Beautiful Machine*, 348) coincides with the views of Elizabeth McKinsey, "An American Icon," in *Niagara: Two Centuries of Changing Attitudes, 1697–1901*, ed. Jeremy Elwell Adamson (Washington, DC: The Corcoran Gallery of Art, 1985), 83–101.

7. Colden, *Memoir*, 65.

8. Ibid., 93.

9. Ibid., 77–78.

10. "Completion of the Grand Canal," *The Emporium*, Buffalo, October 29, 1825.

11. Seelye, *Beautiful Machine*, 330.

12. William L. Stone, "Narrative," in Colden, *Memoir*, 295–96.

13. Colden, *Memoir*, 88.

14. Seelye, *Beautiful Machine*, 331; Mordecai Manuel Noah, "Proclamation to the Jews," September 15, 1825, in *Passport to Utopia: Great Panaceas in American History*, ed. Arthur and Lila Weinberg (Chicago: Quadrangle Books, 1968), 39–48; "Ararat," *Niagara Gazette*, January 18, 1976.

15. Ronald B. Shaw, *Erie Water West* (Lexington: University of Kentucky Press, 1966), 186.

16. Nobel Whitford, *History of the Canal System of the State of New York*, 2 vols. (Albany: State of New York, 1906), 124.

17. Colden, *Memoir*, 198.

18. "Address delivered by Sheldon Smith, Esq. at the Celebration of the Completion of The Erie Canal, in Buffalo, October 26, 1825," *The Buffalo Journal and Mercantile Advertiser*, November 1, 1825.

19. Orasmus Turner described the celebration in the Lockport *Observatory*, Saturday, October 29, 1825; and also in his *Pioneer History of the Holland Purchase* (Buffalo: Jewett, Thomas and Co., 1849), 632–35, 633.

20. Lockport *Union-Sun & Journal*, July 17, 1965.

21. Turner, *Pioneer History*, 635.

22. Turner, Lockport *Observatory*, October 29, 1825.

23. Turner, *Pioneer History*, 635.

24. Ibid.

25. Lockport *Observatory*, October 29, 1825; Stone "Narrative," in Colden, *Memoir*, 297.

26. Stone "Narrative," in Colden, *Memoir*, 293.

27. Ibid., 297–98.

28. During the festivities, "the company were introduced to the memorable Enos Boughton, of Lockport, the pioneer of the Western District—the man who planted the first orchard, and built the first framed barn west of Utica!" Stone, "Narrative," in Colden, *Memoir*, 298. Boughton had kept a tavern at Schlosser on the Niagara River, the northern terminus of the portage around Niagara Falls that began at Lewiston, and, along with Benjamin Barton of Lewiston, was named ceremonial vice president of the Lockport celebration. Barton was a partner in the Porter Barton Portage Company. "It was rather ironical," historian Clarence Lewis suggests, "that he should be celebrating the occasion, because the canal was the beginning of the end of the lucrative portage business," "Celebrations Marked Completion of Canal," Lockport *Union-Sun & Journal*, July 17, 1965.

29. Shaw, in *Erie Water*, argues that the Lockport locks have "almost invariably been chosen by authors from that day to this to illustrate any work on The Erie Canal," 131; this trend has continued in recent works such as Carol Sheriff, *Artificial River: The Erie Canal and the Paradoxes of Progress, 1817–1862* (New York: Hill and Wang, 1996), Russell Bourne, in *Floating West: The Erie and Other American Canals* (New York: W.W. Norton, 1992), and Peter L. Bernstein, *Wedding the Waters: The Erie Canal and the Making of a Great Nation* (New York: W.W. Norton and Company, 2005).

30. Colden, *Memoir*, 366–67.

31. Ibid., 367.

32. Ibid., 369.

33. John F. Sears, *Sacred Places: American Tourist Attractions in the Nineteenth Century* (New York: Oxford University Press, 1989), see photograph "Screen Room of a Coal Breaker," 205.

34. Both of Catlin's images have a western vanishing point; this means that the *Seneca Chief* must be depicted on its return voyage to Buffalo; the location of the towpath cut into the side of the bank also confirms this.

35. Stone, "Narrative," in Colden, *Memoir*, 298.

36. Turner, Lockport *Observatory*, October 29, 1825.

37. Lockport *Observatory*, October 29, 1825.

38. Stone, "Narrative," in Colden, *Memoir*, 303.

39. Jefferson's letter of June 8, 1826, is reproduced in the final section of Colden's *Memoir*; it is in response to the reception of a commemorative medal and copy of the *Memoir*; this exchange occurred in the final year of Jefferson's life.

40. Stone, "Narrative," in Colden, *Memoir*, 299.

41. Ibid., 304.

42. Ibid., 305–6.

43. Ibid., 311.

44. Turner, *Pioneer History*, 632, 636.

45. Ibid., 636, Stone, "Narrative," in Colden, *Memoir*, 321.

46. Turner, *Pioneer History*, 636; the entire volume would take only another six months to complete.

47. Stone, "Narrative," in Colden, *Memoir*, 322–23.

48. Ibid., 337; C. Y. Turner's mural *The Marriage of the Waters* (1905) makes a similarly aggressive statement visually. As Governor Clinton slowly pours the Lake

Erie water, the smokestacks of an American steamship rise up parallel to the falling water, drowning the British warship—with its Union Jack—in thick smoke; the mural was housed in New York City's DeWitt Clinton High School; reproduced in Whitford, *History of the New York Canals*, volume I: 125.

49. Colden, *Memoir*, following p. 352, 122.

50. Colden, *Memoir*, 155; Seelye, *Beautiful Machine*, 330.

51. Colden, *Memoir*, 155, 262.

52. Apparently, women were relegated to the role of spectators; during the "Aquatic Display," two vessels were outfitted specifically for "ladies," Colden, *Memoir*, 352; and Stone, "Narrative," in Colden, *Memoir*, 325; for a discussion of the meaning of the laborers' procession, see Sean Wilentz, *Chants Democratic: New York City and the Rise of the American Working Class, 1788–1850* (Oxford: Oxford University Press, 2004), 87, 89–90, 96; Ronald Shaw points out that, although good feelings were the order of the day, some criticized the celebration, and when a scheduled balloon ascent failed, "thousands of angry people tore the balloon to pieces," *Erie Water*, 189–91nn. 18 and 22.

53. Colden, *Memoir*, 261–62.

54. "Invitation Card," Colden, *Memoir*, following p. 344; a similar design—showing three locks on the left, New York on the right, the state's coat of arms (the rising sun shield, the rising eagle, and the motto "Excelsior")—appears in the commemorative medal and in the frontispiece for the appendix, although, in these figures, the locks are not explicitly labeled. Colden, *Memoir*, 104; Whitford, *History of New York Canals*, following p. 128.

55 Stone, "Narrative," in Colden, *Memoir*, 331.

56. Colden, *Memoir*, 73–77.

57. Colden, *Memoir*, following pp. 4, 7, 113, and 125.

58. Seelye, *Beautiful Machine*, 293.

59. Sacvan Bercovitch, in *The American Jeremiad* (Madison: University of Wisconsin Press, 1978), for example, traces this notion to the earliest Puritan sermons.

60. Report of the Commissioners, February 1824, *Laws*, 156.

61. Quoted in Colden, *Memoir*, 174.

62. Laurence M. Hauptman, *Conspiracy of Interests: Iroquois Dispossession and the Rise of New York State* (Syracuse: Syracuse University Press, 1998).

63. Seelye, *Beautiful Machine*, 371.

64. Colden, *Memoir*, 250–54.

65. Ibid., 271–79; Seelye, *Beautiful Machine*, 344–48.

66. Samuel Woodworth, "Ode for the Canal Celebration," in Colden, *Memoir*, 250–51.

67. Annette Kolodny, *The Lay of the Land* (Chapel Hill: University of North Carolina Press, 1975).

68. Rochester *Telegraph*, October 18, 1825; in Colden, *Memoir*, 146–47; similarly, Mitchill speaks of the increased "intercourse" between regions made possible by the marriage of the "Lord of the Seas" and the "Lady of the Lakes," in Colden, *Memoir*, 278.

69. *Webster's Ninth New Collegiate Dictionary* (Springfield, MA: Merriam-Webster, 1984).

70. Turner, *Pioneer History*, 631.

71. Seelye, *Beautiful Machine*, 348–49.

72. Obituary of David Thomas, Lockport *Daily Journal and Courier*, April 23, 1862; J. J. Thomas, "Memoir of David Thomas: A Paper Read Before the Cayuga County Historical Society, December 17, 1878" (New York: 1878).

73. Biographies of all of these engineers can be found in Whitford, *History of the New York Canals*, II: 1145–71.

74. Shaw, *Erie Water*, 232–34; Whitford, *History of the New York Canals*, 115, 786–807; one might argue that West Point included some engineering in its curriculum before 1824.

75. Roberta M. Styran and Robert R. Taylor, with John Jackson, *The Welland Canals: The Growth of Mr. Merritt's Ditch* (Erin, ON: Boston Mills Press, 1988), 43, 45.

76. Quoted by Janet D. Larkin, "The Canal Era: A Study of the Original Erie and Welland Canals within the Niagara Borderland," *The American Review of Canadian Studies* (Autumn 1994): 299–314, quoted on page 304; original: J. P. Merritt, *Biography of the Honorable William Hamilton Merritt* (St. Catherines, ON: E. S. Leavenworth, 1875), 67.

77. December 14, 1826; Larkin, in "The Canal Era," reports that advertisements inviting workers to the Welland Canal also appeared in other Niagara Frontier papers: Buffalo *Journal*, June 19, 1827; Black Rock *Gazette*, September 24, 1827; 307nn. 34 and 35.

78. Peter Way, *Common Labour: Workers and the Digging of North American Canals, 1780–1860* (Baltimore: Johns Hopkins University Press, 1993), table 17, "Riots, Faction Fights and Civil Unrest on Canals, 1780–1860," 287–95.

79. Clarence O. Lewis, "History of Lockport, New York" (Lockport: Niagara County Historical Society, 1962; revised 1964), 8; and his "The Erie Canal, 1817–1967" (Lockport: Niagara County Historical Society, 1967), 30.

80. Jesse Hawley, "An Essay on the Enlargement of the Erie Canal" (Lockport: Printed at the Courier Office, 1840); in *Souvenir History of Niagara County New York* (Lockport: Niagara County Genealogical Society, 1986, 2nd edition; original: Lockport: Pioneer Association of Niagara County, 1902) mentions that Hawley moved to Lockport in 1836, 160.

81. N. B. Holmes, *Facts and Observations in Relation to the Origin and Completion of the Erie Canal* (New York: N. B. Holmes, 1825).

82. Ibid., "Profile of the Erie Canal from Lake Erie to Utica . . ." showing three separate configurations (the current, with an inclined plane, and with levels and locks), 31.

83. There were also several townships and secondary settlements; e.g., Lockport Heights, Louisiana; Lockeport, Nova Scotia, is not included because of the variant spelling; United States Geological Survey's Geographic Names Information System, http://geonames.usgs.gov/pls/nis/.

84. Oliver G. Steele, *Steele's Western Guidebook and Emigrant's Directory* (Buffalo: Oliver G. Steele, 1849); lists the stops and villages along each of these routes including six Lockports.

85. Thomas A. Becnel, *The Barrow Family and the Barataria and Lafourche Canal: The Transportation Revolution in Louisiana, 1829–1925* (Baton Rouge: Louisiana State

University Press, 1989), 49; Jeffery J. LeBlanc, "The Story of Lockport Louisiana" (New Orleans: n.d.). The canal today forms part of the Intercoastal Waterway.

86. *Souvenir History*, 151–62.

87. *Church Book of the First Presbyterian Church of Lockport*, handwritten session records, January 1838; Michael P. Conzen and Adam R. Daniel, *Lockport Legacy: Themes in the Historical Geography of an Illinois Canal Town*; Studies on the Illinois and Michigan Canal Corridor, no. 4 (Chicago: University of Chicago Press, 1990).

88. *Souvenir History*, 163.

89. "I & M Canal History: Agriculture, Industry and the Waterways"; a website sponsored by the Canal Corporation, http://www.canalcor.org/hisag.htm; "History of the Illinois and Michigan Canal," a National Service sponsored website http://www.ops.gov/ilmi/canalhistory.html; Gooding House, circa 1845," part of the Will County, Illinois, website http://www.willcountylanduse.com/hpc/gooding.html; "Historical Map and Guide to the Illinois and Michigan Canal National Heritage Corridor," published by the Illinois and Michigan Canal National Heritage Corridor, 1993, Michael P. Conzen, editorial consultant.

90. Way, *Common Labour*, appendix II, table 17, 287–95.

91. "Wasteland II Site," Fact Sheet 1, December 2001, Lockport; the Illinois EPA website http://www.epa.state.il.us/community-relations/fact-sheets/wasteland-2/wasteland-2.html; "History of the Illinois and Michigan Canal" (National Parks Service); "Historical Map and Guide to the Illinois and Michigan Canal National Heritage Corridor," Conzen and Daniel, *Lockport Legacy*.

92. "Historical Map and Guide to the Illinois and Michigan Canal National Heritage Corridor."

93. Anne Royall, *The Black Book* or *A Continuation of Travels in the United States*, 2 vols. (Washington, DC: Printed for the author, 1828), 50, 58–59.

94. Sears, *Sacred Places*.

95. Jonathan Culler, "The Semiotics of Tourism," in his *Framing the Sign: Criticism and Its Institutions* (Oxford: Basil Blackwell, 1988), 153–67.

96. William W. Stowe, *Going Abroad: European Travel in Nineteenth-Century American Culture* (Princeton, NJ: Princeton University Press, 1994), 222.

97. Linda L. Revie, *The Niagara Companion: Explorers, Artists, and Writers at the Falls, from Discovery through the Twentieth Century* (Waterloo, ON: Wilfrid Laurier University Press, 2003), 88; see also Sara Mills, *Discourses of Difference: An Analysis of Women's Travel Writing and Colonialism* (London and New York: Routledge, 1991).

98. Alexis de Tocqueville, *Journey to America*, trans. George Lawrence (Garden City, NY: Anchor Books, 1971), 288.

99. Sheriff, *Artificial River*.

100. Washington Irving, "Rip Van Winkle: A Posthumous Writing of Diedrich Knickerbocker," from *The Sketch Book*, in *The Works of Washington Irving* (New York: A. L. Burt Company, 1919), 29–45; Seelye provides an extensive discussion of the role of Hudson River writers in *Beautiful Machine*.

101. Daniel K. Richter, *Facing East from Indian Country: A Native History of Early America* (Cambridge, MA: Harvard University Press, 2001); Malino Johar Schueller and Edward Watts, eds., *Messy Beginnings: Postcoloniality and Early American Studies* (New Brunswick: Rutgers University Press, 2003).

102. Sheriff, *Artificial River*, 53; Patrick McGreevy, *Imagining Niagara: Meaning and the Making of Niagara Falls* (Amherst, MA: University of Massachusetts Press, 1994), 6; it was also referred to as the "Northern Tour"; sometimes the return trip included Quebec.

103. Nicholas A. Woods, *The Prince of Wales in Canada and the United States* (London: Bradbury and Evans, 1961), 235–52; John Disturnell, *Travellers' Guide through the State of New York* (New York: published by the author, 1836), also proclaims that "Niagara Falls is the great point of attraction for all travelers for pleasure or information, going west," 54.

104. Books for travelers include Disturnell, *Travellers' Guide; Thomas F. Gordon, Gazetteer of the State of New York* (Philadelphia: 1836); O. L. Holley, *The Picturesque Tourist; Being a Guide through the Northern and Eastern States and Canada; Giving an Accurate Description of Cities, and Villages, Celebrated Places of Resort, etc.; with Maps and Illustrations* (New York: John Disturnell, 1844); Horatio Gates Spafford, *A Pocket Guide for the Tourist and Traveller, Along the Line of the Canals, and the Interior Commerce of New York State* (New York: T and J. Swords, 1824); Oliver G. Steele, *Steele's Western Guidebook and Emigrant's Directory* (Buffalo: O. G. Steele, 1849); *The Fashionable Tour, an Excursion to the Springs* (third edition, Saratoga Springs: 1828); *The Tourist, or Pocket Manual for Travellers on the Hudson River, the Western Canal* (fourth edition, New York: Harper and Brothers, 1835); in addition, many of the Niagara Falls guidebooks also covered the routes to the falls in detail.

105. Nathaniel Hawthorne, "My Visit to Niagara," in *Tales, Sketches, and Other Papers* (1859; reprint edition, Boston and New York: Houghton, Mifflin and Co., 1883), 42–50.

106. Spafford, *Pocket Guide*, 44–45.

107. Gordon, *Gazetteer*, 559.

108. Holley, *Picturesque Tourist*, 172.

109. "Lockport, in Niagara County, N. Y., Its Rise, Present Condition, and Prospects," *National Magazine* (December 1845): 623–28, 628; see also J. H. Mather and L. P. Brockett, *A Geographical History of the State of New York . . .* (Utica: H. H. Hawley and Co., 1848), 221, 353.

110. Sibyl Tatum, "Account of the Journey of Sibyl Tatum with Her Parents from N. Jersey to Ohio in 1830," unpublished manuscript, 1830, University of Georgia, 13 pp., 11–12.

111. These included James Boardman, *America and the Americans* (original, London: Longman, Rees, Orme, Brown, Green and Longman, 1833; reprint, New York: Arno Press, 1974), 131–32; James Silk Buckingham, *America, Historical, Statistic, and Descriptive*, 2 vols. (New York: Harper and Brothers, 1841), II: 136–37; John Galt, *The Canadas: Comprehending Topographical Information, Concerning the Quality of the Land, in Different Districts; and the Fullest General Information: For the Use of Emigrants and Capitalists* (London: Effingham Wilson, 1836), 261–62; Basil Hall, *Travels in North America*, vol. I (Graz, AT: Akademische Druck, 1965), 174–76; James Lamsden, *American Memoranda by a Mercantile Man, During a Short Tour in the Summer of 1843* (Glasgow: Bell and Bain, 1844), 28–29, he also discussed the notion of modeling the enlarged canal locks on those of Lockport by doubling them.

112. Hall, *Travels in North America*, vol. I: 174–76.

113. Archibald M. Maxwell, *A Run Through the United States During the Autumn of 1840*, 2 vols. (London: Henry Colburn, 1841), I: 269; Jacques Milbert, *Picturesque Itinerary of the Hudson River and the Peripheral Parts of North America*, trans. Constance D. Sherman (New York: The Gregg Press, 1968), 138; Thomas Wharton, "From England to Ohio, 1830–1832: The Journal of Thomas K. Wharton," ed. James H. Rodabaugh, *The Ohio Historical Quarterly* 65, no. 1 (January 1965): 1–27, and continued in the following issue, 24; Thomas Woodcock, Deoch Fulton, ed., "New York to Niagara, 1836: The Journal of Thomas S. Woodcock," *Bulletin of the New York Public Library* 42, no. 9 (September 1938): 675–94, 684–85; John Fowler, *Journal of a Tour through the State of New York in 1830* (New York: A. M. Kelly, 1970, reprint edition; original 1831); O. L. Holley, *Picturesque Tourist*, 172; Thomas L. Nichols, *Forty Years of American Life*, 2 vols. (London: John Maxwell and Co., 1864), I: 121; William Stone, "Narrative," in Colden, *Memoir*, 293; Mrs. Basil Hall, *The Aristocratic Journey: Being the Outspoken LETTERS OF MRS. BASIL HALL Written during a Fourteen Months' Sojourn in America 1827–1828* (New York and London: G. P. Putnam's Sons, The Knickerbocker Press, 1931), 56; Maximilian, prince of Wied, *Travels in the Interior of North America, 1832–1834*, in *Early Western Travels: 1748–1846*, ed. Reuben Gold Thwaites (New York, AMS Press, 1966), XXIV, 177–78.

114. Shaw, *Erie Water*, 131; this trend has continued in recent books by Sheriff, *Artificial River*, and Bourne, *Floating West.*

115. Blake McKelvey, "The Erie Canal: Mother of Cities," *New York Historical Society Quarterly* 35 (1951): 55–80, 55, 64.

116. Nathaniel Hawthorne, "Sketches from Memory—The Canal Boat," *New England Magazine* 9 (1835): 398–404, 398.

117. McKelvey, "The Erie Canal," 63.

118. Mrs. Basil Hall, *The Aristocratic Journey*, 54.

119. Boardman, America and the Americans, 131–32.

120. Henry Tudor, Narrative of a Tour In North America. . . . In a Series of Letters, Written in the Years 1831–2, 2 vols. (London: James and Duncan, 1834), I: 230.

121. E. T. Coke, *A Subaltern's Furlough: descriptive of scenes in various parts of the United States, Upper and Lower Canada, New-Brunswick and Nova Scotia, during the summer and autumn of 1832*, 2 vols. (New York: J. & J. Harper, 1833), 23.

122. Moses C. Cleveland, "Journal of a Tour from Riverhead, Long Island, to the Falls of Niagara in June 1831," *New York History* 27 (1946): 352–64, 359; similar comments were offered by O. L. Holley, *Picturesque Tourist*, 172; Maxwell, *A Run Through the United States*, I: 269; Charles Augustus Murray, *Travels in North America During the Years 1834, 1835 and 1836*, 2 vols. (New York: Harper and Brothers, 1839), 65; Karl Bernhard, Duke of Saxe-Weimar Eisenach, *Travels Through North America, During the years 1825 and 1826*, 2 vols. (Philadelphia: Carey, Lea and Carey, 1828), I: 70.

123. Anonymous, *Journal of a Wanderer; Being a Residence in India and Six Weeks in North America* (London: Simpkin, Marshall, and Co., 1844), 209; Buckingham makes similar comments in *America, Historical, Statistic, and Descriptive*, II: 136–7.

124. Theodore Dwight Jr., *Things as They Are: or, Notes of a Traveller* (New York: Harper and Brothers, 1834), apparently reprinted as *The Northern Traveller* (1841), 48.

125. Boardman, *America and the Americans*, 132.

126. Winifred Lovering Holman, ed., "Diary of the Rev. James-Hanmer Francis, 1837–1838," *The Ohio State Archaeological and Historical Quarterly* 51 (January, 1942): 41–60, 43.

127. Nathaniel Hawthorne, "Sketches from Memory—The Canal Boat," 400.

128. Hall, *Travels in North America*, I: 172.

129. Edward S. Abdy, *Journal of a Residence and Tour in the United States of North America from April 1833 to October 1834*, 3 vols. (London: John Murray, 1835), I: 317.

130. The quote is from Wharton, "From England to Ohio, 1830–1832," 24; other similar comments: Boardman, *America and the Americans*, 132; Frederick Marryat, *A Diary in America: With Remarks on Its Institutions* (New York: Alfred A. Knopf, 1962), 91–92.

131. William Lyon Mackenzie, *Sketches of Canada and the United States* (London: Effingham Wilson, 1833), 2.

132. Hawthorne, "Sketches from Memory—The Canal Boat," 407.

133. Sheriff, *Artificial River*, 64.

134. Spafford, *Pocket Guide*, 47.

135. Carl David Arfwedson, *The United States and Canada in 1832, 1833, and 1834*, 2 vols. (original 1834; reprinted: New York and London: Johnson Reprint Corporation, 1969), II: 304–307.

136. Adam Fergusson, *Practical Notes Made During a Tour in Canada and Parts of the United States in MDCCCXXXI* (Edinburgh: William Blackwood, and London: T. Cadell, Strand, 1833), 171–2.

137. David Nye, *The American Technological Sublime* (Cambridge, MA, and London: MIT Press, 1994), 35.

138. David E. Nye, *Narratives and Spaces: Technology and the Construction of American Culture* (Exeter: University of Exeter Press, 1997).

139. Patrick McGreevy, Review of *The American Technological Sublime* by David Nye, *Annals of the Association of American Geographers* 86, no. 2 (1996),: 354–55.

140. Maximilian, *Travels in the Interior of North America*, 177–78.

141. James M. Jasper, *Restless Nation: Starting Over in America* (Chicago: University of Chicago Press, 2000); Jean Baudrillard, *America*, trans. Chris Turner (New York: Verso, 1988).

142. Revie, *Niagara Companion*, 82–84.

143. Caroline Gilman, *The Poetry of Traveling in the United States* (original: New York: S. Coleman, 1838; reprint: Upper Saddle River, NJ: Literature House, 1970), 103.

144. Captain Oldmixon, R.N., *Transatlantic Wanderings, or A Fast Look at the United States* (London: Geo. Routledge and Co., 1855), 81.

145. Caroline Westerley, *The Young Traveller from Ohio*, Boys and Girls Library, XVI (New York: J. and J. Harper, 1833), 123–24.

146. Herman Melville, *Moby Dick, or The White Whale* (New York: Library Publications, 1930), 277.

147. Tyrone Power, *Impressions of America: During the Years 1833, 1834 and 1835*, 2 vols. (New York: Benjamin Blom, Inc., 1971), I: 413–17.

148. Willis Gaylord Clark, quoted in Lewis Gaylord Clark, ed., *The Literary Remains of the Late Willis Gaylord Clark* (New York: Burgess, Stringer and Co., 1844).

149. Nathaniel Hawthorne, "My Visit to Niagara," in *Tales, Sketches, and Other Papers* (1859; reprint edition, Boston and New York: Houghton, Mifflin and Co., 1883), 42–50; see McGreevy, *Imagining Niagara*, 96–97.

150. Quoted in Clark, ed., *The Literary Remains*, 179.

151. Nathaniel Hawthorne, "Sketches from Memory. By a Pedestrian," *New England Magazine* 9 (December 1835): 398–409, 402.

152. Jasper, *Restless Nation*; Baudrillard, *America*; Antoine de Saint-Exupéry, *Wind, Sand and Stars*, trans. Lewis Galantiere (New York: Yeynal and Hitchcock, 1941), 183–88; Jack Kerouac, *On the Road* (New York: Viking, 1957).

153. Hawthorne, "Sketches from Memory—The Canal Boat," 402, 403, 399.

154. Hawthorne, "My Visit to Niagara," 49.

155. Gilman, *The Poetry of Travelling in the United States*, 103; Royall, *The Black Book*, 58; Stone, "From New York to Niagara," 236; Tatum, "Account of the Journey of Sibyl Tatum," 11–12.

156. Tudor, *Narrative of a Tour*, I: 233–34.

157. Woods, Prince of Wales in Canada, 235–52.

158. Shaw, *Erie Water*, 131.

159. Seelye, *Beautiful Machine*, 332.

160. Marvin A. Rapp, *Canal Water and Whiskey* (Buffalo: The Heritage Press, 1992), 49; Rapp's wording is remarkably similar to Shaw's; his volume was published twenty-six years later but he states in the foreword that "the first publication of this material in a single volume came out in 1965," the year before Shaw's book.

161. Thomas Cole, "Essay on American Scenery," *American Monthly Magazine* (January 1836): 1–12, 6–8.

162. Nichols, *Forty Years of American Life*, I: 121.

163. Maxwell, *A Run Through the United States*, I: 269; Wharton, "From England to Ohio, 1830–1832," 24; John Fowler, *Journal of a Tour*; William Stone, "Narrative," in Colden, *Memoir*, 293; Mrs. Basil Hall, *The Aristocratic Journey*, 54.

164. Woodcock, "New York to Niagara, 1836," 675–94, 684–85.

165. Auguste Levasseur, *Lafayette in America in 1824 and 1825*, or, *Journal of a Voyage in the United States*, 2 vols. (Philadelphia, 1829; reprint, New York: Research Reprints, 1970), II: 213.

166. "Stewart Scott Diary, 1826," unpublished, Manuscript Collection, New York State Library, Albany, 62–63.

167. William Leete Stone, "From New York to Niagara: Journal of a Tour, in Part by the Erie Canal, in the Year 1829," *Publications of the Buffalo Historical Society* 14 (1910), 236–38, 236.

168. Thomas Cole, "Niagara," a poem, quoted in McKinsey, "An American Icon," 92; Margaret Fuller, *At Home and Abroad, or, Things and Thoughts in England and America*, Part One: *Summer on the Lakes* (originally published alone: Boston: Little and Brown, 1844; the whole: New York: The Tribune Association, 1869), 48; see also Elizabeth McKinsey, *Niagara Falls: Icon of the American Sublime* (Cambridge: Cambridge University Press, 1985), 220.

169. Seelye, *Beautiful Machine*, 348.

170. McKinsey, "An American Icon," and *Niagara Falls*; Thomas Cole also makes nationalist use of Niagara Falls in his "Essay on American Scenery," 8.

171. Raymond F. Gates, *The Old Lockport and Niagara Falls Strap Railroad*, Occasional Contributions of the Niagara County Historical Society, no. 4 (Lockport: Niagara County Historical Society), 1950; Alexis Muller Jr., *Looking Back: So That We May Move Ahead*, produced to celebrate the 150th Anniversary of the Grand Erie Canal (Lockport: Lockport Canal Sesquicentennial, Inc., 1975), 30.

172. John C. Calhoun, quoted in Paul F. Boller Jr., *Not So! Popular Myths About America from Columbus to Clinton* (New York: Oxford University Press, 1996), 57.

173. Nichols, *Forty Years of American Life*, I: 50–57, 56.

174. Ramsay Cook, "Cultural Nationalism in Canada: An Historical Perspective," in *Canadian Cultural Nationalism*, ed. J. L. Murray (New York: New York University Press, 1977), 20.

175. George Wurts, "Niagara River July 8th, 1851," in "Journal of a Tour to Niagara Falls, Montreal, Lake Champlain, & C.," (1851) *Proceedings of the New Jersey Historical Society* (1951): 342–62, 347–48.

176. Caroline Westerly, *Caroline Westerly or The Young Traveler from Ohio* (New York: J. and J. Harper, 1833), 81, 115; anonymous, *Notes on a Tour Through the Western Part of the State of New York* (Philadelphia: 1829–30; reprinted, 1916 from the *Ariel*, Philadelphia).

177. Quoted in Charles Mason Dow, *Anthology and Bibliography of Niagara Falls*, 2 vols. (Albany: State of New York, 1921), II: 1206.

178. Janet Larkin, "The Canal Era: A Study of the Original Erie and Welland Canals within the Niagara Borderland," *The American Review of Canadian Studies* (Autumn 1994): 299–314.

179. Hall, *Travels in North America*, I: 193, 248–49.

180. William Howard Russell, *Canada: Its Defences, Condition, and Resources* (London: Bradbury and Evans, 1865).

181. Alexander Mackay, *The Western World, or Travels in the United States in 1846–47*, 3 vols. (New York: Negro University Press, 1968; original: London: Richard Bentley, 1849), III: 117.

182. W. L. Morton, *The Canadian Identity* (Toronto: University of Toronto Press, 1961), 59.

183. Henry A. S. Dearborn, *Letters on the Internal Improvements and Commerce of the West* (Boston: 1839), 22–23; quoted in Seelye, *Beautiful Machine*, 337.

184. Reverend Isaac Fidler, *Observations on Professions, Literature, Manners, and Emigration, in the United States and Canada, Made During a Residence There in 1832* (London: Whittaker, Treacher, and Co., 1833: reprint, New York: Arno Press, 1974), 303.

185. Revie, *The Niagara Companion*, 100.

186. Galt, *The Canadas*, 68, 78, 261–62.

187. Frederick Gerstaecker, *Wild Sports of the Far West* (Boston: Brosby and Nichols, 1864), 48.

188. Gerald M. Craig, *Upper Canada: The Formative Years, 1784–1841* (London and New York: Oxford University Press, 1963), 210–55.

189. Mackenzie, *Sketches of Canada and the United States*, 144–45.

190. Mrs. Frances Trollope, *Domestic Manners of the Americans* (original 1832; reprint: New York: Vintage Books, 1960), 404; Linda Revie discusses Trollope's life and visit to Niagara in depth, *The Niagara Companion*, 70–76.

191. Mrs. Trollope, *Domestic Manners*, 378.

192. Melville, *Moby Dick*, 462–63.

193. Raymond Williams, *Culture and Society, 1780–1950* (New York: Columbia University Press, 1983; original: 1958); Thomas Cole, *The Course of Empire*, a series of five oil paintings: *The Savage State* (1834), *The Pastoral or Arcadian State* (1834), *The Consummation of Empire* (1835–36), *Destruction* (1836), *Desolation* (1836); New York Historical Society; Perry Miller, "The Romantic Dilemma in American Nationalism and the Concept of Nature," *The Harvard Theological* Review 48, no. 4 (October 1955): 239–53; Leo Marx, *The Machine in the Garden: Technology and the Pastoral Ideal in America* (New York: Oxford University Press, 1964).

194. Mary Keys, *Lockport on the Erie Canal New York*, watercolor, 1832, Lockport, NY, signed and dated; Munson-Williams-Proctor Institute, Utica, NY.; included on the cover of Bourne's *Floating West*.

195. *View of the Upper Village of Lockport*, W. Wilson, 1836, drawing, New York Historical Society; lithographed by Bufford; The New-York Historical Society; Reproduced in McKelvey, "The Erie Canal: Mother of Cities," 65; also on the cover of Carol Sheriff's *Artificial River*, and in numerous local history publications in Niagara County.

196. Included in the *Illustrated History of Niagara County* (New York: Sanford and Co., 1878) is a sketched copy of the Wilson print with fifteen buildings and other features labeled.

197. *Northeastern View of the Locks of Lockport, N. Y.*, engraving accompanying descriptive article on Lockport in the *Rural Repository* 20, no. 10, Saturday, December 30, 1843, Hudson, NY, 1.

198. "View Showing the progress of the work on the Lock Section, Taken Sept. 1st, 1839, Drawn by T. Evershed," a separate drawing; the volume is labeled "Working Plans of Enlarged Lockport Locks"; single copies of both of these survive only in the New York State Archives.

199. "Lockport Locks," *Maps and Profiles of New-York State Canals*, designed under direction of Van Rensselaer Richmond, State Engineer and Surveyor, to accompany his report for 1859, Albany, 1959.

200. F. G. Mather, "Water Routes from the Great Northwest," *Harper's* 63 (1881): 414–35, 429.

201. Buckingham, *America, Historical, Statistic, and Descriptive*, II: 136–37.

202. William Henry Bartlett, *Lockport, Erie Canal*, in *American Scenery*, ed. N. P. Willis, 2 vols. (London: 1840), I:160.

203. Seelye, *Beautiful Machine*, 363.

204. Willis, *American Scenery*, 160.

205. Raphael Beck, *Opening of the Erie Canal, 1825*, Lockport Exchange Trust Company, 1927.

206. Briggs, "Lockport," in *The United States Illustrated*, ed. by Charles A. Dana, 162–64.

207. Scott Trafton, *Egypt Land: Race and Nineteenth-Century American Egyptomania* (Durham: Duke University Press, 2004); Burke O. Long, *Imagining the Holy Land: Maps, Models, and Fantasies* (Bloomington: University of Indiana Press, 2003); Timothy Marr, *The Cultural Roots of American Islamicism* (New York: Cambridge University Press, 2006).

208. *Erie Canal at Lockport, NY,* by Hermann J. Meyer, following p. 162, in Briggs, "Lockport," in *The United States Illustrated*, ed. Charles A. Dana. Print courtesy of The New-York Historical Society.

CHAPTER 5. LOSING LOCKPORT: AFTERLIFE OF THE MOUNTAIN RIDGE CONQUEST

1. Catherine B. Burroughs, "Of 'Sheer Being': Fitzgerald's Aesthetic Typology and the Burden of Transcription," *Modern Language Studies* 22, no. 1 (winter 1992): 102–9, 1–9.

2. Alexis de Tocqueville, *Journey to America*, trans. George Lawrence (Garden City, NY: Anchor Books, Doubleday, 1971), 288.

3. *Illustrated History of Niagara County, New York* (New York: Sanford and Company, 1878), 171.

4. Clarence O. Lewis suggests that about half of the canal laborers remained; "History of Lockport, New York" (Lockport, NY: Niagara County Historian's Office, 1964; original 1962), 8.

5. Enlargement details are outlined in the Annual Report of the Canal Commissioners 1839–1843 and 1848–1860; Raymond F. Yates, "The Old Lockport and Niagara Falls Strap Railroad," Occasional Contributions of the Niagara County Historical Society, no. 4 (Lockport, NY: Niagara County Historical Society, 1950), 40; George R. Worley, *Old Home Week: Celebrating of the Hundredth Anniversary of the Erie Canal* (Lockport, NY: Old Home Week Committee, 1925), 22.

6. Carol Sheriff, *The Artificial River: The Erie Canal and the Paradoxes of Progress, 1817–1862* (New York: Hill and Wang, 1996), 138–71; Peter Way, *Common Labour: Workers and the Digging of North American Canals, 1780–1860* (Baltimore: Johns Hopkins University Press, 1993), 163–99; Paul E. Johnson, *A Shopkeeper's Millennium: Society and Revivals in Rochester, New York, 1815–1837* (New York: Hill and Wang, 1984), 37–61; Charles Sellers, *The Market Revolution: Jacksonian America, 1815–1846* (New York: Oxford University Press, 1991), 3–33.

7. Thomas Evershed, "Plan Showing the Relative Position of the Old and New Locks, Race and Culvert, Lockport," Engineering Plans for the Enlargement of the Erie Canal, 1836–1837; W. Wilson, *View of the Upper Village of Lockport*, The New-York Historical Society, 1836; apparently, many of the groceries were rebuilt after enlargement, for "a small-beer grocery" was accidentally pulled in to the canal, Lockport *Daily Courier*, 1853, quoted in Lewis, "The Erie Canal," 7.

8. Reverend Isaac Fidler, *Observations on Professions, Literature, Manners, and Emigration, in the United States and Canada, Made during a Residence There in 1832* (Reprint: New York: Arno Press, 1974; original: London: Whittaker, Treacher and Co., 1833), 203–4.

9. Rochester *Observer*, July 7, 1831; quoted in Ronald E. Shaw, *Erie Water West: A History of the Erie Canal, 1792–1854* (Lexington, KY: University of Kentucky Press, 1966), 221.

10. *New York Evangelist*, December 5, 1836; quoted in Roger E. Carp, "The Limits of Reform: Labor and Discipline on the Erie Canal," *Journal of the Early Republic* 10, no. 2 (summer 1990): 191–219, 202.

11. Sheriff, Artificial River; her subtitle is: *The Erie Canal and the Paradoxes of Progress, 1817–1862.*

12. *Niagara Courier*, July 13, 1836; see also Shaw, *Erie Water*, 215.

13. Sellers, *The Market Revolution*, 237–40; Stephen John Hartnett, *Democratic Dissent and the Cultural Fictions of Antebellum America* (Urbana: University of Illinois Press, 2002) prefers to speak of "cultural fictions" rather than ideologies to describe the same phenomena.

14. William Leonard, "Diary of Thomas William Leonard, 1815–1879," Manuscript, Niagara County Historian's Office, November 29, 1842.

15. Ibid., November 30, 1842.

16. Christopher Carlin, "Douglas Hanging Last Execution for Niagara," Lockport *Union-Sun & Journal*, April 29, 1995.

17. Leonard, "Diary," July 22 and July 23, 1843.

18. Ibid., July 9, 1843.

19. Ibid., August 1 and August 2, 1843.

20. Gertrude Strauss, *Our First One Hundred and Fifty Years*, in commemoration of the 150th Anniversary of the First Presbyterian Church of Lockport, New York, 1823–1973 (Lockport, NY: First Presbyterian Church, 1973), 8.

21. W. J. Rorabaugh, *The Alcoholic Republic: An American Tradition* (New York:1979),10.

22. Sellers, *Market Revolution*, 259–68, 260.

23. "Lockport Hotbed in 'Patriot War,'" Lockport *Union-Sun & Journal*, Saturday, July 12, 1965.

24. William Lyon MacKenzie, *Sketches of Canada and the United States* (London: Effingham Wilson, 1833), 145.

25. "A Secret Military Document, 1825," *American Historical Review* 38, no. 6 (November 1969): 295–300, 299.

26. *Illustrated History*, 172.

27. *Illustrated History*, 179; Lyman Spalding, the manager of the mill, was a strong abolitionist, perhaps involved in the Underground Railroad; this might make him more sympathetic to McLeod, and so perhaps the mill was torched by the same people who wanted to lynch McLeod.

28. *Illustrated History*, 172.

29. Samuel G. Goodrich, *Peter Parley's Pictorial History of North and South America* (Hartford, CT: The Peter Parley Publishing Company, W. W. House, 1858), 301.

30. R. Y. Jennings, "The Caroline and McLeod Cases," *The American Journal of International Law* 32, no. 1 (January 1938): 82–99; Sir Robert Phillimore, "Commentaries on International Law," *Annual Register* 83 (1841): 316; "Lockport

Hotbed in 'Patriot War,' " Lockport *Union-Sun & Journal*, July 17, 1965; *Souvenir History of Niagara County*, second edition (Lockport, NY: Niagara County Genealogical Society, 1986), 170.

31. *Illustrated History*, 172.

32. Leonard, "Diary," September 30, 1843.

33. C. W. Olmsted, city clerk, from a poem delivered on September 1, 1886, at the Big Bridge dedication; quoted in Lewis, "The Erie Canal," 42.

34. Edward Giddins, quoted in Lewis, "The Erie Canal," 41.

35. Leonard, "Diary," January 14, 1845.

36. Honorable Richard Crowley reported this story during an address delivered on September 1, 1886, at the Big Bridge dedication; in Lewis, "The Erie Canal," 42; there were complaints about this situation as early as 1826, when the Lockport *Observatory* reported that there was an unfinished canal bridge and "unprotected banks, twenty-five or thirty feet perpendicular," *Souvenir History*, 154.

37. *Souvenir History*, 154–58; Lyman A. Spalding, "The Journal of Lyman A. Spalding," Syracuse University Manuscript Collections, September 12, 1852.

38. A mill death of a member of the Sons of Temperance is reported by Spalding, "Journal," February 1, 1849; *Lockport Journal*, January 4, 1839; July 18, 1839; and March 25, 1840; quoted in Lewis, "The Erie Canal," 35.

39. Way, *Common Labour*, 299.

40. Whitney R. Cross, *The Burned-Over District: The Social and Intellectual History of Enthusiastic Religion in Western New York, 1800–1850* (Ithaca: Cornell University Press, 1950), 71–76, 72.

41. Johnson, *Shopkeeper's Millennium*, 139.

42. Ibid., 138.

43. Ibid., 137.

44. Ibid., 140–41.

45. Bradford King Diary, quoted in Johnson, *Shopkeeper's Millennium*, 96, November 9, 1830.

46. Mary P. Ryan, "Women's Awakening: Evangelical Religion and the Families of Utica, New York, 1800–1840," *American Quarterly* 30, no. 5, Special Issue: Women and Religion (winter 1978): 602–23, 622; see also her *Cradle of the Middle Class: The Family in Oneida County, New York, 1790–1865* (Cambridge and New York: Cambridge University Press, 1998, original 1981).

47. Sellers, *Market Revolution*, 237.

48. Ibid., 266.

49. Ibid., 216, 214, 230.

50. Ibid., 235–36.

51. Amy Kaplan, "Manifest Domesticity," in *The Anarchy of Empire in the Making of U.S. Culture* (Cambridge: Harvard University Press, 2002), 23–50.

52. *Niagara Courier*, June 7, 1831.

53. Reverend James H. Hotchkin, *A History of the Purchase and Settlement of Western New York, and of the Rise, Progress, and Present State of the Presbyterian Church in That Section* (New York: M. D. Wood, 1848), 140, 142, 160.

54. "Church Book for the First Presbyterian Church at Lockport," session records from January 1823 to August 10, 1838; "Session Record—October 1838 to June 1841—December 1846"; quotations from "Presbyterian 75th Anniversary,"

January 20, 1898, in *Seventy-Five Years: Anniversary Proceedings of the Founding of the First Presbyterian Church, Lockport, New York*, January 20, 1898, a booklet printed in Lockport, 1898; Gertrude Strauss, *Our First One Hundred and Fifty Years*, published by the First Presbyterian Church of Lockport, New York, 1973, in commemoration of our 150th Anniversary year, 1823–1973, 18.

55. John Marsh, *Temperance Recollections, Labors, Defeats, Triumphs: An Autobiography* (1866), 104.

56. *Seventy-Five Years*, 13.

57. "175th Anniversary, 1816–1991, First Baptist Church, a Retrospective History," compiled by Mary L. Newhard, 1991.

58. Davis W. Clarke, *The Life and Times of Rev. Elijah Hedding: Late Senior Bishop of the Methodist Episcopal Church* (1855), 440.

59. Leonard, "Diary," February 18, 1844.

60. "Zeal and Duty," a brochure issued on the 100th anniversary of the movement to organize the Protestant Episcopal Diocese of Western New York, Buffalo, 1934, 5–6.

61. Temperance organizations proliferated; *Illustrated History* provides a list, 197; we can find over forty drinking establishments extant on the *Insurance Maps of Lockport, New York* (New York: Sanborn Map Company, 1914).

62. Marsh, *Temperance Recollections*, 104; in 1879, the Women's Christian Temperance Union was formed at Lockport, "Alcohol Had an Effect on Country's History," Niagara Falls *Gazette*, September 17, 2000.

63. Henry Shaft, *Shaft's Complete Lockport City Directory for 1861–2* (Lockport, NY: S. S. Pomroy, 1861), 17; *Illustrated History of Niagara County*, 170; Johnson, *Shopkeeper's Millennium*, 83–88.

64. Emma Hardinge Britten, *Modern American Spiritualism: A Twenty Years' Record of the Communion between the Earth and the Spirits* (New York: The Author, 1870); Cross, *Burned-Over District*, 345–52.

65. Sellers, *The Market Revolution*, 203–4; Cross, *Burned-Over District*, 286–321, 317, 319.

66. Sylvester Bliss, *Memoirs of William Miller, etc.* (Boston: J. O. Himes, 1853), 172.

67. *Niagara Democrat*, April 19, 1843, quoted in Clarence O. Lewis, "Meteors and Balloonists Thrilled People in 1800's," Lockport *Union-Sun & Journal*, November 16, 1957.

68. E. C. Galusha to William Miller, October 16, 1843; quoted in Cross, *Burned-Over District*, 299–300.

69. Cross, *Burned-Over District*, 300–1.

70. Leonard, "Diary," November 25, 1843.

71. James White, *Sketches of the Christian Life and Public Labors of William Miller, Gathered from His Memoir by the Late Sylvester Bliss, and from Other Sources* (Battle Creek, MI: Press of the Seventh-Day Adventist Publishing Association, 1875), 269.

72. Newhard, "175th Anniversary, 1816–1991, First Baptist Church," chapters 2 and 3; Mary Shaw Parker, *History of the Baptist Church of Lockport, New York: 1816–1928*, published by the First Baptist Church, Lockport, November 1928, 9–10 (Newhard's compilation includes a complete reproduction of Parker's earlier work);

Cross suggests that Galusha was forced to resign against the will of the majority of the church members, and also to give up his denominational positions, *Burned-Over District*, 301–02.

73. Cross, *Burned-Over District*, 301; Clarence O. Lewis, "Settler Kills Indian Who Carried Off Wife," Lockport *Union-Sun & Journal*, December 4, 1968.

74. Leonard, "Diary," March 22, 1844.

75. Ibid., October 21, 1844; the last day was actually Tuesday, October 22.

76. The 1830 U.S. Census lists 44,870 free colored in New York; in 1840, the number was 50,027; New York was finally surpassed by Pennsylvania in 1850.

77. Elinore T. Horning, "The Harris Harassment," *York State Tradition* 26, no. 4 (1972): 26–28; Frank H. Severance, *Old Trails of the Niagara Frontier* (Buffalo: Mathews-Northrup Company, 1899), 227–71; Sabrina Y. White, "The Erie Canal's Role in the Underground Railroad," in *Homefront: The Erie Canal in the Civil War* (Syracuse: The Erie Canal Museum, 1987), 21–23; the farm of Carol L. Murphy of Burt, New York (about seven miles north of Lockport) was the first site named on the National Parks Service "Underground Railroad Network to Freedom" in 2001, see "Freedom Farm," Lockport *Union-Sun & Journal*, April 16, 2001.

78. Fergus M. Bordewich, *Bound for Canaan: The Underground Railroad and the War for the Soul of America* (New York: Amistad, 2005), 159.

79. Spalding, "Journal," April 11 and 23, 1836; January 26, 1838.

80. "Seventy-Five Years," 14.

81. "Church Book of the First Presbyterian Church at Lockport," 1823–1838, April 24, 1838; Reverend Harry H. Bergen, "Congregationalism in Lockport, New York, 1838–1938," Lockport, 1938.

82. Bergen, "Congregationalism in Lockport."

83. Spalding, "Journal," July 4, 1838; "Circuit Riders Started Frontier Methodist Work," Lockport *Union–Sun & Journal*, July 17, 1965.

84. Spalding, "Journal," December 18, 1840.

85. Charles Lenox Remond to William C. Nell, "Letters to Antislavery Workers and Agencies," part 7, *Journal of Negro History* 10, no. 3 (July 1925): 493–519, 501–2.

86. Spalding, "Journal," October 4 and 5, 1849.

87. Arthur O. White, "The Black Movement against Jim Crow Education in Lockport, New York, 1835–1876," *New York History* 50, no. 3 (1969): 265-82, 271.

88. Clarence O. Lewis, "Old Building Once Housed Free Congregationalists," Lockport *Union-Sun & Journal*, December 14, 1966.

89. Clarence O. Lewis, "First Church Built Here Soon After Settlers Came," Lockport *Union-Sun & Journal*, September 8, 1952.

90. John L. Myers, "American Antislavery Society Agents and the Free Negro, 1833–1838," *The Journal of Negro History* 52, no. 3 (July 1967): 200–19, 214; Russell L. Adams, "Black Studies Perspectives," *The Journal of Negro Education* 46, no. 2 (spring 1977): 99–117, 100.

91. *Niagara Courier*, September 9, 1835 see also Arthur O. White, "The Black Movement against Jim Crow Education in Buffalo, New York, 1800–1900,"*New York History* L, no. 3 (1969), 265–82; and Carlton Mabee, "Control by Blacks over Schools in New York State, 1830–1930," *Phylon* 40, no. 1 (1979), 29–40.

92. White, "The Black Movement against Jim Crow Education in Lockport," 270, 272–82.

93. Paul Boyer, *Urban Masses and Moral Order in America, 1820–1920* (Cambridge: Harvard University Press, 1978), 57.

94. Hotchkin, *History of the Presbyterian Church*, 163, 166.

95. Ryan, *Cradle of the Middle Class*.

96. Boyer, *Urban Masses and Moral Order*.

97. C. Stuart Gager, "The School of Horticulture in Perspective," *Science*, New Series, 84, no. 2182 (October 23, 1936): 357–65, 359.

98. Julia Hull Winner, *Belva A. Lockwood*, Occasional Contributions of the Niagara County Historical Society, no. 19 (Lockport: Niagara County Historical Society, 1969), 1–3.

99. Lockport *Daily Advertiser and Democrat*, August 4, 1868.

100. Lockport *Daily Courier*, August 4, 1858; Winner, *Belva A. Lockwood*, 7–8.

101. Lockport *Daily Advertiser and Democrat*, August 6, 1868.

102. Winner, *Belva A. Lockwood*, 8; Lockport *Daily Advertiser and Democrat*, August 3, 4, 5, 6 1868.

103. Susan B. Anthony, Elizabeth Cady Stanton, Matilda Joslyn Gage, eds., *History of Women Suffrage*, 6 vols. (New York: Arno Press, 1969),vol. 3, 64; Winner, *Belva A. Lockwood*, 4–32; Julia Davis, "A Feisty Schoolmarm Made the Lawyers Sit Up and Take Notice," *Smithsonian* (March 1981): 133–51.

104. *Lockport Daily Journal*, May 24, 1872; see also Winner, *Belva A. Lockwood*, 33–78, 20; Davis, "A Feisty Schoolmarm"; Anthony claimed that although the Equal Rights Party did not have a wide membership, the Lockwood candidacy was important for introducing the idea of women candidates for high office, 66–67.

105. Amy Kaplan, "Manifest Domesticity," in *The Anarchy of Empire*.

106. *Illustrated History*, 177; Lyman A. Spalding, *Recollections of the War of 1812 and Early Life in Western New York*, Occasional Contributions of the Niagara County Historical Society, no. 2 (Lockport: Niagara County Historical Society, 1949), 12–19.

107. Annual Report of the Department of Public Works, 1926, 4–5.

108. Spalding, "Recollections," 17, 19; Shaft, *Shaft's Complete Lockport City Directory*, 18–19.

109. Quoted in Ronald P. Formisano and Kathleeth Smith Kutolowski, "Antimasonry and Masonry: The Genesis of Protest, 1826–1827, *American Quarterly* 29, no. 2 (Summer 1977): 136–65, 153; Edward T. Williams, *Niagara County, New York* (Chicago: J. H. Beers, 1921, 337.

110. *Priest Craft Exposed and Primitive Christianity Defended* was a monthly issued between May 19, 1828, and April 6, 1829; published in Lockport by Lyman A. Spalding; The Niagara County Historian's Office has all issues; see also Kathleen L. Riley, " 'A Christian Enterprise of Vast Proportions' on the Erie Canal: Lockport, 1819–1840," The Phi Alpha Theta Lecture at Canisius College, October 30, 1998, Buffalo, New York, manuscript; also see her *Lockport: Historic Jewel of the Erie Canal* (Charleston, SC: Arcadia Publishing, 2005).

111. *Priest Craft Exposed*, July 14, September 1, and October 8, 1828.

112. William Leete Stone, *Letters on Masonry and Antimasonry* (New York: O. Halsted, 1832); Formisano and Kutolowski, "Antimasonry and Masonry," 149–50;

the story of Sheriff Bruce is related in Williams, *Niagara County*, 340; William Morgan, *Illustrations of Masonry, by One of the Fraternity, Who Has Devoted Thirty Years to the Subject; With an Appendix, Containing a Key to the Higher Degrees of Freemasonry; by a Member of the Craft* (date listed as 1826, but actually 1851), from a section entitled "Life, Abduction and Murder of Morgan," 138; the Rochester *Observer* lists charges against four Lockport men, including journalist Orasmus Turner, who was arrested in Canandaigua and released on bail of $2,000, September 8, 1827.

113. John C. Spencer, report of January 1830, state of New York, *Documents of the Senate, Fifty-Fourth Session*, no. 67 (Albany, 1831): 21; quoted in Formisano and Kutolowski, "Antimasonry and Masonry," 152.

114. Formisano and Kutolowski, "Antimasonry and Masonry," 162.

115. Williams, *Niagara County*, 340–41.

116. Cross, *Burned-Over District*, 116.

117. Spalding, "Journal," September 15, 1855.

118. Ibid., March 31, 1842.

119. Spalding to Austin Steward and Benjamin Paul, "Letters of Negroes, Largely Personal and Private," part 5, *Journal of Negro History* 11, no. 1 (January 1926): 160–85, 176.

120. Spalding, "Journal," May 15, 1852.

121. Spalding, "Journal," July 20, 1839, August 12, 1839, February 28, 1840, April 28, 1843, December 26, 1843, August 28, 1844.

122. Patrick McGreevy, "Place in the American Christmas," *Geographical Review* 80, no. 1 (1990): 32–42.

123. Spalding, "Journal," December 25, 1851.

124. McGreevy, "Place in the American Christmas"; Mikhail Bakhtin, *Rabelais and His World*, trans. Helen Iswolsky (Bloomington: Indiana University Press, 1989); Stephen Nissembaum, *The Battle for Christmas: A Cultural History of America's Most Cherished Holiday* (New York: Vintage, 1997).

125. Sellers, *Market Revolution*, 202.

126. *Illustrated History*, 188; "Stone Quarrying Was Big Business," Lockport *Union-Sun & Journal*, July 17, 1965; Donald Zenger, "The Lockport Formation in Western New York," in *Geology of Western New York: Guidebook*, ed. Edward J. Buehler (Buffalo: New York State Geological Association, 1966), 19–23.

127. Clarence O. Lewis, "First Water Tunnel Dug in Lockport in 1824–25," Lockport *Union-Sun & Journal*, June 15, 1960.

128. Richard S. Fisher, *The Book of the World: Being an Account of All Republics, Empires, Kingdoms, and Nations, in Reference to Their Geography, Statistics, Commerce* (New York: J. H. Colton, 1852–53), 282.

129. United States Patent Office, "Subject–Matter Index of Patents for Inventions Issued by the United States Patent Office from 1790 to 1873, Inclusive . . ." (Washington: Government Printing Office, 1874).

130. *Appletons' Cyclopedia of American Biography* (New York: D. Appleton and Company, 1887–89), 380.

131. Morris A. Pierce, "The Introduction of Direct Pressure Water Supply, Cogeneration, and District Heating in Urban and Institutional Communities," PhD dissertation, University of Rochester, 1993; *Illustrated History*, 185–86; "B. Holly's

System of Fire Protection and Water Supply for Cities and Villages," advertising pamphlet (Buffalo: Thomas, Howard and Johnson, 1868), Holly Manufacturing Company, Lockport, New York, with Bignall and McDonald, General Western Agents, Chicago.

132. "B. Holly's System of Fire Protection and Water Supply for Cities and Villages"; *Illustrated History*, 185.

133. *Illustrated History*, 184; Clarence O. Lewis, "First Water Tunnel Dug in Lockport in 1824–5," Lockport *Union-Sun & Journal*, June 15, 1960.

134. Pierce, "The Introduction of Direct Pressure Water Supply, Cogeneration, and District Heating in Urban and Institutional Communities," 1–2, 119–84; "Big City Buildings Owe Warmth in Winter to Lockport Inventor," Buffalo *Evening News*, May 18, 1963.

135. "Big City Buildings Owe Warmth in Winter to Lockport Inventor."

136. "Carlos Holly Had Visions of Modern City Skyscraper," Lockport *Union-Sun & Journal*, December 12, 1939; the name Carlos is a mistake, crossed out on photocopy from Niagara County Historian's Office, and Birdsill is written in.

137. John D'Onofrio, "Birdsill Holly Kept Plugging on Inventions," Lockport *Union-Sun & Journal*, approximately May 10, 1987.

138. Noble E. Whitford, *History of the Canal System of the State of New York*, 2 vols. (Albany: State of New York, 1906), II: 1153; "Can't Steal Niagara," *Lockport Journal*, March 18, 1904.

139. "Engineer Evershed's Water Power Scheme," Lockport *Union*, February 3, 1886.

140. Edward Dean Adams, *Niagara Power: History of the Niagara Falls Power Company, 1886–1918*, 2 vols., I: 115; A. Howell Van Cleve, "Utilization of Water Power at Niagara Falls," delivered before the Buffalo Society of Natural Sciences, March 13, 1903, *Bulletin of the Buffalo Society of Natural Sciences* 8, no.1 (1904): 3–20, 9–10.

141. C. E. Burk, "Lockport N.Y.: Electrical and Industrial," Lockport Board of Trade, 1910; a map showing the power transmission lines out of Niagara Falls in 1910, on both sides of the border, is included in "Information for Visitors," a booklet published by the Niagara Falls Power Company (New York), and the Canadian Niagara Power Company (Ontario), 1910, 11.

142. "Harrison Radiator Has Grown Rapidly Since Founding in 1911," Lockport *Union-Sun & Journal*, July 17, 1965; "Only 121 Radiators Produced the First Year," Lockport *Union-Sun & Journal*, July 17, 1965; "1910–1985: A History of Harrison Radiator Division," *Communicator* (quarterly publication of the Public Relations Department of Harrison Radiator Division, GMC) 3, no. 2 (Summer 1985): 4–11; "Harrison History," *Lock City Lookout* 1, no. 5 (September 1948) official publication of the Chamber of Commerce, Lockport, New York.

143. "Harrison Is Becoming Delphi," Lockport *Union-Sun & Journal*, February 13, 1995.

144. *Revised Charter and Ordinances of the City of Lockport* (Lockport: Common Council of the City, 1913); Clarence O. Lewis, "History of Lockport, New York," (Lockport: Niagara County Historian's Office, 1964; original: 1962), 6; Samuel Woodworth, "Ode for the Canal Celebration," in Cadwallader C. Colden, *Memoir*

Prepared at the Request of a Committee of the City of New York . . . of the Completion of the New York Canals (New York: New York City, 1825); for street names: J. P. Haines, "An Enlarged and Complete Map of the Village of Lockport," Lockport, 1845.

145. See, for example, Elkanah Watson, *History of the Rise, Progress, and Existing Conditions of the Western Canals in the State of New York* (Albany: D. Steele, 1820), 342.

146. Patricia Anderson, *The Course of Empire: The Erie Canal and the New York Landscape, 1825–1875* (Rochester, NY: Memorial Art Gallery of the University of Rochester, 1984), 17.

147. Ibid., 27.

148. N. P. Willis, *American Scenery*, with 121 steel plate engravings from drawings by W. H. Bartlett (Barre, MA: Imprint Society, 1840).

149. James Crawford, "The Erie Canal in Art and Literature," New York State History Conference, Buffalo, June 5, 1998, manuscript, 4; see also his "The Art of the Erie Canal," *Heritage* 12, no. 3 (spring 1996):18–23.

150. James. L. Machor, *Pastoral Cities: Urban Ideals and the Symbolic Landscape of America* (Madison: University of Wisconsin Press, 1987), 123.

151. Sheriff, *Artificial River*, 172.

152. Thomas Cole, "American Scenery," *The American Monthly Magazine* 1 (January 1836): 1–12, 4, 5, 8; for a detailed discussion of Niagara and U.S. nationalism, see Elizabeth McKinsey, *Niagara Falls: Icon of the American Sublime* (Cambridge: Harvard University Press, 1984).

153. Benedict Anderson, *Imagined Communities: Reflections on the Origin and Spread of Nationalism*, revised edition (London and New York: Verso, 1997; original: Verso, 1983).

154. Thomas Cole, *The Course of Empire*, 1844–1834, oil on canvass, The New-York Historical Society.

155. Quoted in Mrs. S. S. Colt, *The Tourist's Guide Through the Empire State* . . . (Albany, NY: Mrs. S. S. Colt, 1971), 23.

156. Angela Miller, "Everywhere and Nowhere: The Making of the National Landscape," *American Literary History* 4 (1992), 207–29.

157. *Illustrated History*, 183–84; "Map of Glenwood Cemetery, Lockport, N.Y.," drawn by Julius Frehsee C.E., 1892.

158. David Schuyler, *The New Urban Landscape: The Redefinition of City Form in Nineteenth-Century America* (Baltimore: Johns Hopkins University Press, 1988); Machor, Pastoral Cities.

159. Lockport *Daily Journal*, August 14, 1864.

160. "Beautiful Glenwood: The Most Handsome Cemetery in This Part of the State," Lockport *Daily Review*, July 28, 1897; see also *Illustrated History*, 184; Bev Chavers, "Glenwood Cemetery: A Walk through Historic Lockport," Lockport *Union-Sun & Journal*, October 27, 1988.

161. Clarence O. Lewis, "Century Ago Lockport Tax Roll Totaled $5,350," Lockport *Union-Sun & Journal*, March 31, 1960; "Civic Benefactor Dies in Oakland," Lockport *Union-Sun & Journal*, March 26, 1953.

162. "The Rollin T. Grant Gulf Wilderness Park, Lockport, New York," pamphlet prepared by the Citizens' Advisory Committee, April 1985, revised April 1986.

163. C. E. Burk, "Lockport, N.Y.: Electrical and Industrial" (Lockport, 1910).

164.George R. Worley, "Old Home Week: Celebrating the Hundredth Anniversary of the Erie Canal," Lockport Old Home Week Committee, Lockport, New York, July 19–25, 1925.

165. Clarence O. Lewis, "Will Lockport Decide to Preserve Landmarks?" Lockport *Union-Sun & Journal*, August 10, 1966; "Interview with Captain James McKain," Lockport *Daily Advertiser*, April 1, 1861.

166. Lockport *Daily Advertiser*, October 20 and 22, 1858.

167. Edwin R. Long, Lockport *Chronicle*, September 3, 1859.

168. Lockport *Daily Journal*, February 1, 2, 3, 7, 10, 12; Jerry Allan, "Lockport's 'Mammoth Cave' Urged for CD Shelter," Buffalo *Evening News*, September 7, 1957; reprinted in Thomas P. Callahan, "The History of the Lockport Caves," mimeograph, 1970.

169. Lewis, "Will Lockport Decide to Preserve Landmarks?"; see also Lockport *Daily Journal*, February 15, March 10, 12, 19, 21, 1883.

170. "Old Cave Never Fully Explored; Company Formed to Exploit the Natural Wonder Petered Out—Tales of Trips into the Darkness," Lockport *Union-Sun & Journal*, October 8, 1921; "Tourist Lure Likely to Be 'Open Sesame' for Old Cave," Lockport *Union-Sun & Journal*, April 21, 1951; Jerry Allen "Buffalonians Agree to Explore Lockport's Large Cave," Buffalo *Evening News*, April 30, 1958; Ilbert O. Lacy and Adelheid Z. Lacy, *Dusty Lockport Pages*, Occasional Contributions of the Niagara County Historical Society (Lockport: Niagara County Historical Society, 1952), 28–36; Allen, "Lockport's 'Mammoth Cave' Urged for CD Shelter"; "The Lure of the Cave," Buffalo *Evening News*, c. April 29, 1958; Thomas P. Callahan, "The History of the Lockport Caves," mimeograph, 1970.

171. Lewis, "Will Lockport Decide to Preserve Landmarks?"

172. Mary Anne Southhard, "Drillers Hunt Lost Cave Sealed in 1886," Buffalo *Evening News*, September 17, 1970; Callahan, "History of the Lockport Caves."

173. "Second Natural Cave Found Beneath City of Lockport, "Lockport *Union-Sun & Journal*, January 20, 1970; Bill Nelson, "Cave Find Sparks Lockport Search," Lockport *Union-Sun & Journal*, February 2, 1970.

174. Bill Nelson and Byers Bachman, "Long Lost Cavern Located by Searchers," Lockport *Union-Sun & Journal*, April 7, 1970; "Exploration of Lockport Caves Drew Much Attention in 1970," Lockport *Union-Sun & Journal*, August 28, 1971.

175. Scott A. Ensminger, *The Caves of Niagara County New York*, Occasional Contributions of the Niagara County Historical Society (Lockport: Niagara County Historical Society, 1987), 64–66.

176. "Lockport Youth Urges Building Cave," Lockport *Union-Sun & Journal*, September 7, 1971.

177. "Explorers Voyage through Tunnel Beneath Lockport," Lockport *Union-Sun & Journal*, January 29, 1970; Boles's team included Peter Smith and Joseph McGreevy.

178. "Lockport Cave Raceway Tour: America's Only Cave Raceway Tour," brochure of the Lockport Hydraulic Race Company, Lockport, 1975; "Lockport Cave & Underground Boat Ride," brochure of the Hydraulic Race Company, Lockport, 1996.

179. "Oh! Rock!" *Niagara Democrat*, November 1, 1843; "Canal Terminus Picturesque Place: End of Power Waterway at Lockport Is Known as Indian Falls," *Buffalo Enquirer*, April 21, 1904.

180. Clarence O. Lewis, "French and Indians Found Large Fish in Abundance," Lockport *Union-Sun & Journal*, May 6, 1954.

181. Orasmus Turner, *Pioneer History of the Holland Land Purchase* . . . (Buffalo: Jewett, Thomas & Co., 1849), 312–15.

182. *Illustrated History*, 146, 167.

183. Edna Deane Thomas (Smith), "Recollections of an Early Settler," transcribed by Laura P. Colton, and published in five parts in the *Lockport Daily Union* beginning with part I (August 11, 1873), part II (August 14), part III (August 21), part IV (August 28), and part V (September 6), quote is from part III.

184. Turner, *Pioneer History*, 29–30; also mentioned in N. P. Willis, ed., *American Scenery*, 2 vols. (London: 1840), I: 160.

185. Sketch of a box canyon, Holland Land Company Field Books, vol. 666, New York State Archives; these surveys were used to prepare the Holland Land Company Townships Maps (1800–1819), the Mountain Ridge site is on the maps of T.14.R.7. and T.14.R.6., New York State Archives. James Geddes Survey Maps of 1817, Original Maps of Surveys for the Erie Canal, New York State Archives.

186. John H. Eddy, Map of the State of New York, 1818; the falls appears on many maps of this period, but perhaps from an initial confusion it is placed too far to the west.

187. Clarence O. Lewis, "Great Glacier Covered County 12,000 Years Ago," Lockport *Union-Sun & Journal*, May 11, 1966; *Illustrated History of Niagara County*, 148.

188. Lockport *Daily Union*, February 7, 1867.

189. *Atlas of Niagara and Orleans Counties, New York* (Philadelphia: Beers, Upton and Company, 1875), Township Map of Lockport, 25; Lockport *Daily Advertiser*, March 22, 1886.

190. "A Gulf Timeline," in "The Rollin T. Grant Gulf Wilderness Park," Lockport, 1985; Lewis, "Great Glacier Covered County 12,000 Years Ago"; Clarence O. Lewis, "Natural Lockport Park Ruined by Use as Dump," Lockport *Union-Sun & Journal*, December 14, 1961.

191. Clarence O. Lewis, "Ship Canal Between Lakes First Envisioned in 1784," Lockport *Union-Sun & Journal*, April 19, 1956; Lewis, "The Erie Canal," 3–5; Joe Higgins, "Complex Questions Await Answer on Practicality of American Canal," *Niagara Falls Gazette*, October 9, 1960; Clarence O. Lewis, "Ship Canal Proposed by Frenchman in 1708," Lockport *Union-Sun & Journal*, February 21, 1962; Clarence O. Lewis, "Lake-to-Lake Canals Started Two Times," Lockport *Union-Sun & Journal*, March 1, 1962; "Proposed Ship Canal Between Lakes Erie and Ontario," U.S. Army Corps of Engineers Buffalo Office, August 30, 1960.

192. "Contract Between the Power Canal Companies," *Revised Charter and Ordinances of the City of Lockport*, Lockport, 1913, 565–70.

193. "Can't Steal Niagara," *Lockport Journal*, March 18, 1904; "Lockport and Cheap Power," *Buffalo Courier*, March 27, 1904.

194. "Another Hearing: Senate Committee to Give It Tuesday, on the Power Canal Bill," *Lockport Journal*, March 11, 1904; "Can't Steal Niagara."

195. "Harrah for Lockport Now," *Buffalo Courier*, April 6, 1904.

196. "Power Canal Bill Passed: Senate Approved It by a Vote of 36 to 12," *Lockport Journal*, April 8, 1904.

197. "Canal Terminus Picturesque Place: End of Power Waterway at Lockport Is Known as Indian Falls," *Buffalo Enquirer*, April 21, 1904.

198. Marcus Ruthenberg, quoted in "Lockport Power May Build Up Great Steel and Iron Industries," *Lockport Journal*, June 3, 1904.

199. "Vanderbilt-Andrews Trolley Plan and the Lockport Power Canal," *Lockport Journal*, August 17, 1904.

200. "Lockport and Cheap Power," *Buffalo Courier*, March 27, 1904.

201. Clarence O. Lewis, "Lake-to-Lake Canals Started Two Times," Lockport *Union-Sun & Journal*, March 1, 1962.

202. Alexis Muller Jr., *Looking Back: So That We May Move Ahead* (Lockport: Lockport Canal Sesquicentennial, Inc., 1975), 21.

203. *Revised Charter and Ordinances of the City of Lockport* (Lockport: Common Council of the City, 1913), article 3, sec. 62, part 26, pp. 56–57; Sanitary Code article 11, sec. 70, p. 379; also pp. 265, 376.

204. Recra Research, Inc., "Lockport City Landfill: New York State Superfund Phase I Summary Report" (Amherst, NY, 1983), 7.

205. "This Is Your Lockport," pamphlet, Lockport City Planning and Zoning Commission, c. 1948.

206. Clarence O. Lewis, "Lewiston, Lockport Gulf Noted Ideal Park Sites," Lockport *Union-Sun & Journal*, January 11, 1961; also by Lewis: "Natural Lockport Park Ruined by Use as Dump"; "Preservation of Gulf as Natural Park Urged," Lockport *Union-Sun & Journal*, May 8, 1963; "Indian Massacre Occurred on Site of Proposed Park," June 13, 1963; "Great Glacier Covered County 12,000 Years Ago."

207. "1910–1985: A History of Harrison Radiator Division," *Communicator* 3, no. 2 (Summer 1985): 4–11, 8–9.

208. Muller, *Looking Back*, 21.

209. Recra Research, Inc., "Lockport City Landfill: New York State Superfund Phase I Summary Report," 7.

210. Letter from URS Consultants to Niagara County Environmental Management Council, December 6, 1989.

211. Recra Research, Inc., "Lockport City Landfill: New York State Superfund Phase I Summary Report," 7–10.

212. "Lockport Chemistry Reported in National Chemical Trade Magazine," August 19, 2001; and "VanDeMark Chemicals Sells Out to French," November 10, 2000, in "Lockport Industrial Index," an online compilation of business and industrial news (http://www.lockport-ny.com/industry.htm).

213. "Lowertown Hit by Chemical Spill," Lockport *Union-Sun & Journal*, March 13, 1998.

214. Bob Kostoff, "No Health Risks from Blaze," Lockport *Union-Sun & Journal*, August 13, 1997.

215. Tom Hartley, "Eighteenmile Creek Feeds Stream of Questions," *Business First* 10, no. 42 (Week of August 1, 1994).

216. Cindi Wittcop, "Widow Wonders about Plant, Early Death of Husband," Lockport *Union-Sun & Journal*, September 18, 2000.

217. John D'Onofrio, "Canal '87 Flotilla Day Ushers in New Season," Lockport *Union-Sun & Journal*, May 11, 1987.

218. Ann Whitcher, "An Afternoon with Joyce Carol Oates," *UB Today* (Fall 1987).

219. Margaret Sullivan, "Oates' Parents Point to the Writer's Roots in Millersport," Buffalo *Evening News*, September 9, 1980.

220. "Canal Terminus Picturesque Place: End of Power Waterway at Lockport Is Known as Indian Falls."

221. Joyce Carol Oates, *(Woman) Writer: Occasions and Opportunities* (New York: Dutton, 1988), 379–80.

222. Joyce Carol Oates, "American Gothic," *New Yorker*, May 8, 1993.

223. Joyce Carol Oates, "City of Locks," *Angel Fire* (Baton Rouge: Louisiana State University Press, 1973).

224. Joyce Carol Oates, *Wonderland* (New York: Vanguard Press, 1971), 91–93.

225. Joyce Carol Oates, letter to the editor, Lockport *Union-Sun & Journal*, June 14, 1995, in response to an article by Brandon Stickney, "Oates Says Area Lacks 'Community,' " same newspaper, May 30, 1995.

INDEX